EVOLUTION AND BELIEF

Confessions of a Religious Paleontologist

Can a scientist believe in God? Does the war of words between some evolutionists and evangelicals show that the two sides are irreconcilable?

As a paleontologist and a religious believer, Robert Asher constantly confronts the perceived conflict between his occupation and his faith. In the course of his scientific work, he has found that no other theory comes close to Darwin's as an explanation for our world's incredible biodiversity.

Recounting discoveries in molecular biology, paleontology, and development, Asher reveals the remarkable evidence in favor of Darwinian evolution. In outlining the scope of Darwin's idea, Asher shows how evolution concerns the cause of biodiversity, rather than the agency behind it. He draws a line between superstition and religion, recognizing that atheism is not the inevitable conclusion of evolutionary theory. By liberating evolution from its misappropriated religious implications, Asher promotes a balanced awareness that contributes to our understanding of biology and Earth history.

ROBERT J. ASHER is a vertebrate paleontologist, specializing in mammals. After finishing his Ph.D. from Stony Brook University in 2000, he studied paleobiology at the American Museum of Natural History, the Berlin Museum of Natural History, and since 2006 has been the Curator of Vertebrates in the Museum of Zoology, Cambridge. Over the past two decades, his work in paleobiology has taken him to Argentina, Britain, Canada, Kenya, Madagascar, Mongolia, South Africa, Spain, western USA, and Venezuela.

EVOLUTION AND BELIEF

Confessions of a Religious Paleontologist

Robert J. Asher

Department of Zoology, University of Cambridge

CAMBRIDGE UNIVERSITY PRESS
Cambridge, New York, Melbourne, Madrid, Cape Town,
Singapore, São Paulo, Delhi, Mexico City

Cambridge University Press
The Edinburgh Building, Cambridge CB2 8RU, UK

Published in the United States of America by Cambridge University Press, New York

www.cambridge.org
Information on this title: www.cambridge.org/9780521193832

First published 2012

Printed in the United Kingdom at the University Press, Cambridge

A catalogue record for this publication is available from the British Library

Library of Congress Cataloging in Publication Data
Asher, Robert J.
Evolution and belief : confessions of a religious paleontologist / Robert J. Asher.
p. cm.
Includes bibliographical references and index.
ISBN 978-0-521-19383-2
1. Evolution (Biology) 2. Science and religion. I. Title.
QH366.A84 2012
576.8–dc23 2011038242

ISBN 978-0-521-19383-2 Hardback

To the memory of Charles A. Lockwood (1970–2008)

CONTENTS

Tulsa City-County Library
Martin Regional Library

Customer ID: **********8990

Items that you checked out

Title:
Behave : the biology of humans at our best and worst / Robert M. Sapolsky.
ID: 32345077560614
Due: 8/14/2018 12:00 AM

Title: Evolution and belief : confessions of a religious paleontologist / Robert J. Asher.
ID: 32345062832895
Due: 8/14/2018 12:00 AM

Total items: 2
Account balance: $1.20
7/31/2018 2:14 PM
Checked out: 2

To renew:
www.tulsalibrary.org
918-549-7323

We value your feedback.
Please take our online survey.
www.tulsalibrary.org/Z45

ACKNOWLEDGMENTS

A number of family members, close friends, and colleagues have provided tremendously valuable input on the text of this book during its gestation. In particular, I wish to thank Lilian, Bernard, Liz, and Kathy Asher. I am grateful to Marcelo Sánchez-Villagra not only for his insightful comments on the book, but also for blurring the distinction between family and friend.

For discussion and/or helpful critiques of the text (which I undoubtedly failed to fully address in at least a few cases), I also thank Denis Alexander, Ken Angielczyk, Brian Beatty, Mike Bell, Stephen Benedetti, Roger Benson, Jimena Berni, Faysal Bibi, Brenda Bradley, Jenny Clack, Simon Conway-Morris, Farhan Feroz, Adrian Friday, Jonathan Geisler, Alan Gentry, Jerome Jarrett, Zerina Johanson, Tom Kemp, Dieter Lukas, Zhe-Xi Luo, Christian Mitgutsch, Nick Mundy, Kevin Padian, Marie Pointer, Bill Sanders, Greg Sutton, and Abbie Tucker.

I am grateful to Tom Kemp, Gea Olbricht, and the staff at the University Museum of Zoology, Cambridge for facilitating access to specimens used in some of the figures, and to Uwe Dörmann and Young-Ok Kim for access to Dr. Kim's medical dissertation on Karl Reichert. I thank Anjali Goswami and Bjarte Rettedal for encouragement and help with photography. Jorge Morales and many others associated with the Museo Nacional de Ciencias Naturales in Madrid served as wonderful hosts during fieldwork in the Miocene of Spain. I am also grateful to Gema Alcalde, Juan L. Cantalapiedra, and Adriana Oliver for their willingness to appear in Figure 8.1.

Thanks to Mike Richardson and Ingmar Werneburg for providing images of vertebrate embryos, and to Dominic Lewis, Megan Waddington, Abigail Jones, and their colleagues at Cambridge University Press for their help and encouragement.

PROLOGUE

I believe in God; therefore, I'm religious. My father is Jewish, my mother Christian, and I was raised in a Presbyterian church in western New York state, USA. At present, I often go to Anglican church services (or "Evensong") at various colleges within my university, and the music is excellent. I'm not a fundamentalist or evangelical of any denomination, and I do not believe that every word in the Bible is an unfiltered indication of His Divine Will. However, for all of its human-caused mistakes, I believe the Bible has a lot going for it. It encourages humility and love, and it asks you to recognize your imperfections, put the needs of others ahead of your own, and as a general rule, treat others as you would like to be treated yourself. In my own intuitive, unscientific way, I think this core message is divinely inspired.

I'm also a paleontologist. That is, I'm an academic who studies evolutionary biology for a living, and I'm particularly interested in the fossil record of mammals. This profession has enabled me to observe firsthand just how right Charles Darwin was about how all mammals share a biological history among themselves and with other forms of life on this planet. At no point has this observation led me to a spiritual "crisis," or to the feeling that God and Darwin are somehow antagonistic. It would have if I equated "God" with superstition or a literal reading of Genesis, but I don't. Biblical literalism was nonsense long before Charles Darwin came on the scene.

The subtitle of this book is "confessions of a religious paleontologist." Actually, all I'd like to confess is the fact that I don't see a contradiction between my profession as an evolutionary scientist and my belief in God. Some of you will disagree. One kind of skeptic thinks it's only mild schizophrenia that enables me to embrace rationality and the scientific method during most waking hours, yet lull my insecure mind to sleep with the comfort that Father looks down from the heavens with special concern for my existence. On the other hand, a stereotypically fundamentalist partisan may think that evolutionary biologists like me deliberately lie to the public—for example, by claiming that religion and science are compatible—so we can go on foisting our privately favored, atheist philosophy on schoolchildren, and in the meantime get paid for it as members of publicly funded university departments. If either perspective represents what you think, then I sincerely hope you'll keep reading this book. I hope even more that, eventually, you'll change your mind.

TRUTH

Certain people have different standards for recognizing "truth."[1] Given access to the same facts, two individuals can look at an issue and reach utterly different conclusions, to the point where they believe those with a different opinion belong somewhere on a spectrum from stupid to perverse. Take evolution, for example. In 2001, the science journalist Carl Zimmer published *Evolution: The Triumph of an Idea*, a popular book setting out the historical context of evolutionary biology and describing the evidence for its continued success as a scientific theory today. Nearly all of the reviews I've seen of his book are very positive. The few negative reviews I've seen are written anonymously as comments on websites selling the book, objecting to its content for reasons including its recognition of an ancient Earth and the common ancestry shared by living things. The presentation of facts behind these phenomena given by Zimmer are characterized by some as "outright lies." One anonymous reviewer writes, incredulously, that only with such "deception" could anyone believe that (as they put it) humans somehow descended from fish, amphibians, and rodents. Other anonymous online reviewers are much more positive about Zimmer's book, but rather disparaging about their anti-evolutionary neighbors. Two possible reasons for this difference of opinion are as follows:

(A) Online reviewers who disagree about Zimmer's book are zealots, people drawn to controversy and for some reason unable to disentangle their personal tastes and preferences from an honest appraisal of the evidence. An accurate verdict on Zimmer's book, and on the matter of evolution generally, lies somewhere in-between their extreme opinions.

(B) Zimmer's book is either right or wrong on the subject of evolution. Life history either happened that way or it didn't. Hence, some of the book's reviewers are wrong, and the others have made an accurate appraisal of the book and more generally about evolution.

To me, the choice is clear: "B" is correct and the reviewers who malign evolution are wrong. Carl Zimmer wrote an excellent book and put at these people's fingertips evidence that, for some reason, they refuse to accept. There is no golden mean to draw between Zimmer's book and the creationists who dislike it; rather, the latter group is misled. By itself, this is no big deal. People make mistakes all the time. More disturbing are the probable reasons why a given creationist clings to her/his opinions on evolution. If I had to guess, I'd say that it is highly unlikely that another book on the subject—one that again sets out some of the vast body of evidence from paleontology, development, and molecular biology in support of evolutionary hypotheses such as common ancestry—would change her/his mind. The creationist has something at stake, some worldview or allegiance, that makes a fair, honest view of the data behind Darwinian evolutionary biology impossible. Why?

Ninety years ago there were better reasons to be an anti-evolutionist. Young Earth creationism as we now know it didn't really exist,[2] but eugenics did. This was an idea about deliberately manipulating human class and race by methods such as forced sterilization of prisoners and the mentally disabled. Unfortunately, this idea was taken quite seriously by many nineteenth- and early twentieth-century biologists, including the author of the high-school textbook used by John Scopes of the infamous "monkey trial" of 1925 in Dayton, Tennessee. In *A Civic Biology*, author G.W. Hunter not only outlines his (crude) understanding of Darwinian natural selection, but also writes about human social engineering:

> Hundreds of families ... exist to-day, spreading disease, immorality, and crime to all parts of this country. ... They not only do harm to others by corrupting, stealing, or spreading disease, but they are actually protected and cared for by the state out of public money. Largely for them the poorhouse and the asylum exist. ... If such people were lower animals, we would probably kill them off to prevent them from spreading. Humanity will not allow this, but we do have the remedy of separating the sexes in asylums or other places and in various ways preventing intermarriage and the possibilities of perpetuating such a low and degenerate race.[3]

If my child were given this passage to read approvingly in a class, I'd be outraged too. Contrary to various reenactments of the Scopes trial, its well-known chief prosecutor, William Jennings Bryan,[4] did not hate evolution because it contradicted his literal belief in Genesis. He didn't have one. Instead, he viewed the Judeo-Christian creation story as most Jews and Christians do, both then and now: as a narrative, entirely compatible with a geologically ancient Earth.[5] His antagonism toward evolution derived from his understanding that evolutionary theory was being trotted out in public schools as a potential justification for massive social inequality.[6] In this context it's entirely reasonable that Bryan, a populist, pro-suffrage, anti-corporate former US Secretary of State and three-time Democrat Party nominee for president, came to regard evolutionary theory with such scorn. He associated the entire enterprise with "might-makes-right," advocated primarily by those who benefited from economic inequality.[7] If that's what Darwinian evolutionary biology were all about, I'd want it banned from schools too.

In the twenty-first century, creationists do not have this excuse to hate evolutionary biology. No serious biologist today advocates the long-discredited notions epitomized in the preceding quote from Hunter's 1914 textbook. On the contrary, it is an evolutionary understanding of human biological history that makes it clear just how closely interrelated all modern humans are to one another, showing that "racial purity" is baseless.[8] To blame evolution for some modern wacko's racial bigotry would be like blaming gravity for the firebombing of Dresden. In fact, it would be worse: physicists still accept that gravity plays a key role in dropping bombs, whereas contemporary evolutionary biologists do not hold that Darwinian natural selection justifies racism. If you want Darwin banned from your school due to his theory's historical connections with eugenics, you may as well try to jettison vector physics too, since it comprises the principles by which nuclear warheads reach their destination when perched atop a ballistic missile. If only the physicists and engineers of the world had never been taught the evil principles of force, acceleration, and the gravitational constant, we might live in a world free of bombs and bullets!

But seriously, there is an obvious explanation for antipathy toward Charles Darwin among the various anti-evolutionist groups of the last 150 years, groups that are often connected to one kind of intense religious creed or another: they think Darwin threatens their worldview. Contributing to this conviction are those biologists who portray evolution as tied to atheism, who help convince the devout that a natural connection of humanity with other organisms is incompatible with their religion. Compounding things further is the fact that adherence to many religious worldviews is not flexible, and any scientific theory or philosophy that seems to threaten certain beliefs *must* be wrong, whatever some scientist may say about evidence.

Let's consider an example of potential conflict between religious doctrine and what science tells us about history. The Book of Mormon makes reference to the use of elephants, horses, chariots, and steel among pre-Colombian civilizations of the New World.[9] While elephants and horses have a long history throughout geological time on the North American continent, members of these groups became locally extinct long before any of the technologically advanced cultures of Central America began to thrive roughly 3500 years ago.[10] The horses present in North America during historical times derive from *Equus caballus* introduced from Spain at the close of the fifteenth century.[11] Furthermore, archeological evidence does not support the use of steel or wheels for transport or warfare in any New World civilization.[12] This is not to say that mainstream Mormons read their scriptures as a literal guide to New World history, or that some interpretive license cannot be applied in understanding these texts. Perhaps the English word "horse" does not accurately convey the meaning of the original term from reformed Egyptian, the language the founder of the Mormon religion, Joseph Smith, was reported to have read when he translated the Book of Mormon.[13] In any event, those Mormons who do interpret their holy books literally cannot simultaneously embrace many well-established facts of paleontology and archeology as they pertain to the last 10 000 years in the Americas. Literalists of the Mormon tradition, and those of many other faiths, have to reject at least some scientific interpretations of Earth history in order to maintain their beliefs.

Incompatibility between some forms of religion and science is not news, and you do not need me to point it out to you. However, this tension highlights the way different people come to regard something as "true." On the one hand, "truth" may derive from experience, from the community in which one is raised, and from an inner conviction that relates to that experience. Potentially different is the "truth" deduced from observation by people with nothing personal at stake beyond an explanation of the phenomenon at hand. An awareness of this dichotomy is essential in making the public case for any rational, scientific explanation of a natural phenomenon, such as evolution.

I don't know if the anti-evolutionist reviewer of Carl Zimmer's book paraphrased above is Muslim, Mormon, Protestant, Jewish, or some other faith. To many anti-evolutionists, their religious beliefs are just as "true" and integral to their identity as any empirical observation about the natural world. Let's say for a moment that I really did want to convey the "truth" of evolution to that person. One approach is to lay out the evidence for it yet again, as University of Chicago biologist Jerry Coyne recently did in his appropriately titled book, *Why Evolution is True*.[14] Coyne's book is excellent; it has one of the best all-round discussions of the plurality of evidence in favor of Darwinian natural selection as the engine behind biodiversity. Will modern creationists be swayed by another account of the hard data behind evolution? I doubt it. In fact, Coyne's book—while well-written and very accessible—doesn't really try to reach people not already in the realm of the strictly empirical.[15] Elsewhere in his talks and writings, Coyne says there is one way to be rational, and any of this stuff about alternative "truth" is relativist nonsense not worth the flatscreen monitor on which it's written:

> What, then, is the nature of "religious truth" that supposedly complements "scientific truth"?... Anything touted as a "truth" must come with a method for being disproved—a method that does not depend on personal revelation. ... It would appear, then, that one cannot be coherently religious and scientific at the same time. That alleged synthesis requires that with one part of your brain you accept only those things that are tested and supported by agreed-upon evidence, logic, and reason, while with the other part of your brain you accept things that are unsupportable or even falsified.[16]

I disagree, and would argue that there are many things in life that deserve the descriptor "truth" but are not amenable to rational disproof. Coyne is absolutely correct to say that coddling the irrational—those for whom "religious truth" means stoning adulterers or drinking poisoned Kool-aid—is incompatible with science and, more generally, civil society. However, while science is a-religious, it is not anti-religious, at least in the important sense that it does not (indeed, cannot) concern itself with phenomena beyond what we rationally perceive. It is not only possible to portray science as lacking fatal consequences for those religious tenets that concern things we cannot empirically observe (such as purpose or agency in life), but it is precisely what scientists have *got* to do to make a compelling case to the public. Coyne tosses "religion" into the same dumpster as any passing superstition, and actively encourages the perception that science is corrosive to any religious sentiment. Yes, there are religious claims that are demonstrably wrong in an empirical sense, such as the horse-drawn chariots of ancient Central America mentioned in the Book of Mormon. However, such specific claims do not do justice to the religion integrally tied into the identity of many laypeople and scientists alike, an identity that by any meaningful definition is worthy of the name "truth."

IRRATIONAL TRUTH

Although I've lived in the United Kingdom for the past several years, I'm originally from the United States—western New York state, to be specific. I have an affinity for my hometown ice hockey team, the Buffalo Sabres. As a metaphor for the point I'm trying to make, ice hockey is not much better than the caricature of superstition that makes up Coyne's view of religion. However, it does show something about "truth" in what I hope is an accessible way.

Rick Jeanneret has been the play-by-play announcer of the Buffalo Sabres since 1971, a tenure spanning nearly the team's entire existence. The Sabres are a professional ice hockey team from maybe the only town in the United States to host multiple big-league sports teams over the last half-century without ever winning any of the national titles of major sports leagues. No superbowls (although we lost four in a row in the early 1990s), no Stanley Cups, ever.[17] We do have a championship lacrosse team (Bandits), and a solid minor-league baseball team (Bisons), which should have gone pro in the 1990s, but the league opted for Denver instead. We used to have a good indoor soccer team (Blizzard), and I have a vague memory of the Buffalo Braves professional basketball team, but they didn't win any titles either. In any event, I've been a Sabres fan all my life, and for some reason I find an extraordinary level of comfort in the fact that in nearly every game I've ever heard on TV or radio, the same voice has been doing play-by-play for 40 years: Rick Jeanneret. With a transistor radio tuned to WGR550 stashed under my pillow, way after my bedtime, I cried with joy when Buffalo thrashed the Leafs 14–4 back in 1981. I shouted with almost as much gusto as Rick did when Chris Drury (who's since abandoned us like so many other mercenaries in the NHL—why can't players sacrifice a million for loyalty?) banged in the tying goal with seconds left in the third to force overtime versus the Rangers in the 2007 playoffs. Buffalo won that series but was eliminated by Ottawa in the next. Chris Drury now plays for the Rangers, and the Sabres have not made it to the playoffs since.[18] I am a Buffalo sports fan!

Hearing Rick Jeanneret shout in ecstasy with a Sabres goal brings me close to uncounted moments of euphoria during childhood. I'm sorry to sound mushy, but I love this guy, and I live in fear that he will retire without ever having heralded my boys through four wins in a Stanley Cup final. He was there when we lost in 1975 versus Philly, and in 1999 when Dallas "won" the series with an illegal [NO] goal in overtime of game six. Just like me, and thousands of other true Buffalo fans, Rick has stuck with the Sabres through thick and thin, and if ever a professional sports team, an announcer, or a city, deserved to win, it's Rick Jeanneret and the Sabres of Buffalo.

Now let's look at professional ice hockey from a rational, environmentalist perspective. To get to their 82 games per year (not including the playoffs), many of which are in places like Atlanta and Dallas (where kids don't realize that ice can also form outside of rinks) or Edmonton and Montreal (where, as in Buffalo, they still admire how a Zambonied ice surface completely lacks frozen twigs), the Sabres as a team emit tons of CO_2 into the atmosphere by flying all over the place. It's safe to say that if the NHL or other pro-sports leagues didn't fly, and if NASCAR shut down altogether, carbon emissions in North America would decrease by a small but measurable amount. Given the environmental change related to global warming that most scientists seriously consider to be threatening humanity's current standard of living, it would be entirely rational for society as a whole to restrict carbon emissions related to non-essential, recreational activity. This is the scenario anticipated by rational thought: burning fossil fuels contributes directly to global warming and will likely lead to major environmental degradation. Broadly speaking, the feeding frenzy of billions of humans during this century will spell the end of the quality of life that we in North America, Eurasia, Australia, and much of the rest of the world enjoy. Rational thought tells me in no uncertain terms that, even given some ambiguity as to the timing, it is in the best, long-term interest of future generations that I, right now, stop contributing to abuse of the Earth's environment. Reducing my consumption of fossil-fuel-derived resources is a part of that, and, as petty as it may sound, I should immediately cease my patronage of consumerist fluff like professional ice hockey, and encourage all of my friends and colleagues to do the same.

Huh? No way! I have to admit there have been times (e.g., the 1999 and 2007 playoffs) when my dedication to this team led me to reject rationality in the lengths I would go to support them. For example, in the midst of a promising playoff run, I cannot rule out offering substantial sums of my own limited wealth, maybe even endurance of some pain or discomfort (sacrificing a toe?), if it would help the Sabres win. My attachment to this team is probably not unlike that of the much maligned (in New York) construction worker Gino Castignoli for his beloved Boston Red Sox. In 2008, Castignoli secretly placed a Boston Red Sox jersey in concrete mix which was then poured into the foundation of the then-new stadium for the New York Yankees. The secret didn't last for long, but long enough for the concrete to dry. When it became clear what had happened, the loyal workers of the New York based construction firm searched for and found the offending talisman under 2.5 feet of concrete, hammering through the newly solidified barrier to remove the Red Sox jersey at considerable time and expense.[19]

My desire to see the Buffalo Sabres to victory is absolutely true, just like that of Gino Castignoli for his Boston Red Sox. However, it is not the product of reason, but an emotional attachment to a group of large, violent men wielding wooden/fiberglass/aluminum sticks who chase a heavy, rubberized disc at high speeds over an artificially frozen playing surface with sharp metal blades strapped to their feet. Worse, the actual players are (with a few exceptions) not particularly dedicated to this team (witness Chris Drury of the New York Rangers). Nearly all of them are from Canada, Finland, Sweden, Russia, Michigan, Minnesota, etc. This doesn't necessarily mean that none of them is loyal; in fact, along with Rick, we've got the longest-serving active coach (former winger Lindy Ruff) and equipment manager (recently retired Rip Simonick) in the NHL. However, even here I lose perspective, as if a dedicated bunch of local hockey players from western New York would make the pointless consumption of resources in this frivolous sport any less environmentally damaging.

As a human, much of how I define myself, including tribal affinities toward sports teams, is not particularly rational by any empirical standard. Multiplied by 6.5 billion, this is a very bad thing for planet Earth, at least insofar as we expect this place to keep supporting our eclectic tastes. However, there is nothing illusory about human devotion to obscure pastimes, such as my attachment to the Sabres. I wouldn't rule out attempts to make ice hockey more environmentally friendly. However, if this is your "goal," you will not get anywhere by telling fans they're idiots for enjoying the sport, or by claiming that their emotional attachment to it is irrational and stupid. For better or worse, we've got the attachment, which is no less rational than our taste for wings, bleu cheese, and canned beer. Consumed in excess they may be damaging, and you may prefer something else. However, we're talking about identity, rather than some purely rational choice. When the Sabres eventually bring home the cup (the very thought of which makes me religious), the sweet taste of victory will exceed even that of our wings, and it will be just as "true" to western New Yorkers as any scientific advancement of recent memory.

RELIGIOUS IDENTITY

In a 2003 lecture in the United States,[20] Richard Dawkins explained why he regards the religious labeling of children to be a highly immoral act. Few would take seriously a five-year-old child who called her/himself a Maoist or a disciple of Freud, and most would regard with some concern the parents of that child. In the same vein, says Dawkins, it is irresponsible to call your five-year-old a Lutheran or Muslim. The state of being a Freudian or Maoist implies certain political sympathies that, if properly understood, require a fairly adult level of political sophistication. Of course one could argue that to the extent anyone properly understands Freud or Mao, they would cease subscribing to their largely discredited socio-political philosophies. Many would say the same for the world's major religions, and that all of these labels have weighty socio-political implications, not for uninformed children. Only adults take upon themselves these labels, voluntarily, after careful consideration as to what the attached belief system behind each label actually means.

Or do they? One of the questions asked of Dawkins after his lecture was this: what can an average church-going adult say about the content of her/his creed that a young child cannot? Do adults really "carefully consider" the content of their religion prior to taking upon themselves its label? The member of the audience who asked this question implied (rightfully, I think) that they generally do not, and that Dawkins was making an inappropriate comparison between socio-political philosophy (e.g., Maoism) and cultural identity (e.g., Islam). Of course we don't call young children Freudians, Libertarians, Jazz fanatics, or other such labels which require some level of experience as an adult to acquire. But "Muslim" as a cultural concept is not the same kind of label as "Maoist." It entails much more of a social identity and not necessarily a political one. A child can be "Muslim" in much the same way as he or she is Montenegran, Minnesotan, or Malagasy, and an individual can grow up to hold a very broad range of socio-political sympathies within any of these identities. Hence, with the important qualification that many conversions into a given religion do happen for carefully premeditated, adult reasons, the majority of a given creed's adherents classify themselves as such for no other reason besides the cultural heritage into which they're born. This fact is important to consider for those who claim that their particular religious worldview is the most "rational," but that is another issue.

For now, the point is that most everyone develops deep connections with one or more aspects of the socio-cultural milieu in which they're raised, and these connections are not intrinsically wrong. To the extent that biologists want to educate people about the natural world, it would be wise to respect this fact. This doesn't mean that people like Jerry Coyne should feign admiration or respect for what he regards as superstition. However, it does mean that if he wants to make a positive difference in the public discourse about evolution, a scientist like him (in the words of astrophysicist Lawrence Krauss[21]) has to reach out to people where they are, not where he thinks they should be. A lot of people out there recognize "truth" without necessarily distinguishing between the empirical and the personal, between factual common sense and the gravitas of how they define themselves as Mormons, New Yorkers, or hockey fans. The challenge for a new book on evolution, one which will at least try to reach an audience that in recent years has been hostile to the idea, I think lies in making this distinction. Camps on either side of the culture wars deserve to be reminded that while nonsense dressed up as religion is still nonsense, not everything about our existence is equally accessible as a subject of scientific inquiry. There probably is at least something comprehensible in the vast majority of what surrounds us, but from my perspective as an evolutionary biologist, scientific inquiry is limited by human rationality and our capacity to observe.

Throughout these pages, I will try to make two points: first, that evolution is "true" as a mechanism that explains how living things on our planet have been derived from similar living things that came before them, and furthermore is not about explaining that very first living thing. Second, I will try to make the case that understanding how evolution works does not address the potential "who" or "why" behind it. This leaves pursuit of such matters to other fields, but not because I want to shield religion or theology from the acid of rational scrutiny. Rather, it is because I want you, the public, to understand that evolutionary biology is not a limitless enterprise for explaining everything. The challenge of comprehending how the Earth's biodiversity has come about, and how species are interconnected, is hard enough without throwing God into the mix. You'll understand biology better if you're aware of this fact.

If you're an atheist, incredulous at the idea that a paleontologist like me is failing to go the whole hog, retaining some sentimentality for the "myths" by which I was raised, then let me start by proposing that evolutionary biology is not the best means by which you or anyone else can decide what is moral or seek to understand the purpose or meaning behind life. If you're deeply religious and skeptical of our atheist colleague almost as much as you are of me—trying to have my spiritual cake and eat it too—then let me suggest that understanding the mechanics of biology does not concern the agency behind it, just as understanding how a lightbulb works does not concern the existence of Thomas Edison. Unless you're superstitious, such an understanding does not have to impinge upon your identity. In both cases, if you'll bear with me, I want to chart a course through the false notion that evolution rules out religious belief. If you come with me that far, then appreciating the facts behind the evolutionary interconnectedness among living things will be much easier.

ONE

SCIENCE AND RELIGION

A lot of disagreements between people are due to honest emphasis on mutually exclusive propositions, both of which have clear value. Examples include social responsibility versus personal liberty, or freedom of speech versus the protection of minorities. In other cases, one party to a debate is just plain wrong, misinformed, or invested in error for extraneous and/or personal reasons. This includes the "divine right of kings" and "separate but equal" racial segregation. Society makes its way along the centuries by recognizing, and dispensing with, the erroneous (e.g., divine right) while building up institutions that can justly scrutinize the real debates, hopefully reaching the right decision more often than not.

The current debate between science and religion, in particular discussions of evolution and public education in the United States, is mostly a phenomenon of the erroneous sort. Some opinions of the partisans, while zealous in their delivery, are just plain wrong. However, making things more complicated is the fact that the "plain wrong" errors are committed by more than one party. On the one hand, the idea that the natural world around us does not teem with evidence in support of Darwin's theory of evolution, that humanity does not share common ancestry with other forms of life on Earth via the mechanism of descent with modification, is profoundly mistaken. You, I, and the rest of humanity are so part of the Tree of Life, this book is to help you count the ways. Equally wrong is the notion that acceptance of this evidence of biological evolution spells doom to a religious worldview (it doesn't), or that Darwin told us anything definitive about the origin of life or how we could be good atheists (he didn't).

Speaking as someone with an academic perspective, who studies evolution for a living and is daily amazed at its explanatory prowess, the hard part of this debate is understanding why so many decent people out there find biological evolution so threatening. What is it about Darwinism that keeps otherwise honest, open-minded people from embracing the remarkable principle that explains—biologically—how they got their five fingers, arched foot, and hearing bones, as opposed to their pet frog's short backbone, long legs, and quiet demeanor? We—meaning not just card-carrying evolutionary biologists but anyone who's read the literature attentively—know with some precision the details behind many of the biological changes between groups of animals. We know not just in terms of the actual record of intermediate forms, both living and fossil, but also in terms of development and genetics. We know this so well that a geneticist who's never seen a fossil before in her life can come up with roughly the same estimate of the pattern of shared ancestry among animals with a backbone (vertebrates) as a paleontologist, using a completely different body of evidence.

I want to spend time in this book sharing with you in a simple, straightforward way a few of the well-documented cases that make Darwinian evolution so compelling. These concern facts and ideas that every biology teacher in every school should know, and they (you) should also know that such ideas do not have fatal implications for a principled, religious worldview. You've probably heard at some point that evolution and religion are fundamentally incompatible. Ironically, the extremists who advance opposite viewpoints on God versus Darwin agree entirely with one another on this point.[1] Nevertheless, it is a matter of simple, empirical fact that practicing evolutionists are not necessarily atheists. Even some who are atheistic recognize that pitting God against Darwin is a mistake and that the two do not comprise an either–or proposition. There have been tensions between religion and science in many quarters, but there have been many first-class evolutionary biologists who are agnostic (like Stephen Jay Gould) or very religious (like Francis Collins) who do not see a necessary conflict between a religious worldview and a materialist orientation of modern science.

In the coming pages, I'll describe in some detail what I mean by the phrase "materialist orientation of modern science." I'll also describe in this initial chapter how Darwinian natural selection does not address questions of ultimate meaning and purpose in existence, nor does it have to. But it does explain a lot of the *how* relating to life's diversity once it began, a beginning that Darwin himself attributed to the Creator (yes, God). Do you vote, pay taxes, or educate young people? If so, then you must know at least some of the factual basis of the Darwin–Wallace theory of evolution, and how it has stood up to thousands of discoveries and scientific tests since it was first published in 1858. This includes a description in Chapter 2 of how evolutionary biology qualifies as a historical science in which hypothesis testing plays a central role, and how evolution by natural selection logically leads to a variety of specific predictions concerning common descent, geological time, the fossil record, the development of living organisms, and molecular biology—topics explored in later chapters.

The fossil record is an important line of evidence supporting the mechanisms of Darwinian evolution. However, it is entirely possible to make a strong case for evolution without recourse to the data from extinct life, and we'll do so in Chapter 3. Of course, when one does examine the fossil record, as any reasonable observer is obliged to do and as we'll do in Chapters 4–8, the case for natural selection as a mechanism behind life's diversity becomes even stronger. Chapters 9 and 10 detail some of the molecular biological agreement with other lines of evidence, agreement that fits specific predictions made by the theory of evolution. Chapters 11 and 12 conclude the book with a discussion of probability, and how claims of impenetrable complexity in biology are now, as they always have been, an inadequate substitute for understanding the mechanisms responsible for the emergence of biological diversity.

DARWINISM, AGENCY, AND CAUSE

When the United States celebrated its two-hundredth birthday, my first-grade teacher, Miss Lee, sent me to deliver a message to her colleague down the hall, Mrs. Sanfrantello. When I realized that in order to deliver this message I had to enter the second-grade homeroom, I was dumbstruck. I could barely imagine how these advanced, civilized creatures would receive a puny first-grader like me into their midst. Fortunately, when I arrived they were occupied elsewhere and their room was empty except for Mrs. Sanfrantello, working at her desk. "Come in, honey," I heard as I poked my six-and-one-half-year-old face through the open doorway. The empty room (I didn't realize anything in that school could be so quiet) accentuated the sense of awe with which I entered; it seemed to me as a huge, cavernous cathedral, echoing with my every footfall. Mrs. Sanfrantello sat at her desk opposite the entrance, and walking up to her with Miss Lee's message seemed to take an eternity as I marveled at the pupils' desks, which came with integrated rulers—further proof of the heady intellectual climate that characterized this center of elementary-school culture. For days after this experience, I could hardly believe that I had actually been in the same room as second-graders!

A few months later, when I sat in this room as a second-grader myself, the integrated rulers really were no big deal, and the room turned out to be the same size as my first-grade classroom. Somehow the sense of awe had completely vanished.

It's possible that some people out there view "science" in a similar fashion as I saw the second grade at the age of six. This is not due to their child-like naïveté, but only because they've been busy with other things and haven't yet taken the time to look into this subject. Prior to being in it myself, I knew little about the second grade, and I attributed to it all sorts of capacities that it didn't actually have. I certainly could not envision myself as part of it. Yet this institution was host to the same sort of kids as me. Any differences between the first and second grades were of degree, not kind.

"Science" differs in a similar fashion from the plain and simple common sense that most people apply regularly in life. It is the fundamental, principal ingredient in the scientific pursuit of knowledge. To turn common sense into science, one simply adds a variety of accessory facts, a more rigorous way of framing questions, careful data collection and assessment of probability, and availability of relevant data in a coherent format to others, and voilà. You're a scientist. Now this is not to minimize the training involved. After all, just using our common sense, the Earth does look flat when looking around while standing in a parking lot. Such a perspective does not allow for collection of enough "accessory facts."

There's nothing magical about the scientific study of biological history or about tracing the evidence of life's evolution. Strains of flu virus can evolve just as speedily among first-year business students as among zoology undergrads—and both are capable of observing, learning about, and suffering from such phenomena. In fact, an appreciation of even just a few of these readily available facts is enough to make a very compelling case for Darwinian evolution—meaning descent with modification over long periods of time.

The phrase "descent with modification" encapsulates Darwin's idea. Attributes of plants and animals have the capacity to be inherited across generations; these attributes may change slightly from one generation to the next; more offspring are produced than can actually survive; some members of one generation may be particularly good at contributing their offspring to successive generations. Over the vastness of time, this process has yielded the biological diversity we see today. Much has been learned about natural selection during the past two centuries, and indeed there are many complicating factors involved. However, the fundamental contribution of Darwin's *On the Origin of Species*[2] remains essential to evolution: heritable genetic variation and differential survival over the course of many generations can lead, eventually, to significant changes in biological populations. This is the Darwin–Wallace theory of descent with modification, or natural selection.

Please note that this process explains *how* biological change occurs. It does so in the same way that you might explain *how* a steam engine works, or the process by which its action is caused: water heated to 100 °C boils into steam, which rises and powers the rotation of a turbine, which then generates electricity at the local power plant, and spins the wheels of your nineteenth-century train, Mississippi riverboat, etc. As an analogy this is a bit dated, but the point should be clear: both explanations are about natural processes responsible for something we observe. It is equally valid to note that Thomas Savery designed the first steam engine, or that James Watt (among others) later improved it. However, the latter is an explanation of a different sort: it is one of agency, not cause. Riverboat passengers at some point may have expressed great admiration for Savery and Watt, the "creators" of their momentum. How does the engine work? Savery did it, helped by Watt. Such an interpretation is true in the sense that Savery and Watt deserve credit as the agency behind the steam engine. However, it says nothing about how the steam engine actually works. There is a materialist, or naturalistic, cause behind the function of their steam-propelled craft which is not changed by recognizing the agency of Savery and Watt in the development of its engine. This kind of natural causation is what I meant earlier when I referred to the "materialist orientation" of science.

The same distinction between agency and cause is very relevant to the current debate on evolution. Darwinian natural selection is a very specific set of ideas about the naturalistic basis by which animals across many generations may evolve. However, to quote the 1980s actor/comic Pee-Wee Hermann, there is a very "big but": exactly how the first organism appeared, or if a higher consciousness was somehow the agency behind biological replication, inheritance, and selection, is not part of the theory of evolution. Darwin himself made this clear in all six editions of the *Origin*. Consider the quotes Darwin listed on the reverse of his title page. In the first edition he cited William Whewell and Francis Bacon; starting with the second, he added another, from Joseph Butler:

> The only distinct meaning of the word "natural" is *stated*, *fixed*, or *settled*; since what is natural as much requires and presupposes an intelligent agent to render it so, i.e. to effect it continually or at stated times, as what is supernatural or miraculous does to effect it for once [italics in the original].

This quote is very important. It recognizes that the human distinction between "miraculous" and "natural" is a relative one, one which seems obvious to us only because we do not fully understand certain things—such as the generation of life—which happen constantly and all around us. As early as 1860, as the second edition of the *Origin* was hot off the press, one of Darwin's regular American correspondents, the botanist Asa Gray, did not fail to observe this addition: "We notice with pleasure the insertion [into the second edition] of an additional motto on the reverse of the title page, directly claiming the theistic view which we have vindicated for the doctrine."[3] Gray's "theistic view" is the idea that natural selection is the means (or cause) by which God (the agent) has effected evolution. Darwin repeated this sentiment in the second through sixth editions:

> I see no good reason why the views given in this volume should shock the religious feelings of anyone. ... A celebrated author and divine [meaning another of Darwin's correspondents, Rev. Charles Kingsley] has written to me that "he has gradually learnt to see that it is just as noble a conception of the Deity to believe that He created a few original forms capable of self-development into other and needful forms, as to believe that He required a fresh act of creation to supply the voids caused by the action of His laws."[4]

Whether or not Darwin himself actually believed in supernatural agency is irrelevant to this point. Furthermore, it doesn't matter at all if you personally believe that there is a God-like agency behind biological diversity. The point is that Darwin's mechanism does not concern the subject of who did it, or why, and that Darwin recognized that his mechanism could not rule out a creator.[5] Rather, however life may have first appeared, he outlined a mechanism that humans can observe and understand. Once started, it allowed life to unfold into the diversity we see today. Whatever his personal beliefs may have been, based on his writings in the *Origin*, Darwin was a "theistic" evolutionist, i.e., one who permitted a divine agency behind the mechanism of biological evolution.[6]

To be sure, his process differs from a naïve interpretation of religious creation stories because natural selection is not a process of a human-like "god" tinkering with organisms as if they were organic Barbie dolls with lots of different outfits, each requiring manual (un)buttoning. But agency is most certainly not ruled out. Whatever the origins of his mechanism, Darwin identified a cause by which species evolve.

Philosophers and theologians have known for ages that scientific and theological explanations of the natural world do not have to be fundamentally at odds with one another. Based in part on the writings of St. Thomas Aquinas in the thirteenth century,[7] and Aristotle centuries before, the Roman Catholic Church has long recognized the distinction between two levels of causation—primary and secondary—which have parallels to what I've called agency and cause.[8] Augustine of Hippo (AD 354–430), a north

African citizen of Rome, is another major historical figure in Christian theology. He famously exhorted his fellow Christians not to force literal belief in the creation story of Genesis at the expense of rational thought.[9] The Catholic Church has followed in this tradition and (beginning with Pope Pius XII in 1950) has recognized that evolution is not fundamentally at odds with Christian faith. Its positive view on the compatibility of evolution and Christianity was made more explicit by John Paul II in 1996, again by Cardinal Ratzinger (now Pope Benedict) in 2004, and most recently in a March 2009 conference at the Vatican.[10] These statements have their roots in the works of philosophers and Christian theologians who lived centuries ago.[11] For example, consider chapters 69 and 70 from *Summa Contra Gentiles*, written around 1260 by Thomas Aquinas:

> [Some] men have taken the opportunity to fall into error, thinking that no creature has an active role in the production of natural effects. So, for instance, fire does not give heat, but God causes heat in the presence of fire, and they said like things about all other natural effects. (book 3, chapter 69)
>
> It seems difficult for some people to understand how natural effects are attributed to God and to a natural agent. ... So, if the action whereby a natural effect is produced proceeds from a natural body, it does not proceed from God. ... However, these points present no difficulty ... In every agent, in fact, there are two things to consider: namely, the thing itself that acts, and the power by which it acts. ... [The] power of a lower agent depends on the power of the superior agent, according as the superior agent gives this power to the lower agent whereby it may act ... as the artisan applies an instrument to its proper effect, though he neither gives the form whereby the instrument works, nor preserves it, but simply gives it motion. So, it is necessary for the action of a lower agent to result not only from the agent by its own power, but also from the power of all higher agents; it acts, [therefore], through the power of all. (book 3, chapter 70)

Creationists and some atheistic biologists have not carefully read Aquinas. They conflate agency and cause by thinking that our understanding of evolution's cause excludes an agency behind it. In fact, as a way of explaining things, agency and cause do not necessarily exclude or compete with one another.[12] To argue that they do in the case of evolution would be just as ridiculous as saying that steam-powered rotation of a turbine cannot be the mechanism behind riverboat thrust, because I know Savery and Watt did it, or to say that since a steam-powered turbine is involved, these Savery and Watt people are only a myth. It is possible to be completely oblivious to the agency (or cause) behind the steam engine, yet know quite well how it works (or who developed it). As I've just observed, ancient theologians, Darwin, and many subsequent authors have recognized this distinction.

Others have not, including many participants in the creation versus evolution debate. Consider this fairly typical passage from a 1999 book on theism and evolution:

> Before Darwin, theists could point to natural objects like the eye and then challenge their philosophically inclined critics to provide a better explanation than theism. Darwin provided a purely naturalistic account for apparent design in the natural world. ... In the *Origin of Species* he challenges his critics, "It is so easy to hide our ignorance under such expressions as the 'plan of creation' or 'unity of design,' etc., and to think that we give an explanation when we only restate a fact." Darwin would have none of that kind of "sloppy thinking." Instead, he proposed a mechanism—natural selection—that would do the work of providing for the patterns in nature that others had only passively described. The results of the debate over design in nature for theism were very great. In the words of ... Richard Dawkins, "Darwin made it possible to be an intellectually fulfilled atheist." ... God is, at best, unemployed in the new cosmology.[13]

Despite citing the *Origin* where Darwin explains how "design" does not provide a cause for biodiversity but only asserts an agency behind it, these authors imply (noting Richard Dawkins' agreement) that natural selection has replaced theism, leaving God "unemployed." But this does not follow: neither atheists nor fundamentalists know the extent to which God as an agent is "employed" or otherwise occupied with a mechanism such as natural selection. Their assertion that he is not is similar to denying the agency of Thomas Savery due to the cause of steam driving a turbine, or, to draw on Aquinas' example, that an artisan is not responsible for nails in a table because it was the hammer that delivered the force. Unfortunately, other examples in which authors have conflated agency and cause are not hard to find.

THE MOST GIFTED BIOLOGY WRITER OF OUR TIME

Stephen Jay Gould died on May 20, 2002. I am one of many biologists who, by reading his engaging prose in books such as *Ontogeny and Phylogeny*, *Mismeasure of Man*, and *Wonderful Life*, was drawn into this field in which he played such a major role.

I found Gould's review[14] of Phillip Johnson's book, *Darwin on Trial*,[15] very entertaining, much in the same way that spectators gawk at blood spilled at a prize fight. Gould had no patience for the misrepresentations and half-truths that filled Johnson's book, and was downright mean in his portrayal of the author himself. Johnson sought to paint evolution as a "theory in crisis," repeating claims of alleged scientific and moral inadequacy that have been rebutted over and over again since 1859. Gould saw through this verbiage and critiqued Johnson's book as a misinformed, quasi-political document, dependent on the reader's ignorance of biology and the history of science to have any effect. Like the Roman public who witnessed Christians thrown before a hungry lion, I enjoyed Gould's shredding of Johnson's book. This lion's name was Steve Gould, and I was cheering him on.

Creationists and atheists alike may read this and come away with the conclusion that by sharing Gould's disdain for Johnson's book, I am an atheist, hostile to religion itself. Such a portrayal attracts some level of blood-lust, identifying me as one of "us" or "them" depending on your orientation as a reader. However, if our goal here is to understand why two academics are shouting at each other, this reaction is not really productive. The disagreement between Gould and Johnson, as evident in Gould's[16] review of *Darwin on Trial* and in Johnson's reply to this critique,[17] is one minor footnote in a long and rancorous debate on the boundaries between science and religion in Western society.

To make clear my reaction to this debate as a religious paleontologist, I'd like to briefly restate its content. The author of *Darwin on Trial*, Phillip E. Johnson, is widely regarded as a founding member of what is now called "intelligent design" (or ID). This idea is derived from some of the same protagonists who had previously advocated creationism,[18] connected in one form or another to the religious, anti-evolutionist movement whose public form has tracked decisions made by the US Supreme Court in 1968 and 1987.[19]

Despite its religious roots, ID asks questions of "design" in nature without an overt appeal to Christianity or other major religions. Descriptions of ID, for example in publications by Michael Behe,[20] state that ID proposes to search for evidence that life was designed, much in the same way that the Search for Extraterrestrial Intelligence (SETI) program searches for signs of extraterrestrial life. SETI researchers sift through vast quantities of cosmic radio transmissions, searching for a pattern that might identify an intelligent source. It is assumed that natural, non-intelligent sources of radio waves, such as a pulsar, would lack patterns, like the series of prime numbers that Jodie Foster, in her role as a data-hungry scientist, dramatically deciphered in the 1997 movie *Contact*. The SETI researcher played by Jodie Foster was not looking for anything supernatural; most ID advocates claim they're not either.

We'll discuss the extent to which ID is about a supernatural intelligence in Chapter 2. For now, it's worth noting that scientists would deny from the outset that a supernatural force could figure at all in their list of possible explanations for something biological. Given the fact that evolutionary biologists are looking to explain the mechanisms behind biodiversity, *how* it has come to exist, this is a legitimate restriction. This is not to say that reasonable scientists object to discussion of agency or purpose in biology, only that doing so is peripheral to understanding the mechanisms of evolution. Nor does it mean that a natural agency is beyond the scope of evolutionary science; paleontologists consider the products of natural agents all the time (e.g., animal trackways, bite marks of predators, or stone tools). However, this restriction does mean that theories proposing to explain natural phenomena should be grounded in the natural realm; any theory about nature should be subject to testing based on the collection of data from nature.

Of course, it's possible that some kind of extraterrestrial being, perhaps supernatural, seeded the Earth with life and its complex accouterments. Because this possibility exists, say advocates of ID, it should be a legitimate target of scientific investigation. They propose to do this by searching for patterns of complexity in the make-up of life—for example, among natural "machines" of varying scales from a protein-transport biomolecule to the *Tyrannosaurus rex* locomotor apparatus. Their search is for units of "irreducible complexity" among these machines, units that could not have been assembled by the randomness they attribute to Darwinian natural selection, just as a pulsar would probably not emit a series of radio signals in units of prime numbers (i.e., 3, 5, 7, 11, 13, 17, etc.).

Setting aside for a moment the fact that the randomness of galactic radio waves is not at all comparable to the essentially *non*-random process of natural selection,[21] this kind of search for agency in nature is what I mean when I say "intelligent design." Taken at face value from its proponents, ID is not specifically about the Judeo-Christian story of Creation. ID is about the search for human-like "intelligence" in the origins of biological diversity. What IDers do not emphasize is the distinction between agency and cause, and the fact that their endeavor is entirely concerned with the former.

Some in the Darwinist camp have also neglected the importance of distinguishing between agency and cause. Returning to Gould's critique of Johnson's *Darwin on Trial*, in that review he contradicts some of his own statements made elsewhere in which he advocates the incompatibility of Darwinism and religion. For example, in a 1977 collection of his essays, Gould famously stated that "Darwin ... was vindicated in his cardinal contention: Cambrian life did arise from organic antecedents, not from the hand of God."[22] A similar quote was part of a February 1982 article in *Discover* magazine (although in fairness its parenthetical allows for a sort of divine beginning): "No intervening spirit watches lovingly over the affairs of nature (though Newton's clock-winding god might have set up the machinery at the beginning of time and then let it run). No vital forces propel evolutionary change. And whatever we think of God, his existence is not manifest in the products of nature."[23]

Anti-evolutionists are attracted to this corner of Gould's technicolor banner, as demonstrated in a 1992 article in *Creation* magazine: "I'm glad [Gould] and I are on the same side about one thing at least—the real meaning of 'Darwin's revolution'. And we both agree that ... Darwin's theory is inherently anti-plan, anti-purpose, anti-meaning."[24] And it's not just a few articles from Gould that they can quote. Statements to this effect in numerous publications from venerable and important evolutionary biologists such as George Simpson, Ernst Mayr, and E.O. Wilson, not to mention entire books by Richard Dawkins, have done much to shape the anti-evolution philosophy of commentators such as Phillip Johnson.

In contrast, in his review of *Darwin on Trial*, Gould notes the independence of religious belief and scientific inquiry: "If some of our crowd have made untoward statements claiming that Darwinism disproves God, then I will find [my third-grade teacher] and have their knuckles rapped for it. ... Science can work only with naturalistic explanations; it can neither affirm nor deny other types of actors (like God) in other spheres (the moral realm, for example)."[25] Of course, this is not the spirit of Gould's writing that Johnson or other creationists quote when discussing his view on Darwinism and religion. In contrast to what Darwin himself wrote, characterizations of "evolution" made from this camp depend on those writings from atheistic evolutionists that ignore the distinction between agency and cause, and that appear to lend a scientific basis to their philosophical denial of any kind of agency or purpose behind the mechanism of evolution.

In my opinion, George Simpson (1902–84) was the most brilliant paleontologist of all time. Nevertheless, his statement that "man is the result of a purposeless and natural process that did not have him in mind"[26] was his own personal formulation of a premise that does not unambiguously follow from a very large body of data which Simpson otherwise interpreted with great clarity. Simpson was a genius, but even he cannot claim to have identified, or negated, the "purpose" or "mind" of nature. Elsewhere in his writing, he seemed to acknowledge this very point: "Evolution, per se, is not antireligious any more than the roundness of the earth is antireligious, although it was once held to be so. ... There are also atheistic evolutionists, but so are there atheistic bankers, who nevertheless keep honest accounts."[27]

Instances of inconsistency such as these form the raw material with which anti-evolutionists make their caricature of evolutionary biology, stating

> God as a remote First Cause remains a possibility, but God as an active creator is absolutely ruled out by the blind watchmaker thesis.[28]... Darwinism and theism can easily be reconciled by those who, like Asa Gray and Charles D. Walcott, misunderstood Darwinian evolution

> as a benevolent process divinely ordained for the purpose of creating humans. ... On the other hand, Darwinism does give atheists and agnostics a decisive advantage to the extent that belief in God's existence is a matter of logic and evidence. Those who really understand Darwinism, but still have spiritual inclinations, have the option of making a religion out of evolution. Theodisius [*sic*] Dobzhansky—Gould's prime example of a Christian evolutionist ... discarded the traditional Christian concept of God ... and worshipped the glorious future of evolution.[29]

According to Phillip Johnson, who in this passage was undoubtedly egged on by the anti-theological rhetoric of atheistic biologists,[30] acceptance of Darwinism coupled with "spiritual inclinations" yields worship of evolution itself. Clearly, Johnson and many other anti-evolutionists[31] do not believe that it is possible to be Christian without the rejection of Darwinian evolution.

This opinion results not only from the confusion of agency and cause, but also from the naïve expectation that the divine intelligence behind "creation" of biological life has to resemble human intelligence. Ironically, anti-evolutionists like Johnson show much in common with the atheist perspective that "randomness" in evolution is opposed to what Johnson calls "design."[32] Allow me to parody this perspective: a venerable, omnipotent deity, who would physically resemble an ageless Charlton Heston and wear flowing white robes that match his beard, constructed biological diversity by a mechanism that has never been specified by any creationist or ID advocate, but which was certainly *not* differential survival of gene frequencies and their resultant phenotypes following recombination, mutation, or other such processes. Rather, each act of creation (and given what we know about the fossil record, there were *lots*), would leave behind unmistakable signs of His intelligence that look, well, like they've been designed—and again no creationist or IDer has ever said *how*. So what do things that have been "designed" look like? Well, like things that we humans design—like watches or mouse traps.

But isn't this just a little bit presumptuous? Why can't the erosional happenstance that carved out the Grand Canyon be regarded as divine design? Couldn't the products of "design" result from an intelligence that is not quite like our own, to the point that the process behind them might seem "random" to us? What Phillip Johnson actually means when he says that the god of a theistic biologist cannot be an "active creator" is that this deity cannot be human-like in Her/His/Its activity. Pardon me if I'm a bit underwhelmed.

You may have heard of Reverend William Paley's fictional watch, found on a heath in his famous 1802 book, *Natural Theology*, in which the essential concepts of the current ID movement were articulated. This watch was clearly the product of an intelligence, said Paley, unlike a stone lying nearby. Paley claimed that biological complexity such as that present in an eye is no less deserving than the watch of the design inference. In a very insightful critique of Paley's argument, the philosopher Elliot Sober points out that Paley's inference of a designer for the eye, and rejection of one for the stone, are inconsistent. Both require initial assumptions about the nature of an alleged designer. This is fine for a watch, something we've seen many times and understand entirely to be within the creative repertoire of humanity. But what about an eye or a stone?

> If Paley gets to help himself to assumptions about the goals and abilities of the putative designer that are favorable to the design hypothesis in the case of the eye, why should he abstain from doing so in the case of the stone? ... The design argument has no more basis for claiming that design is the better supported hypothesis in the case of the eye than it has for saying that chance is the better supported hypothesis in the case of the stone.[33]

In other words, terms such as "designed" and "random" are dependent on our peculiar, human understanding of creative expression. Sober rightfully points out that they are not intrinsic features of objects such as eyes or stones themselves, but require initial assumptions about a particular relationship between randomness and design to a hypothetical creator. At the same time, an objection to Paley's inference also contradicts assertions of the lack of design: "Paley is no more entitled to adopt these favorable assumptions [about design] than [Stephen Jay] Gould is entitled to adopt his unfavorable assumptions."[34]

To cite an unusually morbid example of randomness from a statistics textbook,[35] biometricians Sokal and Rohlf compare the "random and rare" nature of a Poisson distribution to the frequency by which nineteenth-century Prussian soldiers were kicked to death by their horses. Such an event qualifies as random, unlike the regular ticks that mark off the seconds on your wristwatch. The one event by most accounts is a freak accident (although having fled Nazi-dominated central Europe in the 1930s, Sokal's choice of a database consisting of violent deaths among Germans probably wasn't random), whereas the other is meant to precisely measure time—nothing random about it. Yet nodding our human heads in agreement about this difference is one thing; attributing our assumptions about purpose and randomness to the actions of a potential cosmic Creator is completely different. The equations "purpose = God" and "randomness = atheism" are contingent upon some preconceived notion of order, and we humans have no choice but to understand "order" from our own desperately parochial perspective. The fact that we find it difficult to appreciate the creative power of apparent randomness does not mean that God suffers from this problem too. This very simple observation was made by Darwin in all six editions of the *Origin*: "We naturally infer that [complex organs have] been formed by a somewhat analogous process [to human design]. But may not this inference be presumptuous? Have we any right to assume that the Creator works by intellectual powers like those of man?"[36]

Those readers familiar with Gould's review of *Darwin on Trial* might note that I left out a sentence from one of his above quotes.[37] Referring to the knuckle-rapping punishment meted out by his third-grade teacher, Gould adds parenthetically (where I have left only an ellipsis) "as long as she can equally treat those members of our crowd who have argued that Darwinism must be God's method of action." I fall in this camp, and I daresay that in his book *Rocks of Ages* Gould does too. However, neither I nor Gould in his sensitive moments deserve any such knuckle-rapping treatment. Recognizing that the "intelligence" of a creator applied to phenomena such as the origins of life or the cosmos—issues wholly beyond the scope of evolutionary biology—might work through techniques that resemble randomness to a human strikes me as perfectly reasonable, even if it is not amenable to rational (dis)proof. In contrast, requiring that God's style of invention has to resemble our own, as do many creationists and atheists alike, seems extraordinarily presumptuous and vain.

METHODOLOGICAL NATURALISM

Proponents of intelligent design are fond of arguing that, to the detriment of high-school kids everywhere, the capacity to recognize design in biology has been artificially removed from the toolkit of scientific explanation due largely to the intolerance of leftist academics. This is the controversial proposition they want taught in public schools, circumventing the professional scrutiny that scientific ideas must face prior to becoming the stuff of high-school exam questions. Design in biology does not necessarily mean God, they say; it means only something (some *agency*) that derives from an intelligence. ID proponents say this with an air of persecution; some of the brave few who dared to mention the possibility of a Designer in biology have apparently lost their jobs.[38]

While some employment decisions probably have been influenced by an anti-religious bias,[39] this does not diminish the importance of distinguishing between agency versus cause in evolutionary biology. If we ask the question of how life has proceeded over time to diversify into the species we see today, "it was designed" isn't an answer but a repetition of the question.[40] Even if life were designed by a human-like intelligence, the question "how" still stands. Hence, is design a legitimate part of the explanatory toolkit for someone interested in scientifically explaining the cause of biodiversity on Earth over time?

No. Let's imagine that scientists everywhere accepted the idea that aliens with human-like intelligence seeded Earth with life, and have on occasion guided the "machinery" that enables modern biodiversity. We have now agreed on an agency. However, the acceptance of this agency does not change one iota the more profound question: how did they do it? They designed it. Right—we said that already, and never mind that we don't know who designed them. As far as Earth's organisms are concerned, aliens are the agency, but how did they cause our planet's biological diversity? What are the biological mechanisms by which organisms alive today have become so varied? There are several important qualifications, but essentially we have an answer: descent with modification. Speculation about the aliens doesn't change this.

"Methodological naturalism" is a rule of science that says one should not use supernatural phenomena to explain causation in the natural world.[41] This is what I've outlined above: science is all about the *how* behind nature, not the *who* or *why*. One of the first to articulate this phrase as it applies to scientific inquiry was a Christian theologian, Paul DeVries.[42] Over the past few centuries, many scientist-theologians have outlined and endorsed concepts of separating the natural and divine in the search for causation,[43] and would agree that supernatural forces are no more an option to explain the process behind human evolution than the virulence of the 1918 flu epidemic, or the current pace of climate change. This is not to rule out the participation of those agents in one or both phenomena. To be perfectly honest, no one can really know for sure; maybe there was some kind of supernatural agency behind the "Spanish flu," which killed more humans over the course of two years than, probably, both World Wars put together. However, whether or not there was such a supernatural force does not affect our capacity to understand how viruses foil immune systems and replicate to the detriment of their hosts.

Analogously, whether or not the Mafia, Fidel Castro, or the Freemasons played a role in the assassination of John F. Kennedy does not change the lethality of the bullet that ripped through his head. Methodological naturalism can help us comprehend the pathology of a gunshot wound no matter who pulled the trigger. In short, methodological naturalism is the best means we have to understand and deal with natural phenomena productively and is the backbone of science itself.[44]

Now the skeptics among you may want to proceed a step further: of course we can rule out a role for God and spirits in the 1918 flu epidemic. Those are fairy tales, inventions of superstitious people who don't know any better. This is the view of philosophical naturalism, which regards the scientist's assumption of dealing only with observable nature (and rejecting the supernatural) in explaining the *how* of life as not just a methodology, but a worldview. Philosophical naturalists do not accept "truth" beyond the empirically demonstrable. Hence, my inner conviction that the Judeo-Christian God is somehow behind the Big Bang, or that this God represents the "Creator" behind the breath of life—spelled with a capital "C" in the second through sixth editions of *On the Origin of Species*[45]—or (on a much more banal level) my conviction that the Sabres are the most worthy team in professional hockey, all such feelings amount to intuition, personal taste, and do not deserve the descriptor "truth." Philosophical naturalism says that there is a good reason why scientists assume that what they cannot rationally perceive plays no role in nature: what they cannot rationally perceive does not exist.

That is a very big step, one that separates religious from atheist scientists. Both practice methodological naturalism, but atheists continue to apply the acid of skepticism beyond the realm of what is rationally perceptible. The irony is that reason itself suggests that our own five senses—whether they are stimulated directly or indirectly—do not provide an adequate register for the potential activity of all phenomena. In other words, because we are aware of the extent to which we are limited in our capacity for perception—for example, our limited grasp of audiovisual spectra and our experience of time as linear—it is reasonable to expect that there are, or have been, factors influencing our existence that we cannot perceive. We've already discussed why methodological naturalism is critical to understanding *how* biology works. However, it is not justified to extend this essential scientific practice to questions that are not particularly scientific, i.e., those that concern issues beyond the mechanics of nature.

It would be possible using methodological naturalism to rule out many kinds of agency, such as that of a human-like intelligence behind certain phenomena. Were German Freemasons responsible for unleashing the 1918 flu? Of course not; a moment's reflection combined with a few facts—they suffered from it too, they lacked the technology to manufacture such a thing, and credible witnesses who were in a position to see such a conspiracy have never attested to it—reasonably demonstrates that this was not the case. We might also rule out the agency of more distant protagonists, such as human-like aliens, due to their alleged activity in the same natural sphere as you, I, and the Freemasons. In the same way, say the philosophical naturalists, we can rule out supernatural agency behind pretty much anything.

But wait: we know that German Freemasons existed; we know that they worked, slept, ate, and peed just like anyone else. Their involvement in any given historical event would be subject to the same rules as your or my involvement, or that of any other fundamentally natural protagonist, potentially observable by our five senses. This is not true of a supernatural agent. The act of ruling out human-like agency is not the same as that of a supernatural being,[46] and derives from the unwarranted assumption that the actions of a supernatural agent manifest themselves in nature in the same way as those of a human agent. Passing judgment on an agent potentially beyond human reason, which is a fair characterization of a deity evidently capable of sparking time and the universe into being, is completely different from rationally deducing the irrelevance of a natural agent to specific historical events, such as Freemasonry to the 1918 flu. In order to make a scientific claim about the existence, or not, of a supernatural agency, it is necessary to assume that it resembles human agency. Both intelligent design and philosophical naturalism do this without any justification that I can discern.

The line between methodological and philosophical naturalism can be subtle and is frequently crossed, often by those who conflate agency and cause. This is one of the issues that makes the current debate on evolution versus religion so frustrating. Proponents of ID want to identify agency in biological diversity; evolutionary biologists want to identify the cause or mechanism by which diversity arose. Regardless of your opinion on the scientific status of the search for agency, it cannot replace or preclude the search for natural cause. Conversely, scientifically resolving the specifics of natural selection does not address the who or why behind the biological diversity on our planet.[47] On this basis, I agree with the Christian philosopher Alvin Plantinga, who, in a 2006 lecture, made a similar distinction between science and methodological naturalism on the one hand, and scientific secularism or philosophical naturalism on the other:

> The confusion of science [methodological naturalism] with scientific secularism [philosophical naturalism] is egregious. It is little better than confusing ... stamp collecting with the claim that [stamp collecting] is enough. But I believe this confusion, colossal as it is, is widely perpetrated, and by people from both sides of the divide between science and religion. There are many who enthusiastically endorse science, but they go on to confuse it with scientific secularism. Perhaps this is because they see ... methodological naturalism as essential to science, but then confuse it with [scientific] secularism. ... Others who emphatically reject secularism fall into the same confusion. They are suspicious, distrustful of science because of its association with scientific secularism, or the so-called scientific world-view. But the fact that science is associated with secularism, the fact that some people associate the two, is not a decent reason for suspicion of science. It is no better than being suspicious of music history just because someone thinks it's enough. This confusion I believe is one factor underlying the continuing mutual distrust between science and religion.[48]

I also agree with Stephen Jay Gould's opinion on the subject as published in his 1999 book, *Rocks of Ages*. Correctly understood, evolution (and more broadly, science) and religion occupy "Non-Overlapping Magisteria" (often referred to as "NOMA"), and are basically compatible with one another in the sense that they deal with fundamentally different questions.

THE COMPATIBILITY DELUSION

If you've read any of the recent literature that advocates or implies a scientific basis for atheism,[49] or indeed some of the creationist literature,[50] you'll rightfully have the impression that my endorsement of Gould's "NOMA" depends on a somewhat idealistic conception of these two magisteria, religion and science. It depends at least in part on setting their limits rather differently than the most enthusiastic practitioners of each would accept.[51] For episcopalians and economists, NOMA is fine, but for pentacostalists and paleontologists, no way. "Faith," says the philosophical materialist, makes a virtue out of a scientific cardinal sin: belief in something for which you have little, no, or even contrary evidence. Those who profess to uphold the spirit of scientific rationality are not at liberty to accept any religious dogma or creed simply because they seek to fill their own spiritual void. At least some beliefs of today's many religions have to be recognized as what they are: irrational, destructive, or worse.

I have previously noted progressive strides taken by the Catholic Church in recognizing compatibility between religion and science,[52] such as the opinion on evolution issued by Pope John Paul II in 1996.[53] That's good. Now consider John Paul II's opinion from 1988 that "no personal or social circumstance"[54] could ever justify the use of condoms. Consider also how multiple Roman pontiffs have toured the world, telling their 1.1 billion followers (such as husbands with HIV and wives without) that latex contraception is a sinful practice[55]—one that is not only medically safe, but positively life-saving as condoms demonstrably reduce rates of HIV transmission,[56] not to mention their eventual effect of reducing human-induced environmental degradation by helping to relax population pressure. Combined with evidence that religious, abstinence-only sexual education programs are ineffective in actually promoting abstinence,[57] the idea that we should attribute divine insight into human sexuality to elderly men sworn to celibacy (or "ghastly old virgins" as Christopher Hitchens once put it) seems like a good candidate for religious irrationality.

There are many other candidates. Some religious institutions have presided with great conservatism over centuries of ignorance, intolerance, and belligerence, all with the interests of their own, very human power-brokers in mind. Flagrant examples of greed, hate, or simple absurdity justified by one or another religious doctrine abound: the piracy and enslavement of Atlantic seafarers by the Islamic Caliphate of North Africa (inspiring "... the shores of Tripoli" in the lyrics of the familiar but seldom appreciated *Marine Hymn*), suspicion of foreign ideas (heliocentrism) and persecution of those who dare to defend them (Galileo), open war against "infidels" in Toledo, Kfar Etzion, Belfast, Srebrenica, the Gaza Strip, Hebron, and New York City. Listing such events committed in the name of someone's "god" has recently become the focus of a lot of press,[58] and rightfully so. In this book I will not apologize for the wicked and well-documented legacy of humanity's monotheistic elite, or their enduring malfeasance as perpetrated by religious groups of various stripes, such as the Taliban, the Jewish National Front, or the Moral Majority. The claim that Mullah Mohammed Omar, Rabbi Meir Kahane, and Reverend Jerry Falwell share an ideology may shock some readers; however, in terms of claiming very specific, divine insight into the minutiae of human behavior, and in terms of their willingness to pass severe judgment on others based on this claim, they're quite similar.

Let us start with the Islamic fundamentalists, who promote illiteracy among women and value their legal testimony as that of, at best, "half a man."[59] This is irrational. Condoning the "marriage" of underage girls to adult men is also irrational, as a Saudi court recently did when they rejected a petition by a traumatized mother to annul her 8-year-old daughter's marriage to a 47-year-old man.[60] Claiming propriety over the state of Israel[61] based on millennia-old "documentation" is irrational, along with the belief that religious homogeneity of the occupied West Bank and Jerusalem is necessary to fulfill biblical prophecy,[62] to hell (literally) with the non-Jews who have lived there for generations. Attributing blame for September 11 terrorism perpetrated by Islamic zealots to a Divine Will enraged by homosexual behavior, feminism, and the separation of church and state is irrational, even if that attribution was half-heartedly retracted at a later date.[63] As previously mentioned, telling tens of millions of impoverished slum-dwellers to reject cheap, simple AIDS protection and birth control is irrational. Religious rhetoric applied to contemporary politics is particularly irrational among those who deny climate change and even human population growth.[64]

While many of the above religious claims about the world are fatuous and wrong, they do not have much to do with arguments regarding the existence of a Creator. Here is where a distinction can be made between deism, or belief in God, as a first-cause or ultimate agency, and theism, or belief in God as not only a first-cause but also as active participant in the minutiae of human thought and action. Like Phillip Johnson's caricature of an anthropomorphic "active God" referred to previously,[65] this distinction relies heavily on the naïve attribution of human limitations in the Creator's mode of activity, and the ways in which such activity is evident to us. This makes the distinction between deism and theism rather artificial, but because the terms appear frequently in discussions of religion and science, they're worth mentioning here.

Both deists and theists are moved by what Immanuel Kant[66] described as the starry sky without, i.e., the immensity of the cosmos and its apparent beginning from nothing, and the moral law within, i.e., the fact that across cultures, humans discern right and wrong in a consistent way. The deist attributes such universals to a God who set the universe in motion, but does not have any human-like interest in micromanaging life. The theist, in contrast, goes much further by attributing to God not only the initial spark of life, but also specific human qualities, such as compassion, patience, worry, and vengeance. More importantly, many theists claim the ability to interpret these divine emotions for their fellow humans in great detail. Everything that I labeled fatuous in the previous paragraph derives from theistic beliefs: from virgins (or was it raisins?[67]) awaiting the martyrs of Islam, to a fast-tracked Messianic arrival as more Jews settle the West Bank, to the Pope's disapproval of condoms and birth control. Such specific, theistic beliefs can and do conflict with scientific rationality, but deistic beliefs do not.

Rational science (such as evolutionary biology) is quite compatible with deism. To the extent that one recognizes that theistic, divine action does not have to resemble human action, i.e., that it could manifest itself in our existence via naturalistic mechanisms, rational science would not be able to rule out such theism. A similar kind of observation has been made by many theologians, biologists, and philosophers. For example, the eminent biochemist (and theistic Christian) Denis Alexander notes that "the ID literature gives the impression that there is something inherently 'naturalistic' about certain aspects of the created order and not about other aspects, and such thinking appears to stem from a very inadequate doctrine of creation."[68] This perspective was also clear to nineteenth-century scientists:

> Neither Darwin nor Lyell (nor numerous other classic representatives of the modern view of the world) saw the world as a closed nexus of purely natural causes that totally excluded the hand of God. Intelligent design creationism, then, misrepresents the modernist view of the world as it was known to Darwin and his contemporaries—theorists for whom the initial creation and intelligent design of the world was not in doubt, but who did doubt the existence of naturalistically inexplicable signs of redesign and/or supernatural intervention in the world.[69]

When theistic beliefs slide into superstition, for example by requiring that God's activity be "naturalistically inexplicable," they become incompatible with evolutionary biology or any other rational, data-oriented science. While I am eager in this book to point out that religion and science can be compatible, I do not deny that many contemporary practitioners of the world's big three monotheisms do subscribe to scientifically irrational beliefs. In other words, many Christians, Jews, and Muslims are superstitious.

THEISTIC CHRISTIANITY AND EVOLUTION

In order to make sense of Judeo-Christian scripture, some level of interpretation is necessary. For example, without completely abandoning your reason, it is impossible to believe the distinct creation stories of Genesis chapters 1 and 2 in a modern, literal sense, since they provide contradictory sequences of events. In chapter 1, male and female humans are created at the same time, after other animals; in chapter 2, man precedes non-human animals and woman is created last. Within chapter 1, the sequence of events doesn't make sense. The creation of vegetation (day three) took place prior to the existence of the sun (day four). There was some kind of light since the very first verse—the presence and absence of which God called "day" and "night" (respectively)—but this preceded the existence of the sun, moon, and stars (day four).

If we look again to the writings of Denis Alexander,[70] he notes that the "New International Version" English translation of Genesis 2 departs substantially from the original Hebrew by using the pluperfect tense: "Now the Lord God *had formed* out of the ground all of the beasts of the field..."[71] Alexander suggests this was an attempt by the English translators to make interpretive room for its temporal compatibility with Genesis 1. In fact, the verbs of the original Hebrew relevant to Creation are identical in both. There is no pluperfect in that language and the order of creation in each case is simply different, as accurately reflected in the King James English Bible.

Here is another interesting passage with an element that cannot literally have been true:[72] chapter seven of 1 Kings is a detailed description of Solomon's opulent Palace of the Forest of Lebanon, "the length thereof was 100 cubits, and the breadth thereof 50 cubits." At over a square kilometer, this was a substantial bit of real estate. In charge of the interior decorating (at least the metalwork) was Hiram of Tyre, who fashioned for Solomon a "sea of cast metal" as detailed in verse 23: "circular in shape, measuring ten cubits from rim to rim ... it took a line of thirty cubits to measure around it." My New International English translation adds a couple footnotes, indicating that 10 cubits are equivalent to 15 feet or 4.5 meters and that 30 cubits are roughly 45 feet or 13.5 meters. In the King James translation, this structure is "round all about" with the same dimensions.

Now here's the mistake: because the circumference of any circle has to be the product of its diameter and the constant pi (about 3.141592), no circle can have a diameter of 10 and circumference of 30. Either the measurements are off, or it's not a circle. At 10 cubits, the circumference must have been about 31.4 cubits, not 30. Okay, fine; so the dimensions in 1 Kings were approximate—30 is pretty close to 31.41592—or maybe Solomon's bronzed jacuzzi wasn't perfectly circular. And that's exactly the point. Some interpretation is necessary here because of what we know about the laws of nature (mathematics, in this case), and how they contradict the existence of a circle that measures 10×30. Now does this passage present a real theological problem? No, because (with a few prominent exceptions, like creationists) biblical scholars and laypeople alike *interpret* the text; they don't believe every word literally. If they did, you might find yet more warning stickers in public school textbooks: "Geometry is *just a theory*."

Elsewhere in the Bible, a "firmament" separated water in the sky from that in the seas below, and provided a surface to which stars could be attached.[73] Moses parted the Red Sea.[74] The sun stood still to aid Joshua in battle.[75] Literal belief in any of this runs against the laws of nature as we know them today, such as those that dictate the geometry of circles. This is one reason why most scholars and laypeople in the Judeo-Christian tradition recognize that the authors of the Bible wrote metaphorically, rather than historically. In addition, scholars have pointed out that the original Hebrew of Genesis contains meter and rhyme,[76] not translatable into English, further evidence to regard its content as poetic rather than literal. Since the time of St. Augustine in the sunset of the Roman Empire, most Christians have recognized these apparent inconsistencies as creative license of nonetheless divinely inspired authors.

Consider now the core of the Christian tradition in which the earthly incarnation of God, personified in Jesus Christ, is said to have been born of a virgin, have performed miracles including both production (bread and fish) and transformation (water and wine) of sustenance, have raised others from the dead, and to have been resurrected from the dead himself after his own crucifixion. Do you believe that Christ's mother was a virgin? This is a surprisingly tough question for many modern Christians.[77] More generally, don't all of the major religions demand one or another parcel of belief that flies in the face of the rational basis of science?

When religion demands that we believe in something apparently irrational, and our current understanding of biology places virgin birth in this category, the boundaries of Gould's Non-Overlapping MAgesteria (or "NOMA") or have been transgressed. Hence, Christianity, as an article of faith, appears to violate Gould's principle of harmonious co-existence with science. The domain of asking "who" and "why" (agency, religion) has interfered with the domain of asking "how" (cause, science), since the biological basis for sexual reproduction generally demands that both egg and sperm participate in the conception of human life, and Christianity denies this in one particularly famous case. So how does the Christian get out of this?

Even if we agree that a given religion, say Christianity, has violated NOMA, it is still true that science generally (and evolution in particular) is about *how* and not *who* or *why*. Understanding the meaning behind the first "breath" of life is beyond the scope of evolutionary science, which simply does not concern supernatural agency or purpose. While scientists can and do ask about the conditions under which life began, even if this is no longer within the realm of evolutionary biology but more relevant to chemistry and cosmology, science remains decidedly silent in attributing meaning to life's origins, and does not take up the teleological slack if we are to abandon any given religion.

In addition, the question "can a scientist be religious?" is an empirical one, much like another question answered tongue-in-cheek by the theologian N.T. Wright, who reported the response of a bishop when asked if he believed in infant baptism: "yes, I've seen it done."[78] Sure, scientists can be religious, and I've met some! Unfortunately, this does not yet include Francis Collins, former head of the Human Genome Project (and author of *The Language of God*), but he's worth quoting here: "Can I as a scientist actually hold up my hand and ... say I believe in the literal resurrection of Christ? Yes I can."[79] In the utmost depths of time, In The Beginning one might say, Collins admits what by rationalist standards must be called a miracle. Yet there is no other way to describe it; rationality, at least not human rationality, did not yet exist. Indeed, nothing at all existed, certainly nothing that could have left behind evidence. On this point I would paraphrase Collins as follows: in this rationally unknowable instant, an eternal, timeless God created something from nothing, and from its (God's) perspective outside of time, saw the start, progression, and end of its creation in a single moment. If you go this far in admitting a divine spark to the origin of the cosmos, as Collins does and as Darwin did, one more step to admitting a Christian miracle 2000 years ago is no big deal.

> If you're willing to answer yes to a God outside of nature, then there's nothing inconsistent with God on rare occasions choosing to invade the natural world in a way that appears miraculous. If God made the natural laws, why could he not violate them when it was a particularly significant moment for him to do so? And if you accept the idea that Christ was also divine, which I do, then his Resurrection is not in itself a great logical leap. ... I find absolutely nothing in conflict between agreeing with [Richard Dawkins] in practically all of his conclusions about the natural world, and also saying that I am still able to accept and embrace the possibility that there are answers that science isn't able to provide about the natural world—the questions about why instead of

> the questions about how. I'm interested in the whys. I find many of those answers in the spiritual realm. That in no way compromises my ability to think rigorously as a scientist.[80]

In a 2007 lecture at the Faraday Institute in Cambridge, Collins noted Richard Dawkins' grudging admission of a God that might possibly have existed, in which he said "I accept that there may be things far grander and more incomprehensible than we can possibly imagine ... beyond our present understanding."[81] Noting this, Collins quipped, "I think Richard Dawkins has just found God."[82]

I'm tempted to end this chapter here, leaving the acceptance of miracles to my senior colleague, Francis Collins. If I end now, I can conveniently omit setting forth my own answer to the question, "Do you believe Christ had a biological father?" But if I expect my readers to seriously consider my amateur theological musings, I have to be honest and give you at least some idea of where I stand on this important question, so here goes. Probably from years of conditioning, and definitely from a personal "feeling" that is not defensible from material evidence, I admit that I want to believe in a Creator, and more specifically a Christian one. Importantly, this desire is not completely without factual basis. There are historical components pertaining to the person of Christ himself that are honestly impressive, including scholarship supporting many events of the New Testament.[83] Compared to other historical figures (e.g., Socrates), the historical evidence for Jesus is quite good. For example, the existence of Christ has been documented by non-Christian sources, such as the Roman writers Tacitus and Suetonius, who presumably would not have had any self-serving interest in attesting to his existence.[84] There is a strong case not only for the existence of a person called Christ, but also for the temporal authenticity of the documents we call the New Testament. At least some Christian scriptures (e.g., Paul's letters and Mark's Gospel at around AD 50 and AD 70, respectively) appear to have been written close to Christ's lifetime, well within range of an oral tradition based on eyewitness accounts.[85]

However, does this enable me to believe in an actual human being born of a virgin? No, it does not—at least not in a biological sense, which is how most people understand this question and how, therefore, I should answer it. Female humans do not give birth unless they have been inseminated. As He was a human being, I infer based on what I know of biology that Christ would have developed in His mother's womb, from zygote to morula to embryo to fetus. Christ had organs, a nervous system, bones, muscles, connective tissue, and had cells full of organelles, a nucleus, and chromosomes. Most of this initial growth took place nourished by a chorioallantoic placenta, as in other placental mammals (see Chapter 3), not from a big supply of yolk, like a turtle, or from a long period of drinking mother's milk outside the uterus, like a kangaroo. In the days before pediatrics and prenatal care, He was probably born after about 38 weeks in utero, exiting the womb head-first with His umbilical cord still attached and with the placenta following shortly thereafter. As a child, and adult, He would have sometimes gotten tired, annoyed, giddy, and hungry; He would even have occasionally crapped and bathed (hopefully in that order).

Everything that I understand about human biology indicates that He, too, had a biological father. There is no doubt, however, that this father was perceived as divine by his followers. As a human being, of course Christ had a biological father; it is not rational to believe otherwise. Personally, however, I really do believe that father and son were inspired individuals, worthy of the impressive documentation with which their legacy has been recorded. I don't have concrete evidence for this belief, at least not of the sort that one would need to demonstrate that, for example, pelycosaurs share a closer evolutionary relationship with mammals than they do with birds (see Chapter 5). Simply stated, Christianity is my faith. It is not an unshakable faith, nor do I believe literally in many parts of the Bible. Indeed, much of the text in this chapter disqualifies me as a theistic Christian by most evangelical standards. Nevertheless, Christianity seems to me a legitimate account of the agency behind life, and while the causes behind life's diversity are fascinating, they are not of immediate relevance to this faith.

Let me phrase this differently. Do I believe in miracles? If by "miracle" you mean the spontaneous failure of a natural law due to the contrary influence of some supernatural agency, then no. I don't believe that such things happen—not now, not 2000 years ago. However, this is not at all the same thing as denying the power or existence of divinity, including the Christian sort. For example, God most definitely can turn water into wine: a minute fraction of the rains in southern France end up in vineyards, and some small proportion of that water ends up in grapes, some of which ultimately winds its way into bottles, fermented.[86] There's your miracle (and the process works even without the bottles). Remember the quote from Joseph Butler, given by Darwin on the title page of the second through sixth editions of the *Origin*, mentioned in the beginning of this chapter? "*What is natural as much requires and presupposes an intelligent agent to render it so, i.e. to effect it continually or at stated times, as what is supernatural or miraculous does to effect it for once.*" The "do you believe in miracles?" question assumes an opposition between "nature" and "god" that is wholly our own fabrication, as if the two compete with one another for our attention. This question presumes a philosophy that the two things are independent, even antagonistic—but I don't think they are. Rather, one is an expression of the other. God cannot "intrude" into the normal operation of nature because, the way I see it, nature is a part of God; it represents God's thought, or laws, in action. He cannot intrude upon Himself.[87]

To sum up, I am not superstitious, but I am religious, and I do not base my religious faith on peculiar human myths about some extraterrestrial spirit breaking the laws of nature. The divinity I see in Christian scripture is the author of the Book of Nature, not its critic.

Did Charles Darwin believe in any of this? He did have Christian faith during much of his life and a crisis with it following the death of his daughter Annie in 1851. Following this event, he was apparently more agnostic. In 1879, near the end of his life, Darwin wrote

> It seems to me absurd to doubt that a man may be an ardent Theist & an evolutionist. ... What my own views [on religion] may be is a question of no consequence to anyone but myself. But as you ask, I may state that my judgment often fluctuates. Moreover whether a man deserves to be called a theist depends on the definition of the term ... In my most extreme fluctuations I have never been an atheist in the sense of denying the existence of a God. I think that generally (& more and more so as I grow older) but not always, that an agnostic would be the most correct description of my state of mind.[88]

Darwin's opinion on the veracity of miracles, such as the non-human paternity of Christ, is not particularly important for the purposes of this book, and based on the above quote I think he would agree. Humans can and should operate on the assumption that, to understand cause throughout the vastness of life history, it is not necessary to postulate miracles of any kind. Instead, when faced with uncertainty about how our cosmos works, we should simply acknowledge our as-yet insufficient awareness of natural mechanisms. Stated differently, we should recognize a "miracle" as an artifact of our own limited capacity to understand, rather than a substitute for understanding. Relatedly, we need to remember that developing an understanding of natural cause does not rule out a possible agency behind it.

In The Beginning, long before the origin of life on Earth, there was no understanding, no rationality, no evidence; there was nothing, or at least nothing that is relevant to science. Such a cosmic "Beginning" is far away from anything touching upon evolutionary biology on Earth, which is what this book is really about. Despite the fact that many debates between creationists and evolutionists eventually degrade to arguing about the origin of life, this is completely irrelevant. Evolutionary biology is not about The Beginning. It is about the process that has been going on ever since, one which joins together all of the living, biological points we happen to observe in our present slice of time.

TWO

EVOLUTION AS A SCIENCE

A year before the publication of his *On the Origin of Species*, Darwin co-authored with Alfred Russel Wallace the first publication detailing their view of natural selection.[1] Wallace was the younger of the two and less well-known in the scientific circles of the time. Like Darwin, he was a veteran of serious exploration in remote (for Europeans) corners of the world, particularly South America and southeast Asia. Independently from Darwin, Wallace formulated essentially the same idea that variation and differential survival over many generations provided the mechanism by which biological diversity evolved. The letter Darwin received from Wallace in 1858 detailing this idea finally motivated him to act publicly. With the encouragement of his colleagues Joseph Hooker and Charles Lyell, Darwin jointly published a paper with Wallace in the *Journal of the Proceedings of the Linnean Society, Zoology*, entitled "On the tendency of species to form varieties; and on the perpetuation of varieties and species by natural means of selection." On July 1, 1858, in the absence of both authors and a year before the *Origin* was published, this paper was read publicly at a meeting of the Linnean Society in London by George Busk, a secretary of the society.

The big deal with the Darwin–Wallace theory of evolution is that it provides a mechanism whereby one organism can comprise part of an ancestor–descendant lineage that connects it to other species. This does not mean that you are descended from monkeys alive today, but it does mean that if you go back far enough in time, you'll find the common ancestor shared by you and that monkey. Both of you occupy different, terminal twigs on the Tree of Life, just as you and your cousin occupy different branches on your family tree, but on a much larger scale. This rather unsettling idea differs from traditional accounts of the "origin of species" because of its emphasis on an actual mechanism. Most religious creation stories tend to focus on giving credit to the deity responsible for the appearance of biological species, rather than dwelling at length on the means by which the deity did the creating. Natural selection does the opposite. Darwin was not concerned with describing a deity pulling the strings behind this new mechanism of "creation" because he was interested in *how* this process happened, i.e., its cause. At the same time, neither Darwin nor Wallace ruled out some kind of Creator, or agency. But that was the stuff of the last chapter, and the subject now at hand is cause, not agency.

How is it possible to know that the Darwin–Wallace mechanism behind biodiversity, and its implications for our relatedness to other living things, is true? What are the kinds of data that would lead us to favor one sort of theory behind a species' "cause" over another? Science can help us answer this question.[2] As already discussed, it is glorified common sense that makes it possible to recognize the accuracy of the Darwin–Wallace theory. Some natural sciences (chemistry, applied physics, engineering) use the scientific method to explain phenomena that are active and observable on a day-to-day basis. Others, such as evolutionary biology, also operate on principles that we can observe today, but have in addition a substantial historical component.

HISTORICAL SCIENCE AND NATURALISM

We cannot witness evolutionary change between major groups of animals before our eyes because it has occurred on a scale far greater than a human lifetime, and because for most living biodiversity, it happened long before humans existed. For example, according to the theory, your shared common ancestor with a living monkey existed over 25 million years ago.[3] However, being temporally removed from the object of study is true of many sciences. Astronomists often work with light from stars that in some cases has taken many millions of years to reach Earth. This light tells us about events that are exceedingly distant in the remote past, not unlike fossils uncovered by a paleontologist. Forensic pathology is much closer to home, but the principle is the same. No one alive today was present at the death of the seventeenth-century aristocrat and writer John Wilmot.[4] Although no one was around during his lifetime who could have recorded the results of modern medical diagnostics, it is still possible to blame venereal disease. By scrutinizing his writings and those of his contemporaries, his own bony remains, and what we know of seventeenth-century hygiene, we make a diagnosis. The etiology of syphilis, caused by a bacterium (*Treponema pallidum*) that still infects people today, is relatively well understood, and can leave distinctive pathologies on the skeletal remains of its victims. If advanced syphilis contributed to Wilmot's death, his remains would likely show evidence of bone lesions, and there should be some hint from biographical data that he was sexually active, if not promiscuous.

In contrast, other explanations such as a head injury during combat, suicide by hanging, or old age would leave behind different kinds of evidence, at least some of which would be incompatible with the hypothesis of syphilis as the cause of death. Signs of blunt trauma to the skull would favor combat. Heavily worn teeth, adjacent to empty tooth-sockets resorbed by bone, would favor old age. Given access to the relevant material and a little bit of training, any sensible person would be capable of collecting the data relevant to this case and reaching a conclusion as to how this individual died nearly 400 years ago. After doing so, that person would probably have a better appreciation as to why so many monastic religious orders demanded celibacy of their adherents in an age when venereal disease was often fatal.

Even though a given phenomenon may have occurred long ago, it is still possible to make scientific inferences about it because we can be confident that the phenomena of erosion, gravity, decay, etc. that we observe today would have also been applicable at that time. This draws upon the principle of uniformitarianism, a term made famous by one of Darwin's mentors, British geologist Sir Charles Lyell. He is regarded as one of the founding fathers of modern geology because of his application of this principle in understanding Earth history. In terms of its focus on the past, evolutionary biology is not substantially different as a scientific discipline than astronomy, forensics, or geology. We take the evidence we have, even if removed temporally from our present perspective, and make rational predictions of discoveries that should follow if our proposition is true, assuming that the physics, mathematics, and chemistry that we know today are also relevant to the past.

Uniformitarianism is a kind of naturalism, described in the previous chapter. As understood today, it is not quite the same as the "laws" advanced by Lyell in the early nineteenth century, which included now-outdated ideas about an overly monotonous pace of geologic change. Nevertheless, the naturalistic essence of uniformitarianism, i.e., the notion that processes observable today are applicable to the past, represents the core of the scientific method.[5] Creationists don't like it when science is bound by natural processes, because to them science so defined excludes at the outset their favored "explanation" of a supernatural role in the origin of species—which as we've discussed is not an explanation of cause but an attribution of agency. So they cry foul, and claim that scientific naturalism is a political doctrine. It isn't. Naturalism is a pragmatic methodology that underpins scientific efforts to figure out how things work. Without naturalism, we'd have to admit that maybe water has previously run uphill, since people have not always been around to testify one way or the other.

We can be sure that throughout Earth's history, gravity has had some effect on the action of water because we see the results in both modern and ancient geological deposits. Gravity holds a small vertebrate to the ground as it runs across a muddy surface, leaving tracks. A flood inundating that surface—again driven by the action of gravity—would erase those tracks and leave behind particles of varying sizes, proportional to the energy of the flood (more energy transports larger particles). Low-energy deposition, for example at the bottom of a lake, would yield a deposit with relatively fine-grained, homogeneous particles. Vertebrate tracks and fine-grained deposits are exactly what we see scattered throughout various layers in the rock exposures of the Grand Canyon of northern Arizona. There are substantial outcrops of limestones, shales, and sandstone, not to mention the remains of algal beds and tree stumps, within horizontally widespread deposits throughout the canyon. These things result from gradual deposition of sediment over a long period of time; they do not negate the observation that other phenomena evident at certain places within the several thousand feet of Grand Canyon geological section can occur quite suddenly, such as an intrusive flow of volcanic rock. The explanation of the patterns observed in the rocks of the Grand Canyon[6] is a simple extrapolation from the processes that we know would form them today, requiring, among other things, that water run downhill.

The uniformitarian brand of naturalism says that the present is the key to understanding the past. Importantly, naturalism is not dogmatic about specific physical resemblances between the two and it allows for considerable flexibility in extrapolating about the conditions of the ancient Earth. For example, we know that days now are slightly longer than they were during the Paleozoic;[7] that the oxygen content of the Earth's atmosphere has varied tremendously over time;[8] and that the positions of the continents, with major consequences for ocean currents and climate, were quite different in previous geological eras.[9] Naturalism does not mean that everything we observe today must have always been so. Rather, it enables scientists to make inferences about remote events based on the physical processes that are accessible to us today. This can result in predictions that are repeatedly confirmed about an utterly different kind of planet in the deep geological past. This year, spring in Cambridge was hospitable to cherry blossoms, baby foxes, and long walks. Fifty thousand years ago, this part of Britain was much colder and connected by land to the rest of Europe. About 300 million years ago, the Earth's continents completely lacked mammals, birds, and flowering plants, and were surrounded by oxygen-poor seas.[10] Naturalism enables scientists to understand Earth history, and does not insist that all formative actions of the distant past be "slow and gradual." On the contrary, it forms the basis for inferring sudden events such as asteroid impacts, giant volcano eruptions, and even catastrophic floods.[11]

Scientific disciplines such as medicine, chemistry, and engineering are not typically regarded as controversial. Certain topics within each may be contentious, but each discipline as a whole is not. Importantly, these depend on naturalism just as much as evolutionary biology does.[12] Qualified medical professionals today assume that demons, spirits, and voodoo are irrelevant to human health. They do not treat cholera or bubonic plague by sending crack teams of interns to flail themselves in penitence at the site of the outbreak; yet this was one of the responses of medieval Europe to disease, a community that did not benefit from naturalism as we do. Today, rather than entertaining what sort of supernatural agency might be behind disease, medical science begins with the assumption that there is a natural mechanism behind sickness—something that can be treated by natural means. Most people capable of reading this book can do so thanks to the dependence of contemporary science on naturalism. Perhaps naturalism saved or extended their life through medicine; perhaps their livelihood is dependent on the high urban population density that engineering and sanitation now permit.

Consider carefully: do you want a physician who seriously entertains the possibility that some supernatural force has made you sick, or do you want someone who focuses on the mechanisms behind your symptoms, someone who assumes a natural basis for the ailments of all of the patients in her or his waiting room? Natural agents (viruses, assassins) may very well play a role in your health, but at the end of the day, it is only because medical practice understands the natural basis of your health, *how* your body functions, that it can help you recover. If your doctor started pondering a supernatural agent behind your illness, you would not commend the pursuit of free speech in medical diagnostics. You'd look for another doctor.

So, if you prefer a physician who assumes a natural basis behind the symptoms presented to him or her in the office, and therefore has a shot at prescribing a remedy, why would you accept anything less in your child's science classroom? I think the proof that anti-evolutionists are truly dedicated to rejecting scientific naturalism will be when they stop taking vaccines, driving cars, or engaging in other activities that have resulted from applications of naturalism in science.

UNIFORMITARIANISM AND INTELLIGENT DESIGN

Intelligent design (ID) advocate Stephen C. Meyer professes a low regard for naturalism, but high regard for uniformitarianism. In many of his publicly available talks and lectures,[13] he infers the "design" that we observe today always to be a product of intelligence, and claims to use the uniformitarianism of Charles Darwin to justify his inference. To paraphrase, a very complex device we observe now, such as a wristwatch, computer, or piece of software, has only one source: human ingenuity. According to him, none of these things can be made without it. It follows, he says, that a similarly complex device we observe in the geological past must also have arisen as a result of something like human ingenuity, i.e., intelligence. The processes we know and observe today are relevant to explaining the phenomena of the past, and we know that particularly complicated things we see today have an intelligence behind them. Biology is particularly complicated, ergo, "intelligent design." Meyer repeats an argument articulated long ago by scholars like William Paley,[14] but applies it to areas of complexity about which Paley knew nothing, such as cellular microbiology and DNA.

There are three broad (and overlapping) categories in which portrayals of intelligent design as a mechanism behind biodiversity are wrong: philosophical, theological, and biological. The first two we discussed already in the previous chapter; the last crops up throughout this book and others with information on the natural world. At the risk of some repetition, I think here is a good place to summarize all three.

Philosophically, "design" concerns only a limited explanation of the subject, namely the agency behind the watch, software program, etc., and not how the item actually works. If you want to know how a wristwatch powers each tick (coiled spring versus battery?), or where some bit of software stores information (does it need access to a network?), then "Joe designed it" is not an answer. You might very well think that God, space aliens, or both have directly intervened as human-like tinkerers in shaping the Earth's biota. Fine. Even if you were right, you've not said anything about *how* either agency has affected the actual course of life history. Furthermore, by attempting to replace a causal mechanism (natural selection) with an attribution of agency (design), ID advocates such as Meyer are decidedly anti-uniformitarian. What process of today could possibly lead to this understanding of the past? How could a biological phenomenon, even if designed, be simply willed into existence without an actual mechanism? At least young Earth creationists are honest about their appeal to a supernatural divinity.

Theologically, ID advocates constantly invoke words like "undirected" or "random" when they refer to Darwinian evolutionary processes, perpetuating not only the myth that natural selection is random (it isn't; see Chapter 11), but also the vain supposition that what they deem to be "undirected" cannot be the result of an agency. Conversely, they define "directed" or "non-random" relative to human intelligence. Theologically speaking, it would be an impoverished Creator indeed who suffered from our human limitations concerning perception and time, and could only create like we do.[15] This is the other edge of the theological sword that Meyer has made for himself. If you agree with Meyer that supernatural intelligence equates with what humans perceive as non-random, it follows that what is random (by human standards) is not within the creative capacity of that intelligence. Some deity! The Christian philosopher Alvin Plantinga recently made a similar point regarding the implied irrelevance of the ID "god" to phenomena that we perceive as consistent with natural law:

> One thing about intelligent design that strikes me as a little disturbing is the fact that it seems to concentrate the finding of design in only certain very narrow areas. ... If something happens by way of law then you don't attribute it to design, and if it happens by way of chance you don't. It's only if it doesn't happen by either of those two things [that one infers design]. But the fact is, seems to me, if you're a Christian or a theist you'll think that what happens by virtue of law also happens by virtue of design; it's God who set the laws for us, [who] set the laws for our world.[16]

Biologically, human-like intelligence as an agent is not required for complexity to arise. Those who speak loudest for the ID movement do not accurately portray the capacity of processes such as natural selection to result in high levels of complexity over time. Instead, ID proponents make false generalizations about "randomness" and how it cannot produce "information" (see Chapter 10), or about how the fossil record is somehow lacking or irrelevant in documenting the step-wise accumulation of features integral to complicated anatomical systems (see Chapters 4–8).

Strictly speaking, Meyer is correct to say that some kinds of signals evident to us in the natural world would justify the inference of a human-like intelligence involved in its current state. The absence of such signals is why it is possible to rule out human-like agency behind certain historical events, like the conspiracy theory that German Freemasons engineered the spread of the 1918 flu. Evidence in favor of the participation of a human-like agency in biological evolution might include multiple, independent discoveries of the remains of prehistoric synthetic polymers. Artifacts from monumental architecture from some distant, pre-human age, or the remains of an actual computer and fragments of the manual to use it, would also help. If the "intelligence" of the ID movement actually seeded the Earth with various biological novelties over time (like bipedal apes), it would be reasonable to expect that intelligence to have left behind a record, in the same way that any other intelligent being would leave behind a record.

For starters, we'd expect to find hard organic remains such as bones or teeth, since all known intelligent agents have them. Furthermore, if these agents could engineer a new organism, they'd be way past us in terms of technology, and we should reasonably expect them to leave behind some of the more banal traces of their existence, like infrastructure and waste, beyond simply their finished product, such as a new ape. If they had a long-term commitment to their creation, they could have even arranged a particularly novel pattern of text in some archipelago or constellation in the night sky: "Please don't ruin our rainforests," for example. All of these things—cultural artifacts, manuals, computers, celestial or geographic twitters—result from human-like, intelligent agency.

A phrase frequently used by Meyer in his publicly available lectures[17] is "uniform and repeated experience," particularly regarding how he sees the connection between complexity and intelligence. Based on that experience, he says, intelligence is responsible when marked complexity is observed. Another "uniform and repeated experience" that we have about intelligent agents is that they leave behind a plethora of evidence when and wherever they have existed. Humans have made an extraordinary impact on our planet; we will leave behind many obvious traces that will be evident in Earth's history long after our species goes extinct: cigarette butts, nuclear waste, metal alloys, tires, bunkers, etc. Remains of such things derived from our "intelligence" will be at least as obvious to future geologists as the global traces of an asteroid impact 65 million years ago are to geologists today. This impact probably contributed to the extinction of many large vertebrates like terrestrial dinosaurs, and left behind global traces of an element called iridium, followed by an abundance of ferns. Sixty-five million years later, the K–T boundary asteroid impact is recognized in the geological record by a globally detectable "iridium spike," adjacent to which is a "fern spike."

If we adopt uniformitarianism to ask whether or not intelligent agents could have directly influenced the course of our evolutionary history, we would expect those agents to have left behind the same kinds of traces as other such agents. Humanity is the only one we know of so far, and we make an exponentially greater amount of garbage than we do functional designs. One of the most obvious kinds of evidence that a human-like intelligence in Earth's distant past would have left behind was spelled out with one of the most famous lines, indeed one of the most famous words, ever uttered in twentieth-century film: "Plastics."[18]

Far from being persecuted for a discovery that raises the issue of design, anyone finding a "plastic spike" in deep time, accompanied by one or more additional types of artifacts known to indicate the existence of a human-like intelligence, and corresponding temporally to one or more major evolutionary events, would be assured of a successful, mainstream academic career (to say the least). While such artifacts wouldn't tell us anything about *how* biodiversity actually came about, they would indicate that something out there, significant and human-like, served as an agent behind life history on Earth.

Meyer argues that one such artifact has already been found. It is DNA itself, and he reads it as saying "I'm right" in a software-like, digital code. While the complexity of DNA makes an interesting analogy to human creative expression, the analogy falls short as proof of human-like intelligence as the cause behind biodiversity for the philosophical, theological, and biological reasons enumerated here and elsewhere.[19] We know already that complicated structures can result from mechanisms unrelated to a human-like intelligence, observable within the scale of a single human lifetime. Some of these examples are complex in a repetitive fashion, such as the symmetry of a snowflake. Others are complex in more patterned ways, such as (for example) novel digestive pathways in bacteria,[20] changes in skeletal morphology,[21] reproductive anatomy,[22] or perception of visual spectra[23] in different groups of vertebrates. A meaningful understanding of "information" in these cases means the capacity of some organism to perform a novel task (e.g., eat nylon, give birth to live young, perceive a higher wavelength of light) or exhibit a novel phenotype (e.g., conduct sound via multiple bones in the ear). Many of these cases show that novel "information" has resulted from very specific genetic changes, ones which we can actually observe and ones which are demonstrably the subject of mutation and selection in many organisms over time[24] (see Chapter 10).

An appreciation of animal and plant development, genetic regulation of phenotype, and the expanse of geological time plus the multitude of fossilized intermediates between groups we recognize today, all comprise persuasive lines of evidence supporting a major role for the *non*-random action of natural selection as a mechanism behind modern biodiversity. Compared to this, an appeal to the human-like, cosmic tinkerer of the ID movement as an alternative to Darwinian evolutionary biology robs us of an actual biological mechanism and explains nothing about cause.

When we focus on this attempt to replace the mechanism of natural selection with "design," the supernatural orientation of the ID movement (one which is regularly denied by ID advocates) becomes clear. Supporters of ID invoke the activity of an "intelligence" behind biodiversity which does indeed have some natural analogues. However, the ID god is beyond nature, since they expect or imply that it has been active in this world in ways that are unique and unlike any natural intelligence. ID advocates claim that their human-like designer has left behind evidence in living things, but are silent about its own cultural and organic remains. They say nothing about the absence of these more banal traces of their designer's existence, traces left by every known form of natural "intelligence." Despite claims to the contrary, ID does not use the uniformitarianism of Darwin or other scientists. Rather, it attempts to use the label "science" for their efforts to identify the activity of a human-like, supernatural force.

SCIENCE AND HYPOTHESIS TESTING

As is true for geology and many other scientific disciplines, evolutionary biology is a historical science, one in which it is essential to apply the principles of naturalism (including uniformitarianism). Evolutionary biologists qualify as "scientists" because of this emphasis on the natural processes of the present—phenomena immediately relevant to disciplines such as physics, chemistry, and mathematics—and because we use these processes to make testable inferences about our ideas. Stated differently, evolutionary biologists use common-sense logic based on principles that are now observable to construct hypotheses about ancient life.

A "hypothesis" is simply a precise statement of some event with testable consequences. Ideally, hypotheses should be easily falsifiable. "The Earth is spherical" is an accepted hypothesis that could be rendered false in an instant by looking back from an orbit of a few thousand kilometers and observing our planet as a cube or trapezoid. "Swans in my neighborhood are blue" is another, easily falsifiable (and falsified) hypothesis against which most tourists have collected the data; they keep feeding them and they only seem to come in one color—white. Now consider "all swans are white." By observing that all local swans are white, we have verified our hypothesis as much as we can. With each additional white swan observed, we provide yet more verification; however, in most parts of the world we cannot falsify it, since differently colored populations are absent. Under these circumstances, this hypothesis is harder to test. Nevertheless, most practicing scientists tend not to be too dogmatic about setting up hypotheses that are relevant only to "falsification" and not "verification" and perfectly respectable science involves some of the latter.[25]

Here's another example that's a bit less abstract about framing a question scientifically, again emphasizing that "science" is a souped-up version of common sense. Say I wanted to convince you of the hypothesis that two wayward grade-school students cheated on their spelling test by copying each other. If true, I should be able to predict something about their exam results and where they were seated during the test (the data to be collected). Specifically, if they cheated, I predict that most, if not all, of their mistakes should be the same and also that they sat adjacent to one another. So we find out that in fact they were seated nearby during the exam, and that 99% of their mistakes were identical (e.g., they omitted apostrophes and failed to obey "I before E except after C" in the same places). Pretty compelling evidence that they copied each other, wouldn't you say?

While the degree of verification provided by the data seems sufficiently compelling to accept the hypothesis, especially given the background knowledge we already have about human behavior when taking exams, such verification is not a guarantee that the hypothesis "these students cheated by copying" is actually true. In other words, verification is a weaker means of scientific hypothesis testing than falsification (when, as per our previous example, all you need to see is one black swan). Hence, to be extra careful we should go about collecting more data, particularly if the extent of similarity in their errors were less, or if they had no way to communicate with each other during the exam. They could have independently picked up the same bad spelling habits from the grammatical conventions of mobile phones, or their mistakes could have been due to chance.

Here's one more example about hypothesis testing: Noah's flood. According to verses 21–23 of Genesis chapter 7 of the King James English Bible, quoted on the creationist website *Answers in Genesis* (AiG),[26] "all flesh died that moved upon the earth, both of fowl, and of cattle, and of beast, and of every creeping thing that creepeth upon the earth, and every man. ... Noah only remained alive, and they that were with him in the ark." Both the King James and New International versions describe God's anger with humankind at the time of the flood, to the point where He declares "I am going to bring floodwaters on the earth to destroy all life under the heavens, every creature that has the breath of life in it."[27] The clear implication of these passages, interpreted literally from Genesis chapters 6 and 7, is that during the flood, no animal life survived on Earth, except for that hosted by Noah on his ark. (With plants and invertebrates it's harder to say; they certainly respire but maybe aren't "flesh" and I guess could be interpreted as lacking the "breath of life.") Interestingly, the AiG website states that aquatic and avian life would not have been present on the ark,[28] which seems inconsistent with a literal understanding of Genesis, as quoted above. The same website takes on the extra burden of including "dinosaurs" on the ark, a group which actually includes birds, but they probably mean extinct life in general.

Distilling the Genesis account of the flood into a testable prediction is made easier by the relatively precise dimensions of the ark given in Genesis 6–7. A testable hypothesis following from these texts would go something like this: reproductively capable representatives of all animals on Earth (asexual singletons and breeding pairs for sexually reproducing types) can survive for over 40 days and nights (one also has to include the extra weeks waiting on Mt. Ararat for the floodwaters to recede) on a wooden boat measuring 300×50×30 cubits, a craft characterized by the AiG website as "a mid-size cargo ship."

The process of investigating this question is an exercise in historical science. No one alive today was there at the time, so we cannot vouch for this event via direct testimony. But we assume that the natural laws of Earth as we know them today would also have been applicable at that time. If we can come up with a rational answer to this question based on what we know about the Earth today, then we can infer the extent to which this account is likely to have happened in the past. Importantly, if you reject the applicability of current natural principles to the time of Noah's flood, there's no reason why you should feel bound by the verdict supported by those principles. By appealing to some unknowable, supernatural circumstance (God squeezed them all into the boat, kept them from eating each other, made them comatose for a few months, etc.) you wouldn't qualify as a scientist, nor should you be able to bring your thinking into any classroom and label it "science." You might be correct, but you would not be behaving scientifically.

In any event, a calm, scientific assessment of this hypothesis has only one conceivable verdict: false. By rational standards, the biomass of the requisite animal life and the technology required to sustain creatures from temperate, tropical, montane, desert, subterranean, arctic, and aquatic environments for several months on a ship comprise data that clearly falsify the hypothesis of the ark, not to mention beg the question as to how these organisms attained their current distributions. (And what about all those plants, freshwater fish, and invertebrates?) You can still opt to believe in Genesis as a literal exposition of Earth history, but in doing so you are not abiding by common sense in either its vanilla or scientific flavors. A literal belief in this particular bit of scripture is irrational based on the principles of nature and common sense.

HYPOTHESES DERIVED FROM NATURAL SELECTION

Natural selection as a process behind evolution has fared much better than "flood geology" or intelligent design as a source of scientific hypotheses that have withstood many attempts at falsification. Here are several important kinds of observations about the natural world, divided arbitrarily into four interrelated categories that will comprise the remaining chapters of this book. If the Darwin–Wallace theory about natural selection as a process behind evolution is true, then each class of observation listed below is a logical consequence.

1 *Characters and common descent*: Evolution by natural selection implies that most or all of life shares a common ancestor.[29] As a result, it predicts that characteristics of living organisms (e.g., avian feathers, mammalian milk) appeared on the Tree of Life independently of the groups of species for which we now regard those features as diagnostic. Following from this, natural selection predicts that certain organisms should mix adaptations and morphologies seen in others, comprising what are popularly known as "missing links," or species that stretch the definition of exactly what constitutes a "bird," "mammal," or other such category of living things recognized today.

2 *The fossil record*: Many species formerly alive are now extinct. On the very large scale of the past 3.5 billion years of life history, evolution by natural selection predicts that species deemed to be ancestral to the most living species should be found in relatively older strata of the Earth's geological record. Conversely, those extinct species most closely related to specific, modern groups should be found in relatively younger strata. Importantly, because we know that fossilization does not exactly correspond with the geological start- and end-points of most species, this prediction does not demand a perfect correlation between ancestry and geological age. Nor does it mean that any one species is "on its way" to becoming another, or that ancient organisms cannot be specialized. Rather, it comprises the more general expectation that a common ancestor for a large group of living animals should be older than one for a relatively smaller group of living animals. For example, evolution predicts that the common ancestor of vertebrates should be found in the record at an older date than the common ancestor of primates.

3 *Development*: Natural selection predicts that features of a developing embryo, fetus, and/or juvenile should be conserved in animals that share common ancestry. In other words, even though adult structures in two animals may appear to be very different, such adult structures should still develop from a common embryonic pattern.

4 *Molecular biology*: If evolution by natural selection has occurred, we should expect a number of phenomena regarding animals' genetic make-up. For example, the genetic similarity of close relatives should be more pronounced than that among more distantly related species. Second, we would expect to find a relationship between "genotype," an animal's genetic make-up or DNA, and "phenotype," how the animal looks or behaves. In other words, something particular regarding an animal's morphology or behavior should be predictably affected via a particular change to its DNA. Third, we should expect to find "molecular fossils," or the genetic rudiments of structures that are no longer functional in an adult of a given species (e.g., eyes in a mole), but which were present in its evolutionary ancestors.

Anti-Darwinians vary in the extent to which they believe evidence supporting the above predictions relate to their arguments concerning creationism and ID. Textbooks with a creationist perspective that are aimed at a popular, school-aged audience—including *Of Pandas and People*[30] and *Explore Evolution*,[31] both of which were written at least in part by prominent members of the contemporary ID community—articulate viewpoints that reject common ancestry and other postulates of Darwinian evolution (see Chapters 8 and 9). In contrast, some ID advocates today seem to accept common ancestry and/or aspects of certain other points listed above, but characterize these predictions of Darwinian evolutionary biology as "trivial"[32] compared to their belief in an intelligent designer.

This acceptance of common ancestry combined with an insistence on a guiding intelligence is similar to ideas of certain nineteenth-century authors. Stripped of its anthropomorphic insistence on a human-like "designer," it is not far off from what Darwin himself wrote (see Chapter 1). As far back as 1864, the chemist Josiah Parsons Cooke seemed content with Darwin's account of common descent, but disputed that it could have arisen with what he labeled an "unintelligent cause." He stated "the difficulty to my mind in Mr. Darwin's particular theory is not its developmental feature," by which Cooke meant common descent by natural selection, "but in the fact that [Darwin] refers the development to what I can understand only as an unintelligent cause."[33]

Such arguments for anthropomorphic intelligent design, whether articulated in 1864 or 2010, do not change the fact that no one prior to the Darwin–Wallace theory had devised a cogent mechanism of evolution from which common descent and the other predictions outlined above follow so closely. Without Darwin's mechanism, no one can coherently explain why animal and plant anatomy tend to vary together with their geographical distribution and stratigraphic record, or why we should expect any correlation between genetic and anatomical similarity. Far from being "trivial," predictions derived from evolution by natural selection concerning common descent and anatomical, geographic, stratigraphic, and genetic similarity are core propositions of Darwinian evolutionary biology. As we will explore in some detail in Chapters 3–11, it is precisely the ability of the Darwinian mechanism to predict many facts about biology that, 150 years later, makes his argument so compelling.

Each of the predictions outlined above entails a number of active scientific disciplines composed of thousands of postdocs, PhD students, and professors, scattered among many departments in countless universities, museums, and other research institutions. There is no way that a comprehensive review of scientific research derived from all of these topics can be presented in a single popular book, and I won't describe at all the evidence from a few other fields (e.g., biogeography and behavioral ecology) which have also led to a Darwinian understanding of evolutionary biology. Details from just about all of these fields have been amply described elsewhere.[34] In the coming chapters, I add to the existing literature by discussing a number of specific topics concerning the distribution of anatomical features across living representatives of the Tree of Life, the fossil record and development, and the molecular basis for understanding evolutionary relationships and phenotype.

THREE

CHARACTERS AND COMMON DESCENT

From our time-bound perspective, humans are capable of observing only a snapshot of life. If the Darwin–Wallace theory is true that over time one species of organism has given rise to others (an idea known as "common ancestry") via natural selection, then a full appreciation of species through time should reveal examples of animals that combine features from what we now categorize as other, distinct organisms. A different way of phrasing this very important concept is as follows: the features that we see defining certain animals (like the mother's milk of mammals, the baleen of blue whales, the shells of turtles, or the feathers and beaks of birds) appear on the Tree of Life independently of any specific animal that we would recognize today because of those features.

As a logical correlate of the common ancestry entailed in natural selection, the anatomical structure we recognize as "feather" has an independent origin relative to the first appearance of animals that anyone today would call "bird."[1] Relatedly, the common ancestry required by natural selection predicts that such animals (e.g., a feathered non-bird) possessed characters (e.g., teeth) that we associate with other groups. Such animals may be rightfully dubbed "intermediates" in the sense that they mix features we see as exclusive to animals that seem today to be well defined. No bird alive today lacks feathers or has teeth, just as no mammal alive today has more than a single bone in its jaw or lacks mother's milk. But common ancestry predicts that this was not the case for all of the ancestors of birds or mammals, and that considerable uncertainty may therefore complicate the decision as to when the first "bird" or "mammal" first appeared in the fossil record.

Importantly, a feathered Cretaceous *Velociraptor* is "intermediate" not as half bird, half terrestrial dinosaur, or as a "reptile" trying to become a bird. Rather, it shows some of the features of the common ancestor shared by Mesozoic theropods (the specific group of dinosaurs within which birds have their origin) and modern birds. It fits the prediction of evolution by natural selection[2] that anatomical features have appeared on the Tree of Life independently of the animals in which they predominate today. Animals that possess such combinations of features from distinct, modern groups might legitimately be viewed by us to be intermediates, and they do not have to be extinct. Many are known only by fossils, but this is not so in every case. Neither Darwin nor Wallace formulated their initial ideas based primarily on the fossil record. Rather, they made their most decisive observations about the natural world based on living plants and animals. Fossilized organisms that mix features of different animals alive today can be pretty spectacular, like a feathered dinosaur with a jaw full of teeth.[3] When this happens with a living animal it's even better.

EVOLUTIONARY TREES

Let us consider the following pairs of animals that differ substantially when viewed in isolation from other species: toad and crocodile, crocodile and kangaroo, kangaroo and galago, galago and monkey. Figure 3.1 conveys genealogical information, like a family tree, and depicts these animals according to their degree of relatedness, as determined from multiple lines of evidence such as comparative anatomy, development, fossil history, and molecular biology (as described in Chapters 4 through 9). This evolutionary tree, or cladogram, shows that the animals occupying branches closest to one another at the tips of the tree (e.g., tarsiers and monkeys) share a more recent common ancestor than either does with any of the other animals. Each intersection among branches, or each node, represents an ancestor of all of the animals above that point in the tree. Hence, the tree also implies something about time, although not at any fixed rate. A node farther down the tree represents a common ancestor that preceded in time the nodes and terminal branches above it.

In many cases the common ancestor of two living species may have closely resembled one of its descendants, which may exhibit little or no evolutionary change over a long period of time. Evolution does not demand that change occur at any particular rate, or at a constant one across different branches. Hence, a living chimpanzee is just as much a descendant as is the human with whom the chimp shares a common ancestor. While chimps have

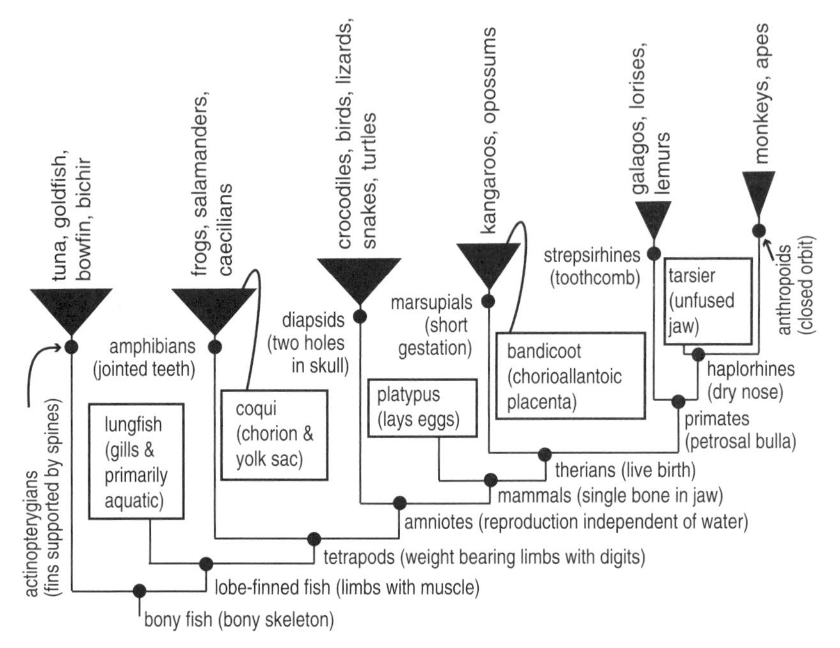

Figure 3.1. An evolutionary tree, or cladogram, depicting the pattern of genealogical relations among vertebrates with a bony skeleton. Each node (black circles) represents a common ancestor from which a group of organisms has evolved (e.g., diapsids and mammals share a common ancestor and are collectively known as amniotes). Note that due to space limitations, this tree does not include all members of the named groups, and that branch lengths are arbitrary. Each node is distinguished by shared features (e.g., mammals have a single bone in their jaw). Animals in boxes belong to the labeled nodes along the base of the tree, but exhibit features present in other groups (e.g., the platypus is a mammal but lays eggs). The existence of such animals that do not have all of the features we regard as typical for a given group supports the idea that those features have evolved independently on the Tree of Life. Mammals, for example, evolved a single bone in the jaw independently of live birth.

a relatively bad fossil record, that of many other anthropoid primates (the group to which humans, chimps, other apes, and monkeys belong), is quite good. In many regards, the chimp–human common ancestor was cranially more like a modern chimpanzee than a human. But that common ancestor was not a chimp.[4]

One can trace common ancestry in Figure 3.1 by taking a given species and moving downwards, that is, from more recent (above) to more ancient (below) shared ancestry. Everything that's connected above a certain node forms an exclusive group, descended from a single, ancestral population of one interbreeding species that existed earlier in time than nodes farther up the tree. For example, the common ancestor of galago, tarsier, and monkey (primates) evolved more recently than the common ancestor of kangaroo, bandicoot, and primates (therians). In addition, although a kangaroo and galago share a common ancestor, one would have to include several other animals as well (bandicoot, tarsier, monkey) in order to account for all of the descendants of that common ancestor in a single group (therians).

With this in mind, you can better appreciate an important principle of biological taxonomy: named groups have to encompass all of the descendents of a single common ancestor. Applying this rule to the tree in Figure 3.1, the group "primates" include galagos, lemurs, lorises, tarsiers, monkeys, and apes; "marsupials" include kangaroos, opossums, and bandicoots; "therians" include primates plus marsupials; "mammals" include therians plus platypus; "amniotes" include mammals plus crocodiles, and so on. It would not be permissible according to this convention to name a node, say mammals, and leave out one of its descendents, say bandicoots. By the way, it should also be clear to you that since this page has limited space, I've not included all of the actual members of these groups (e.g., humans within primates, koalas within marsupials, snakes within amniotes, etc.).

LIVING "INTERMEDIATES"

Understanding the significance of evolutionary trees, we can move on to examine in more detail a few of the species that mix anatomical features seen in other living groups—species that you might therefore be tempted to call living "intermediates" (Figure 3.1). We mentioned a number of very different twosomes above: toad and crocodile, crocodile and kangaroo, kangaroo and galago, galago and monkey. While the members of each pair show major differences in terms of their anatomy, it is not difficult to find yet other living animals that mix anatomical features seen in each. These include the coqui frog, platypus, bandicoot, and tarsier. The coqui frog lays eggs without a hard shell (like a toad), but does so on land with a big yolk sac (like a crocodile). The platypus lays eggs and has multiple bones in its shoulder-skeleton (like a crocodile), but provides milk for its young, shows a single bone in its jaw, and has three ear bones (like a kangaroo). The bandicoot nurses its young after a brief pregnancy in a pouch (like a kangaroo), but shows a placenta constructed from certain embryonic membranes, the allantois and chorion (like a galago). The tarsier has elongate ankle bones and lacks a fully ossified connection between its right and left jaws (like a galago), but has an eye-socket that's walled off from its chewing muscles plus a "dry" upper lip that lacks a midline connection into its mouth and nasal cavity (like a monkey).

The fact that the coqui, platypus, bandicoot, and tarsier show such a mosaic of features found in other groups does not mean that these animals are now in the process of becoming something else, or that they're not legitimate, full-blown critters in their own right. Rather, their existence is consistent with the mechanism of descent with modification as an evolutionary process. As stated above, their anatomical features (embryonic membranes, shoulder and ear bones, jaws, lip philtra) first appeared in the vertebrate Tree of Life independently of the animals that now have these features. The theory of natural selection, with its implication of common descent, predicts that features we deem to be characteristic of major groups (such as amniotes, mammals, and primates) appeared piecemeal over the course of evolution, and that some animals may exhibit only a subset these features, depending on where they are in the Tree of Life.

I've picked these animals because they demarcate intermediates between well-established points on the Tree of Life. Within the larger "phylum" of vertebrates (part of the classification hierarchy established by the famous Swede Carl Linnaeus), they represent some of the more conspicuous, living intermediates among groups that happen to be successively closer to *Homo sapiens*. The coqui shows transitional features between amphibians (like other frogs and salamanders) and reptiles (e.g., a crocodile), the platypus between reptiles and mammals (e.g., kangaroo), the bandicoot between other marsupials and placental mammals (e.g., galagos), and the tarsier between lemurs and anthropoid primates (e.g., monkeys). The evolutionary relationships among these major groups are relatively well understood (Figure 3.1). The fact that there appear to exist living "transitions" between them demonstrates that many anatomical features are independent of the animals in which they occur.

It's important to recognize that my choice of animals showing these intermediate states is arbitrary. One could just as easily say that the galago represents an "intermediate" between platypus and tarsier, rather than viewing the platypus as an "intermediate" between reptiles and mammals. It all depends on your perspective. Usually, well-known, common members of a given animal group (e.g., toads, crocodiles) provide biologists with the features that we now regard as typical for the groups to which they belong (e.g., amphibians, reptiles). Hence, today we regard the platypus as an anomaly since it is rare, and because it mixes a few features seen in more common animals, such as crocodiles and monkeys. It is an accident of history that the platypus is not more common, and that we don't view crocodiles as exhibiting "transitional features" between amphibians and the platypus.

By the way, throughout this section I frequently place the terms "transition" and "intermediate" in quotes because they are potentially misleading.[5] In common English usage they frequently imply something in the process of becoming something else. However, unlike governments, animals are not really ever in "transition." Living things are not trying to become something else. Apes from the early Miocene did not anticipate that some of their descendants would evolve into habitual bipeds. It is only due to the peculiarities and contingencies of natural variation, environmental change, adaptation, and constraint over the last 25 million years that some such populations did contribute to the evolution of our own bipedal, savannah-adapted species.

Terms such as "transition" and "intermediate" are useful because they convey the real sense in which both living and fossil animals mix anatomical and molecular attributes from various parts of the Tree of Life, and because they were used by Darwin himself. However, the sense of active transformation as commonly implied by those words in English is not really applicable in evolutionary biology. Hence, these terms as applied to animals are best avoided, or when used applied to features of species, not to the species themselves.[6] "Transition" evokes the image of a now entirely outdated notion, popular in the eighteenth and nineteenth centuries, of the "great chain of being" in which life shows "progress" (a term even worse than transition) from "sea creature into amphibian into reptile into rodent into primate into human."[7] This is a misguided sentiment we mentioned briefly in the prologue, paraphrased from anti-Darwin critics of Carl Zimmer's 2001 book, *Evolution: Triumph of an Idea*. Evolution is not inherently progressive. The mechanism of natural selection is about contributing heritable characteristics to future generations of your lineage. Features of some species that might seem to us to make a faster, smarter, prettier, or stronger animal will be completely irrelevant if those features do not contribute to the reproductive success of that animal.

With this brief foray into the vocabulary and nature of the Tree of Life behind us, let us now start with the task at hand: considering how certain living animals mix features found in other groups alive today.

REPRODUCTION ON LAND AND IN WATER

One of the most important differences between fish and amphibians on the one hand, and reptiles, birds, and mammals on the other, is the relative independence of the latter from water. Most amphibians lay their eggs in a pond or lake, within which their young hatch and begin life as qualitatively different animals (larvae or tadpoles) compared to their parents. A tadpole shows completely different strategies for locomotion and feeding than a typical adult frog. Some fish and amphibians can survive in arid environments, such as the lungfish and spadefoot toad, because they enclose themselves in a watertight cocoon and/or burrow, and hibernate for extended periods of time. However, such arid-adapted fish and amphibians will eventually require periodic rains or flooding for the water they need to reproduce.

Lizards, snakes, turtles, crocodiles, birds, and mammals are not so constrained. These animals are collectively known as "amniotes" due to important features they all share in the anatomy of their egg, including a structure called the amnion. As the fertilized egg begins to develop, four membranes extend from the embryo, each with specific attributes: the amnion (surrounding the embryo proper), chorion (lining the egg), allantois (storing waste), and yolk sac (storing nutrients). In many amniotes, a desiccation-resistant shell envelopes all of these, and together these features give amniotes something that no other vertebrate has: the ability to reproduce without access to standing water.

Amniotes such as a Grant's golden mole are capable of going through their entire lives without even seeing water, much less getting their toes or tails wet in a pond. They obtain water from their solid diet and are very good at not losing any unnecessarily. For reproduction, they essentially bring the "pond" with them, using the four membranes to preserve the aquatic environment that all organisms require for early development. Non-amniotes such as most frogs and salamanders may possess a yolk sac, but they lack consistent development of the other three membranes and therefore cannot achieve complete independence from water during reproduction. However, some non-amniotes can attain partial independence.

AMPHIBIANS AND THE PUERTO RICAN COQUI

If we contrast two strategies—typical amphibian reproduction in water in which many small eggs hatch into larvae (as in a bullfrog) versus amniote reproduction on land in which few large eggs hatch into juveniles (as in a crocodile)—we can single out a few amphibians that show a remarkable resemblance to the latter strategy. Unlike most other amphibians, Puerto Rican tree frogs, or coqui (*Eleutherodactylus coqui*), not only lay their eggs on land, but also show internal fertilization and lack a larval stage. Upon hatching, out comes a miniature version of an adult coqui, not an aquatic larva. By not laying their eggs in water and by skipping the tadpole stage (also known as direct development), they can avoid aquatic predation in ponds or lakes.

Amphibians such as the coqui help us to understand the early evolution of amniote reproduction. The yolk sac of a coqui egg is relatively large, and during development the embryo inside of that egg encloses both itself and its yolk with a membranous extension of its own body tissues.[8] You did the same thing when you were an embryo, but you had four embryonic membranes, not two like the coqui. Other animals such as most amphibians and sharks have zero or one such membrane. By providing relatively greater independence from standing water in their reproduction, facilitated by their direct development and large yolk sacs, coqui suffer less predation from aquatic organisms during early development, and this comprises a major selective advantage. To the extent that you could shield your offspring from hungry fish and insects at the pond's edge by moving your growing eggs farther onto land, and providing them with a larger built-in store of food (yolk), you would be that much more likely to enable your offspring to survive and reproduce.

Scientists do not yet know the exact genetic and developmental mechanisms that lead to increased yolk mass and larval size, two of the critical features of coqui and amniote eggs. However, we do know that yolk mass can vary substantially between species.[9] Furthermore, another feature of coqui reproduction that differs from that of most other amphibians concerns how males fertilize female eggs. Unlike most other frogs, more than one coqui male may contribute to fertilization; i.e., coqui frogs are polyandrous. Interestingly, polyandry appears to be correlated with yolk mass.[10] That is, when more than one male contributes sperm to a clutch of eggs, each egg ends up with more yolk.[11] This observation provides one intriguing avenue with which to investigate mechanisms behind some of the distinctive features of the coqui egg, and ultimately, ask if it might have been relevant to the evolution of the amniote egg.

There are many other amphibian species, such as the Surinam toad (*Pipa pipa*) and Appalachian salamander (*Plethodon jordani*), that show amniote-like features, including large yolk size, internal fertilization, terrestrial clutches, and direct development.[12] By keeping its eggs on its back (*Pipa*), or hiding them in moist burrows (*Plethodon*), both animals similarly relieve the aquatic predation that would otherwise threaten their young. Like the coqui, these amphibians are also direct developers. That is, rather than producing tadpoles or larvae, the eggs of both amphibians hatch directly into juveniles, or miniature versions of the adult. Direct development in *Pipa* is partial, since the larval phase is condensed to a brief period that takes place entirely within the egg while situated in the mother's back. In both the coqui and *Plethodon jordani* direct development is complete, both animals having lost larval features of their embryos.

To emphasize yet more the inherent diversity of amphibian reproduction, other amphibians skip not the larval stage, but the adult. For example, axolotls (*Ambystoma mexicanum*) and Texas blind salamanders (*Eurycea rathbuni*) don't bother with metamorphosis and keep at least some "juvenile" features their entire lives. Sexual maturity in these animals has been accelerated to the point that it takes place while the animals still have their larval features, such as external gills, a caudal fin, and an aquatic habitat.

Consideration of reproductive strategies among modern frogs and salamanders reveals a tremendous amount of variety, including some that come rather close to strategies observed in amniotes. This variety, including terrestrial eggs, large yolks, internal fertilization, and extraembryonic membranes, has been the raw material on which natural selection has worked over the course of geologic time. Certain living amphibians, such as the coqui, thus present reasonable examples of the kind of reproductive system that might have characterized the common ancestor of amniotes.

MAMMALS AND MONOTREMES

A few years back, when Western Union finally stopped offering a printed telegram service, a major US newspaper commemorated the end of an era by listing the top ten. Among them was a response by Cary Grant to a reporter who wanted to know his age. "How old Cary Grant?" wired the reporter. The actor responded curtly: "old Cary Grant fine. How you?" A notch or two down the list of memorable telegrams printed by this newspaper was one sent by William H. Caldwell in 1884 to the scientists gathered for the meetings of the British Association in Montreal: "monotreme oviparous, ovum meroblastic."

Thus began the global recognition of an extraordinary fact: there lives today a mammal who lays eggs. In fact there are a few: the platypus (*Ornithorhynchus anatinus*) and four species of long- (*Zaglossus*) and short- (*Tachyglossus*) nosed echidnas, all of which are endemic to Australia and/or New Guinea. These animals belong in the group known as "monotremes." Caldwell's telegram, driven by the economics of pay-per-word communications and enriched by a classical academic training, meant that the echidna and platypus lay eggs ("monotreme oviparous") consisting of a large yolk on one side and a nascent embryo on the other ("ovum meroblastic").

As shown in Figure 3.1, monotremes are situated at the very base of living mammal diversity. They are one of the three main divisions of mammals recognized since the nineteenth century, along with placental and marsupial mammals, and all are still supported as distinct groups today. Both echidna and platypus share defining features of living mammals: a single bone comprising the jaw and three small bones conducting sound in the middle ear, lactation as a means of early nutrition, and the use of fur as insulation. Among living organisms, this combination of features is found only among mammals.

Monotremes also have very specialized traits that reflect the nature of their habitats. The platypus is a dedicated aquatic predator, with a flat tail and webbed fore- and hindfeet that make it an agile swimmer. You know already that it has a superficially duck-like bill, but you may not have known that it uses the bill to locate small prey items in freshwater streams and ponds. Unlike a duck's bill, that of the platypus is covered with electrosensitive skin, enabling the animal to detect minute impulses emitted by the muscular activity of its prey.[13] When the animal dives, both its eyes and ears are shut tight; yet it is capable of honing in on bugs a tiny fraction of its body size at distances of several meters. This sensitivity is enabled by the animal using its bill to triangulate the location of prey items, likely associated with the characteristic back-and-forth sweeping of the head while it hunts.

The bill is of course only one of the animal's distinctive features. The male platypus also has a sharp spur on its ankle, through which it can deliver a painful dose of venom. This is apparently used for male–male competition for mates, not for predation or defense. Another specialized feature of the modern platypus is that adults don't have hard, enamel-covered teeth, but only keratinous swellings which, nevertheless, do help the animal to masticate prey items. Juveniles show the initiation of real teeth during development, but these are never functional and do not break the gum. (Incidentally, there are platypus fossils that have enamel-covered teeth in adults.[14])

Both the more common short-nosed echidna (*Tachyglossus*) and the much more rare, montane, long-nosed species from New Guinea (*Zaglossus*), also possess a number of specialized traits that reflect their current habitat. *Tachyglossus* is a dedicated ant and termite eater; *Zaglossus* is more specialized for earthworms and insect larvae. Both have a long, narrow snout which lacks teeth but shows an elongate tongue. *Tachyglossus* uses its powerful forelimbs to break apart bits of decaying wood and vegetation on the forest floor and lap up the bugs inside. Like the platypus, the male echidna also has a spur on its ankle, but one with only a vestigial poison gland, so the spur does not deliver any venom. To protect itself in its terrestrial environment, echidnas are covered with dense spines. Hence, *Tachyglossus* resembles a hedgehog or porcupine, and potential predators cannot greatly disturb an adult echidna as it waddles along, thriving in a variety of different habitats.

While both echidnas and the platypus show several remarkable features related to the ways in which they make their living, they are perhaps most notable for features they share with reptiles. Egg-laying is first on the list (more on that in a moment). The zoological term that encompasses platypus and echidna, "monotreme," hints at another reptile-like character. Unlike most other mammals, the digestive and reproductive tracts coalesce into a single channel inside the animal, known as a cloaca.

Mother's milk is certainly not a reptilian character, and like other mammals, monotremes have this in abundance. However, they deliver this milk to their young in a very unmammalian way: without nipples. In a female echidna, the maternal pouch develops seasonally to correspond with egg-incubation and post-hatchling development. Milk glands are present along either side of the pouch, and when they fill, what was just a depression becomes an actual pouch. At the front of the pouch are two hairy patches which correspond to the openings of the milk glands. Here is where the developing echidna ingests its first meal. Unlike other mammals, it does so without a teat or nipple.

In terms of their skeleton, there are additional bones (interclavicle, coracoid) in the monotreme shoulder skeleton, present as separate elements in reptiles but not in marsupial and placental mammals. The largest element of the shoulder that monotremes do share with other mammals, the scapula, shows an important difference: it lacks a scapular spine, or the "shoulder blade" that, in a human or other therian mammal, divides the external surface of the scapula in two (Figure 3.2). Furthermore, the habitual stance of the echidna and platypus on land is a sprawling posture. Rather than holding their arms and legs underneath the body, these are extended to either side, with the elbows and knees flexed to about 90 degrees to keep the animal's belly off of the ground (Figures 3.2 and 5.1).

Another resemblance to reptiles has to do with the organ of hearing, or cochlea. This is located inside a skull bone called the petrosal. Birds and mammals depend heavily on the sense of hearing, and by making the cochlea longer, both groups increase their sensitivity to acoustic signals. Mammals like you and I make the cochlea longer using a unique strategy: we coil it (discussed further in Chapter 10). In all marsupials and placentals, the cochlea is so tightly coiled it resembles a snail shell. Monotremes also show some curvature in their cochlea, but not much; it is not tightly coiled as in marsupial and placental mammals.

And of course monotremes lay eggs. This is not the hard-shelled variety you see at the supermarket, but a smaller, more spherical, and leathery version. As previously mentioned, female echidnas seasonally develop a pouch and carry the egg within it as they forage. The aquatic habits of the platypus prevent this, of course, since a submerged amniote egg (pouch or not) would inhibit the exchange of oxygen and carbon dioxide that every developing embryo needs. So the female platypus uses a special nursery burrow to leave her eggs while she is off hunting.

The cryptic nature of these burrows is part of the reason why, after European scientists first became aware of this animal at the close of the eighteenth century, it took another 80 years for the question about its mode of birth to be settled. This answer turned out to be rather decisive with the

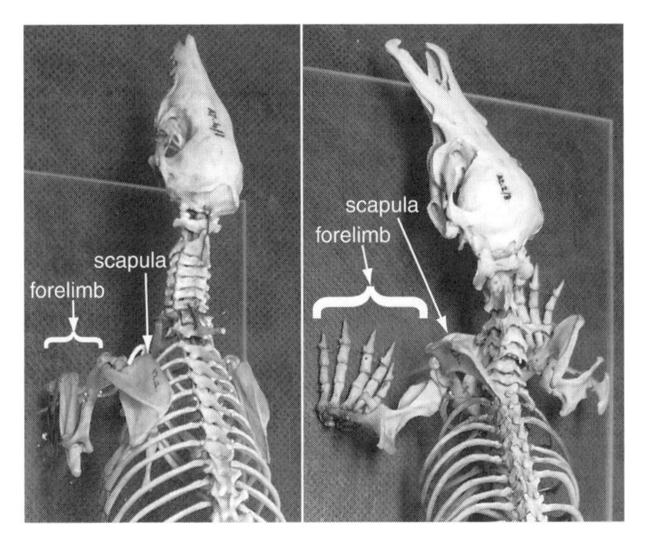

Figure 3.2. Anterior skeletons of a therian (left, the bandicoot *Isoodon obesulus*) and monotreme (right, the platypus *Ornithorhynchus anatinus*). Note the orientation of the forelimbs beneath the shoulder in the bandicoot, and out to the sides in the platypus. Also note the division of the external surface of the scapula by the scapular spine in the bandicoot, compared to the flat external surface of the platypus. These are two of the many skeletal differences that distinguish marsupial and placental mammals (collectively known as therians) from the echidna and platypus (collectively known as monotremes).

famous telegram quoted above. Interestingly, there were several Australian-based naturalists interested in the platypus during the early and mid-nineteenth century, and Australian Aborigines had known about egg-laying in monotremes for centuries. There had even been reports to this effect from European settlers dating to 1818.[15] Nevertheless, perhaps because there was not an actual cadaver involved, and/or because Victorian science demanded proof from one of their own, it took William H. Caldwell, a 20-something, Cambridge-educated Scot to settle the issue. He did not have many scruples about employing aboriginals to kill hundreds, perhaps thousands, of these animals in search for the perfect female, one which would still have an egg within its reproductive tract. He got it, and claimed to have shot the animal himself.[16]

Both the echidna and platypus are well adapted to cope with the demands of their environments. In fact, the short-beaked echidna also does quite well on the suburban frontier in environments shaped by humans, as the handful of "echidna in my garden" videos on YouTube demonstrate.[17] (The long-beaked echidna from the New Guinea highlands is an entirely different story; all species of *Zaglossus* are currently listed as critically endangered on the IUCN Redlist.[18]) But why do monotremes show such a mix of features that, in any other part of the world, are exclusively found in either a reptile (multiple shoulder bones, no scapular spine, uncoiled cochlea, sprawling gait, egg-laying, no nipples) or a mammal (single jaw bone, three ear bones, mother's milk, fur)? Why is it that only in Australia and New Guinea, mammals with a similar habitat as hedgehogs (in the case of the short-beaked echidna) or otters (in the case of the platypus), show so many peculiar, reptile-like features in their skeletons and soft tissues?

Taking an evolutionary view of mammal diversity, there is a perfectly rational answer to this question. Compared to hedgehogs, otters, or any other living mammal, monotremes have evolved from a different part of the mammalian Tree of Life. They derive from a node in that tree which long predated the divergence of not only hedgehog and otter from one another, but also the divergence of marsupial and placental mammals. As such, they carry with them part of the anatomical signature of that distant common ancestor, one which was a real animal that existed over 160 million years ago, based on our current understanding of paleontological and molecular data.[19]

That common ancestor existed prior to any kangaroo, opossum, monkey, shrew, rodent, or whale. While the ancestors of modern placental and marsupial mammals were adapting and evolving in response to their respective environments, the ancestors of monotremes were themselves the subject of considerable evolution. During their long, independent history, monotremes acquired such peculiarities as an electrosensitive bill, ankle spurs, poison glands, defensive spines, or a long tongue in a mouth without teeth. Their long evolutionary past clearly shows that neither echidna nor platypus is simply a throwback to some 160-million-year-old animal. However, it is equally clear that these animals mix anatomical features otherwise found in reptiles and mammals in just the way one would expect if Darwinian natural selection was the mechanism behind their evolution.

MARSUPIALS AND BANDICOOTS

Compared to the *ca.* 5000 species of living placental mammals, monotremes barely register on the scale of pure biomass. There are also far fewer marsupials—about 300 species. However, this comparison masks a surprising diversity. In addition to the iconic kangaroo, koala, and the American opossum, there are living marsupials that resemble mice, raccoons, weasels, pigs, anteaters, gliding squirrels, moles, otters, and shrews. Of course, these placental mammal names don't do justice to marsupials. Despite certain ecological similarities to other mammals, living marsupials share a number of attributes with each other to the exclusion of any placental mammal, including a very brief period of development inside the mother's uterus and an extended one postnatally, nourished by milk. Marsupials don't replace as many teeth as placental mammals; only the last premolar has a "baby tooth" or deciduous precursor. They also typically have more than three incisors; a certain bone called the alisphenoid forms the front of their middle ear; and their ankle and jaw bones are quite distinctive. All such features make marsupials unique, and no zoologist has had difficulty telling the living ones apart from other mammals since the nineteenth century.

Marsupials are naturally found throughout the Americas, Australia, and the southeastern archipelagos of Asia. The American species are impressive, including the very adaptable opossum (*Didelphis*) which is found from Argentina to Canada, the tropical *Caluromys* and *Marmosa*, the semiaquatic *Chironectes*, and a variety of small, shrew-like species from South America. However, in comparison with their Australian relatives, American marsupials occupy a much narrower range of habitats.

The bulk of living marsupial diversity is found in Australia and New Guinea. These islands are host to three main groups of marsupials: diprotodonts (e.g., kangaroos, wombats, koalas, possums, gliders), dasyuromorphs (e.g., quolls, numbats, devils, Tasmanian wolves), and peramelians (e.g., bandicoots), plus the enigmatic *Notoryctes* (marsupial mole). As a brief aside, it's worth mentioning the great difference between an American "opossum" and an Australian "possum." Both are extremely adaptable and can do quite well in human-influenced environments. In Sydney the animal that raids your unlocked garbage can is a brush-tailed possum (*Trichosurus vulpecula*), whereas in Buffalo it's the Virginia opossum (*Didelphis virginiana*). But these animals differ by much more than the letter "o": the former is a diprotodont, the latter a didelphid, and in evolutionary terms the two have had a distinct history within marsupials for about 60 million years.[20]

In the very recent geological past, these groups contained an even greater level of diversity. Diprotodonts were represented by the rhinoceros-sized *Diprotodon* and the hypercarnivorous, sabre-toothed *Thylacoleo* (the Australian diprotodont analogue of the similarly adapted South American hyper-predator *Thylacosmilus*). Both peramelians and dasyuromorphs also have a diverse fossil history, but one that was not quite as speciose as that of the diprotodonts.[21]

This brings us to the living bandicoots. Within the time of European settlement in Australia, there have been at least 22 species classified in the genera *Perameles*, *Isoodon*, *Macrotis*, *Echymipera*, *Peroryctes*, *Microperoryctes*, and *Rhynchomeles*. The so-called pig-footed bandicoot, *Chaeropus ecaudatus*, has not been seen in the wild since the early twentieth century and is likely now extinct. Bandicoots share numerous features of their anatomy and reproduction with all other marsupials, including opossums and kangaroos. Perhaps the most conspicuous of these is the short intra-uterine gestation, followed by a relatively long period of lactation and development in the pouch. However, bandicoots deviate from the typical "marsupial" pattern of development in a particularly interesting way that concerns the morphology of its placenta.

THE MARSUPIAL PLACENTA

"Placenta" is a general term that means an organ present in the mother's uterus that helps transmit nutrients to and waste from the developing fetus. "Placentalia" is a taxonomic name for a category of mammals that includes cows, elephants, sloths, mice, you, and me, for which the phrase "placental mammals" is equivalent. You would understandably be tempted to think since we belong to the group called Placentalia, and marsupials do not, that we're the only ones with a placenta. Indeed, egg-laying animals like crocodiles, birds, and turtles do not utilize the kind of intra-uterine development seen in placental mammals; almost everything a chick needs for nutrients and waste storage (except for a little bit of gas exchange) is contained within its egg. However, lots of other animals have some kind of live birth which requires a placenta, including certain fish, amphibians, snakes, lizards, and, as it turns out, marsupials. Hence, despite being outside of the taxonomic category of Placentalia, marsupial mammals (and many other vertebrates) do indeed have a placenta.

This organ in marsupials and non-mammals with live birth shows many differences from our own. Hopefully you recall from our earlier discussion the four embryonic membranes that all amniotes possess, and that enable them to reproduce without direct access to standing water: the amnion (surrounding the embryo), chorion (lining the egg), allantois (storing water and waste), and yolk sac (storing nutrients). In placental mammals, the placenta comprises the union of two of these membranes, the chorion and allantois, leading to the term "chorioallantoic" placenta. Most marsupials, in contrast, have a placenta comprising the chorion and yolk sac (also known as the vitelline membrane) which forms the "choriovitelline" placenta.

As in any placental organ, the union of the embryonic choriovitelline tissues with the wall of the marsupial mother's uterus allows for exchange of nutrients between fetal and maternal blood supplies. In a marsupial such as a kangaroo, this exchange is very important, but doesn't last for too long. An adult eastern gray kangaroo falls within the size range of a human (*ca.* 40–70 kg), yet has an intra-uterine gestation time roughly one-tenth as long (about 28 instead of 270 days). Development of the fertilized egg inside of the marsupial uterus is very brief. Hence, "birth" in a marsupial results in a barely recognizable little worm, something capable only of crawling from the maternal genital tract to the pouch, without a fully functional brain, hindlimbs, or internal organs. The vast majority of marsupial development takes place external to the uterus, in the pouch where the young is nourished by mother's milk.

Bandicoots depart from the typically marsupial choriovitelline placenta and show instead one formed by the chorion and allantois; they are therefore chorioallantoic like humans, mice, cows, elephants, sloths, and other placental mammals. Overall, they still possess a typically marsupial reproductive pattern, with an intra-uterine gestation lasting only a couple weeks and infant dependence on lactation thereafter. However, they show a few fascinating differences from other marsupials. One of them concerns the fact that when born, their young are among the most developed of any marsupial, and they also have the shortest lactation times among marsupials.[22]

The bandicoot *Macrotis lagotis*, or bilby, reaches about 1 kg as an adult, comparable in size to some small kangaroos (e.g., *Bettongia gaimardi*, or Tasmanian bettong) which, like most other marsupials, and unlike bandicoots, has a choriovitelline placenta. Yet *Macrotis* has a gestation time that's one-third shorter (14 versus 21 days) and a period of pouch-lactation that's almost half (90 versus 160 days) that of *Bettongia*. Furthermore, the newborn *Macrotis* is generally larger than the newborn *Bettongia* (350 versus 320 mg), reflecting the fact that at birth, peramelians are developmentally similar to marsupials with longer gestation times. Peramelians have the fastest growth rates seen in marsupials, despite having the shortest gestation time and period of juvenile dependence on lactation.[23]

One explanation for this difference concerns the efficiency of a chorioallantoic relative to a choriovitelline placenta. For reasons that are not yet entirely clear, it seems that when the allantois joins with the chorion to form the placenta, as in peramelians and placental mammals, the developing young grow faster. The fusion of the allantois with the chorion seems to enable interchange of nutrients in the blood supply of mother and embryo more thoroughly than when the yolk sac fuses with the chorion to form the placenta. Independently of the placenta, the very high fat content of peramelian milk, among the highest of mammals and almost as rich as that of some aquatic carnivorans (e.g., seals), plays a role in the fast postnatal development of these animals.[24]

Why, then, if the chorioallantoic placenta is more efficient, did it not evolve more frequently among marsupials, and why are the bandicoots not the most successful in terms of species diversity? There are far fewer peramelians (*ca.* 22 species) than diprotodonts (well over 100 species), so how good can their placentas actually be? Surely if natural selection were the operative force behind the evolution of the placenta, the efficiency of the peramelian chorioallantoic pattern would have been elaborated and diversified over the course of geologic time, and this group would now be better represented than it actually is.

There are good reasons to recognize how the unique conditions of Australia over its geologic history have not provided an ideal selective framework whereby "more efficient," as we've described above for the bandicoot placenta, equates with "better" in an overall, adaptive sense. With the exception of Antarctica (which has a geologically rather close relationship with Australia), no place on Earth is more inhospitable to animal life than Australia. The organisms on that continent have had to frequently contend with droughts and climatic unpredictability, possibly making the capacity to abandon a pregnancy with relatively small cost to the mother one of the most important factors to long-term survival.[25] The efficiency of bandicoot placentation and lactation is dependent on a steady supply of nutrients, and if the mother herself cannot survive periods of food-scarcity, then it doesn't matter how well she can supply her embryo or joey with nutrients. The ability of choriovitelline marsupials such as kangaroos to jettison an embryo in times of scarcity, without the mother herself losing an inordinate amount of energy in the process, has ensured the success of their reproductive strategy. During the last century there have been populations of human-managed sheep (like us, sheep are placental mammals too) in Australia that have been decimated by droughts. The wallabies, without human minders looking after them, have done much better. [26]

Another possible reason why the chorioallantoic placenta of bandicoots has not led to a more placental-mammal-like pattern (e.g., by prolonging the period of intra-uterine gestation) concerns the constraints of its reproductive anatomy. For example, rather than a midline uterus that connects via a single vagina to the cervix, as in humans and other placental mammals, marsupials have two uteri that connect to the cervix via two lateral vaginas and a central, pseudovaginal canal. Birth in a marsupial typically occurs via the pseudovaginal canal, and the connection between uterus and the outside is more convoluted than in placental mammals. This may constrain the size of the newborn marsupial, peramelians included, regardless of how efficiently a chorioallantoic placenta enables an embryo to develop.

Bandicoots are modern animals, terminal twigs within the marsupial portion of the Tree of Life (Figure 3.1). No contemporary biologist argues that modern placental mammals evolved from a bandicoot. Note, however, the recurring theme emphasized in this chapter: anatomical features that characterize modern groups, such as the chorioallantoic placenta present in every living placental mammal, tend to crop up sporadically elsewhere in the Tree of Life. Although they unequivocally qualify as part of the marsupial radiation, living bandicoots depart from their brethren in a surprisingly "placental" way and, like the coqui frog, platypus, and tarsier (our next example), demonstrate a point that is best understood in a Darwinian context: anatomical features appear on the Tree of Life independently of the animals in which they predominate today.

TARSIERS

All living primates belong in one of two groups: strepsirhines or haplorhines. The former contains lemurs, lorises, galagos, and a number of more obscure animals like indris and aye-ayes, located primarily on the island of Madagascar. The latter contains monkeys, apes, humans, and the tarsier.[27] One of the fairly obvious differences between the two groups has to do with the size of their nose and their dependence on smell: a lemur has a large, more typically mammalian nose, whereas that of most monkeys and apes comprises a relatively smaller proportion of overall skull size. The difference is a slightly less extreme version of that between a human and a dog. Think of the small distance between your eyeball and nostril compared to that of your pet retriever. As clever as you think your dog is, there's more nose in his head than brain, and a big chunk of the brain that is there doesn't do much else besides interpret smell.

In contrast, haplorhine primates are among the most visually oriented of all mammals, and tarsiers are among the most visually oriented of all haplorhines. These small gremlin-like animals inhabit the jungles of the Philippines, Borneo, Sulawesi, and other islands of the southeast Asian archipelago, and consist of roughly a half-dozen species, similar in size to a small rat. You would never mistake one for a rat, however. These animals look like two massive eyes connected to a pair of long legs. Due to their large eyes taking up nearly all of the space in the front part of the animal's skull, the tarsier's nose is relatively small, even by haplorhine standards (Figure 3.3). Tarsiers are active at night in southeast Asian rainforests and depend on visual acuity in a low-light environment to hunt insects and small vertebrates.

For over a century, scientists have noted that tarsiers share a number of features with lemurs, galagos, and lorises (strepsirhines) on the one hand, and monkeys, apes, and humans (anthropoids) on the other. For example, the diet and size of a tarsier closely resemble those of a strepsirhine, as do the two halves of its jaw, loosely connected to one another in front, not solidly fused into a single, horseshoe-shaped bone, as they are in an anthropoid (Figure 3.3). On the other hand, tarsiers resemble anthropoids in having a bony wall in the back of their orbit (Figure 3.3), multiple enclosed spaces within their middle ear, a right angle defining the connection of its astragalus to its fibula (comprising the ankle), and in not having the capacity to make their own ascorbic acid, or vitamin C.[28]

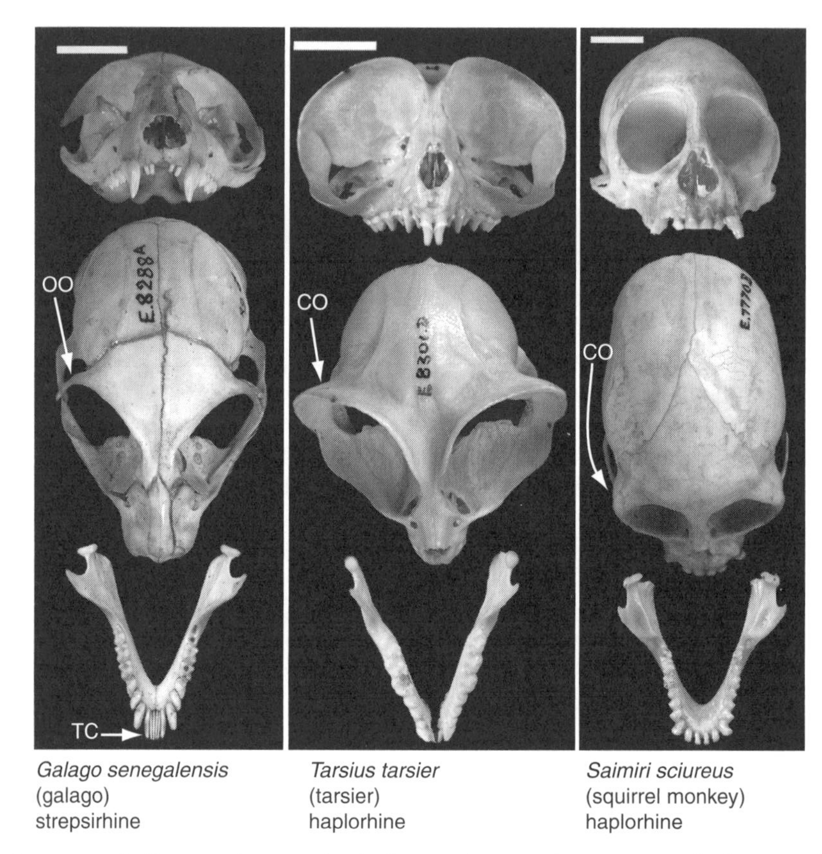

Figure 3.3. Differences between haplorhine and strepsirhine primate skulls. Note the toothcomb (TC) and open orbit (OO) in the galago, as opposed to the relatively short incisors and closed orbits (CO) of the tarsier and the squirrel monkey. The squirrel monkey and tarsier lack a reflective structure inside the eyeball known as the tapetum lucidum. In the tarsier, this means in order to survive as a nocturnal, visual predator, it has to substantially enlarge its eyeballs. The galago has a similar niche, but because of its placement on the Tree of Life near other strepsirhine primates, it has a tapetum and a toothcomb. The scale bars at the top left of each image equal 1 cm.

A further difference between tarsiers and strepsirhines relates to the enormous size of the tarsier's eyes. As noted above, these are large to facilitate their need for hunting at night, dependent as they are on visual acuity to find prey. However, many visually oriented animals are successfully nocturnal with relatively smaller eyeballs. In fact, this is a common niche for other small primates, such as galagos and lorises. Why are tarsier eyes so much bigger? The answer derives from another similarity to anthropoids: tarsiers, like anthropoids, lack a reflective structure within the eyeball called the tapetum lucidum. This serves to make better use of the limited visible radiation at night by reflecting what is there back within the chamber of the eyeball. This results in the "eyeshine" you see when a deer, raccoon, dog, or cat stares at you at night from in front of your car. It's widely distributed in all kinds of mammals, but is absent in anthropoids (including humans) and tarsiers. So to make up for its lack of a tapetum in its nocturnal vocation, tarsiers need really big eyeballs.

Southern Asia and Africa have a number of small, nocturnal, strepsirhine primates that are very similar to the tarsier in their ecology and diet of insects and small vertebrates. These include lorises (*Loris*, *Nycticebus*) in southern Asia and galagos (*Otolemur*, *Galago*) and pottos (*Perodicticus*) in Africa. All of these animals have several specializations not seen in the tarsier. One of the most obvious and important of these is the toothcomb (Figure 3.3), a device that does essentially what you'd expect from the name. It is indeed a comb consisting of elongate, anteriorly pointing lower incisors and usually canines. Nocturnal, insect-eating primates such as lorises, galagos, and pottos can use the toothcomb in social interactions to groom each other, but because these animals tend not to form large groups, more often than not they use their toothcomb to procure food. One kind of galago (*Euoticus elegantulus*) uses its toothcomb to scour tree branches and feed on sap. Other strepsirhines, such as ring-tailed lemurs from the island of Madagascar, live in large groups and use the toothcomb more frequently in social interactions.[29] Just as chimps and gorillas manage their social hierarchy by the patterns of who-grooms-who, many lemurs do the same thing, but use their teeth more frequently than their hands.

With notable exceptions (such as the sap-eating *Euoticus* and gregarious lemurs), several small, strepsirhine primates have a very similar niche as the tarsier. These noctural predators (galagos, lorises, pottos, mouse lemurs) also visually hunt insects at night in a forest, but show important anatomical differences which reveal something about their evolutionary history. From a pure engineering point of view, galagos don't need a toothcomb or a wet nose to make a living, since tarsiers don't have them and yet do quite well at hunting insects at night. Nor do they need an open orbit, for the same reason. Similarly, and again in terms of simple design, a tarsier doesn't necessarily need its huge eyeballs; if it had a tapetum it would get along just fine with galago- or loris-sized eyes. There's nothing essential about its sharply angled ankle joint (a feature shared with monkeys and apes) for nocturnal insect-hunting, either. In sum, many of the features of tarsiers and other small, nocturnal, insect-eating primates do not exist due to the pure functional requirements of making a living in the forest at night, since other animals in this niche lack them and do just fine. So why do these groups have these particular bits of anatomy?

Evolution by natural selection gives us an explanation: tarsiers evolved from one part of the Tree of Life, and galagos, lorises, lemurs, and pottos evolved from another, and each came with a specific set of anatomical baggage. Specifically, tarsiers share a common ancestor with monkeys and apes, one which possessed the specializations of these living animals: a walled-off orbit, sharp ankle joint, lack of a tapetum lucidum, and inability to synthesize vitamin C (to name a few). In contrast, galagos and other strepsirhines share a common ancestor that possessed not only a tapetum, but also a toothcomb, open orbit, sloped ankle joint, and vitamin C metabolism. The suites of characters present in each group of small, nocturnal predators are only partly determined by the requirements of making a living in this kind of environment. Their distinct evolutionary histories, which we represent by their positions on the Tree of Life (Figure 3.1), have had a profound influence in determining their anatomy.

SUMMARY

In the preceding pages I've detailed just a handful of animals that we can better understand thanks to the Darwinian basis of evolutionary biology. Anti-evolutionists can complain that there are unsolved questions and uncertainties, and indeed a careful search of the literature will find qualifications to the generally accepted ideas concerning adaptation and evolutionary relationships that I've summarized above.[30] This is the nature of any vibrant scientific field. Nevertheless, the reality is that the Darwinian process of natural selection is reasonably demonstrable as a major explanatory factor in all of the above cases. No one has a better mechanism to understand how tarsiers, bandicoots, monotremes, or coqui frogs have come to possess their own particular suites of characters. Calling such an animal an act of "design" or "creation" simply repeats the fact that they exist. We knew that. No such claim does the hard work of specifying a mechanism by which their particular suites of characters came about in an individual, living species.

Darwin's mechanism is about how biological species express natural continuity, and is apparent by observing the shape, size, and distribution of organisms such as those reviewed above. Charles Darwin knew a lot about living plants and animals, where they existed, and what they looked like, but there are other lines of evidence about which he knew relatively little by today's standards, such as the fossil record. To this we now turn.

FOUR

THE FOSSIL RECORD

One of the most powerful means by which scientific inquiry can confirm a hypothesis is the congruence of data from different sources toward a common answer. We've just reviewed above some of the comparative anatomy (the study of the skeleton, teeth, limbs, organs, etc. across animal groups) that biologists use to show how vertebrates share common ancestry. Monkeys and humans are closer to one another than either is to a galago, and these three in turn share more recent common ancestry among each other than any does with a kangaroo. All mammals are more closely related to one another than any is to a crocodile, and so on, reflecting the branching structure shown in Figure 3.1. This broad-scale theory of genealogical relations among animals was initially proposed on developmental and anatomical evidence. Now, we know that completely independent bodies of data from the fossil record and molecular biology have also lent support to this theory of interrelationships.

The pattern of genealogical relationships shown in Figure 3.1 has been articulated by many biologists with substantial consistency, in some cases before[1] and more frequently after[2] publication of *On the Origin of Species*. Pre-Darwinian biologists such as the Swedish taxonomist Carl Linnaeus or the French anatomist Georges Cuvier did not have a robust causal mechanism, or process, to explain what they observed in terms of shared features across plants and animals, or pattern. Nonetheless, Linnaeus and Cuvier (and anti-Darwinists such as Richard Owen or Louis Agassiz) were perfectly

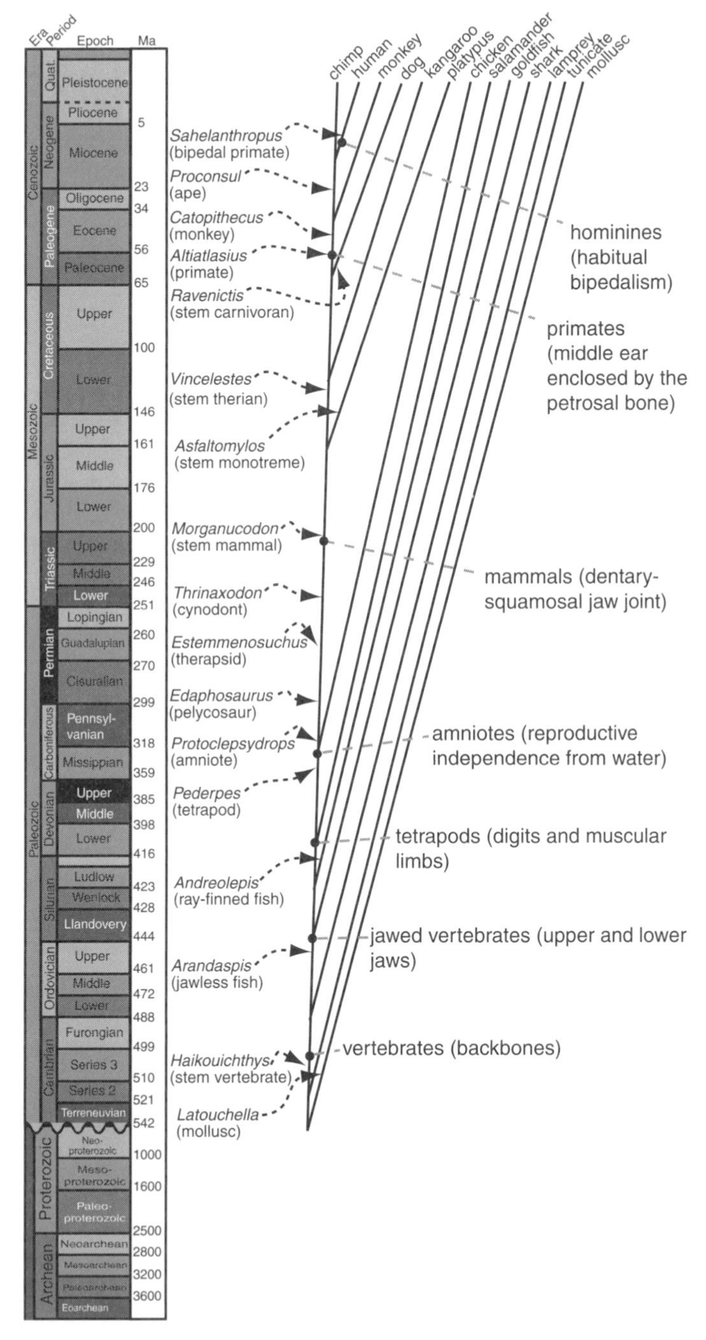

Figure 4.1. Correspondence of stratigraphic first appearance of fossils representative of major animal groups (after Benton *et al.* 2009) with their positions on the Tree of Life, arbitrarily focused on primates. The stratigraphic column is based on Gradstein

capable of observing the pattern left behind by evolution. They knew that some animals resemble each other in many ways despite living in very different environments (such as seals and cats, or hippos and camels). While they did not understand the process behind these patterns of resemblance, non-Darwinian taxonomists were aware of the patterns and even incorporated them into their classifications.

As an important aside, this explains why natural selection as a mechanism behind evolution would have been articulated by someone else had Darwin and Wallace not published on it first. Sir Richard Owen, founder of the Natural History Museum in London, was publicly very critical of *On the Origin of Species*, although in at least one personal meeting with Darwin in 1859 he was very complimentary about the book and its explanatory value.[3] There were even attempts to credit Owen during his lifetime with recognition of the "evolution of species" prior to Darwin's book.[4] In fact, Owen never clearly articulated what kind of mechanism he envisioned behind the obvious change over time that he observed as a veteran paleontologist, but he certainly *did* recognize that animals evolved at some level.[5] Owen followed the intriguing but vague tradition of German and French natural philosophers in which animals were derived from a common, platonic archetype, the imprint of which was evident in fossils and modern organisms alike, consistent with the belief that across geological time there was common ancestry. Natural selection per se was not a part of this tradition. However, anyone with an open, critical mind, awareness of the natural world, and sufficient intellectual courage could have made this connection between pattern and process via natural selection. Indeed, as we previously discussed, that is exactly what Alfred Russel Wallace did. His letter in early 1858 compelled Darwin to publish the first explicit paper on evolution by natural selection a few months later, with Wallace as his co-author.[6]

The taxonomies generated by nineteenth-century biologists show many imperfections by today's standards. However, for animals with a backbone (i.e., vertebrates like fish, frogs, birds, you, and I), they imply a pattern of evolutionary relatedness very similar to that accepted today. This Tree of

Caption for Figure 4.1. continued

et al. (2004). Some of the more widely known nodes are labeled at right with their diagnostic anatomical features. Zoologists of the nineteenth century had inferred a tree very similar to this one (see Figures 4.2, 4.3), but without knowing much about the fossil record of each lineage. The fact that the oldest hominine, primate, mammal, amniote, tetrapod, jawed fish, and vertebrate (to name a few) all follow each other sequentially in the record is strong support for common descent, one of the predictions of evolution by natural selection.

Life is in turn supported by two completely independent bodies of data about which mid-nineteenth-century biology knew relatively little: stratigraphy and molecular biology. The succession of animal forms in the fossil record, or stratigraphy, was beginning to form a sophisticated discipline in the nineteenth century, but the fossil record of vertebrates at that time was vastly inferior to what is known today. Molecular biology, of course, did not develop as an academic discipline until well into the twentieth century.

The comparative anatomical data used in the nineteenth century to build biological classifications is mutually consistent, or congruent, with what we now know from stratigraphy and molecular biology. All provide strong evidence in support of the idea that natural selection is a mechanism behind evolution. Stated differently, this congruence is what we would expect to find if natural selection were the means by which animals alive today evolved from one or few common ancestors in the geological past. Early evolutionary biologists did not know the details of many of the fossil groups now represented in the stratigraphic record (Figure 4.1), and they knew nothing about the molecular biology of these animals. Yet both kinds of data support ideas of vertebrate interrelationships similar to those first derived from comparative anatomy and embryology. The details of how molecular biology fits into this picture are discussed in Chapters 9 and 10. Here, we'll have a look at how the fossil record documents evolution.

STRATIGRAPHY

Figure 4.1 represents the kind of image that has been shown in textbooks for over a century, although the more recent versions have a much better empirical basis than the older ones. It shows that the most ancient vertebrates from fossil-bearing rocks are segmented and aquatic, but lack jaws. Those sediments bearing the oldest fossils of jawed vertebrates—again all sea-creatures—are younger and lack tetrapods, i.e., animals with muscular limbs ending in digits capable of propelling them in shallow water or on land. Those with the oldest tetrapods are younger still and lack mammals or birds. Those with the oldest mammals and birds are again younger and lack primates, whales, ostriches, and hummingbirds. Our own species shows not only anatomical bipedalism, but also an exceptionally large braincase, a relatively small toothrow, and a chin, among other features. Anatomically modern humans don't appear in the fossil record until somewhere between 0.1 and 0.2 million years ago, a minute, utmost upper slice of the last 600 million during which we have a decent fossil record. In sum, fossils discovered in geologically older strata tend to be potentially ancestral to those from younger strata. Relatedly, fossils from younger strata tend to show more similarity to modern groups than do those from older strata.

Think about it: why are there no birds, snakes, or mammals from geological deposits that contain trilobites or placoderm fish? Why are there no human or rabbit fossils from Permian deposits that contain early synapsids such as *Dimetrodon*? The answer is simple: they hadn't yet evolved. This does not mean that every fossil represents a "transition" on the ladder of nature that serves as a connection between groups alive today. As previously discussed here and elsewhere,[7] the fossil record does not represent a "ladder" at all. The animals that paleontologists identify are, broadly speaking, cousins on the "bush" of life connecting various animal groups with which you are familiar. Combined with the fact that extinction and fossilization occur for reasons that have little to do with species' origin, it should not come as a surprise that occasionally, while fossil organisms identified as ancestral generally are found in older strata, some such populations postdate (on a geological scale) others that are more specialized.

Even with this qualification, the overall signal is unambiguous: reconstructions of the vertebrate Tree of Life based on anatomy correspond closely with the fossil record. Figures 4.2 and 4.3 show genealogies of vertebrate animals derived from two nineteenth-century publications, one by Ernst Haeckel[8] and another by Theodore Gill.[9] For the most important branching points, both Haeckel and Gill proposed an arrangement based on anatomy and development that largely corresponds with the succession of vertebrates in the fossil record we know today, even though at the time they had only a very dim notion of what that record actually was.

In his 1863 book, *Man's Place in Nature*, Darwin's close colleague Thomas Henry Huxley also outlined his ideas on how the Darwinian mechanism implied genealogical relationships among animals. Like Haeckel and Gill, Huxley viewed humanity as part of Mammalia, based largely on developmental evidence:

> the more closely any animals resemble one another in adult structure, the longer and the more intimately do their embryos resemble one another: so that, for example, the embryos of a Snake and of a Lizard remain like one another longer than do those of a Snake and of a Bird; and the embryo of a Dog and of a Cat remain like one another for a far longer period than do those of a Dog and a Bird; or of a Dog and an Opossum; or even than those of a Dog and a Monkey. Thus the study of development affords a clear test of closeness of structural affinity. ... Does [humanity] originate in a totally different way from Dog, Bird, Frog, and Fish, thus justifying those who assert him to have no place in nature and no real affinity with the lower world of animal life? Or does he originate in a similar germ, pass through the same slow and gradually progressive modifications, [does he] depend on the same contrivances for

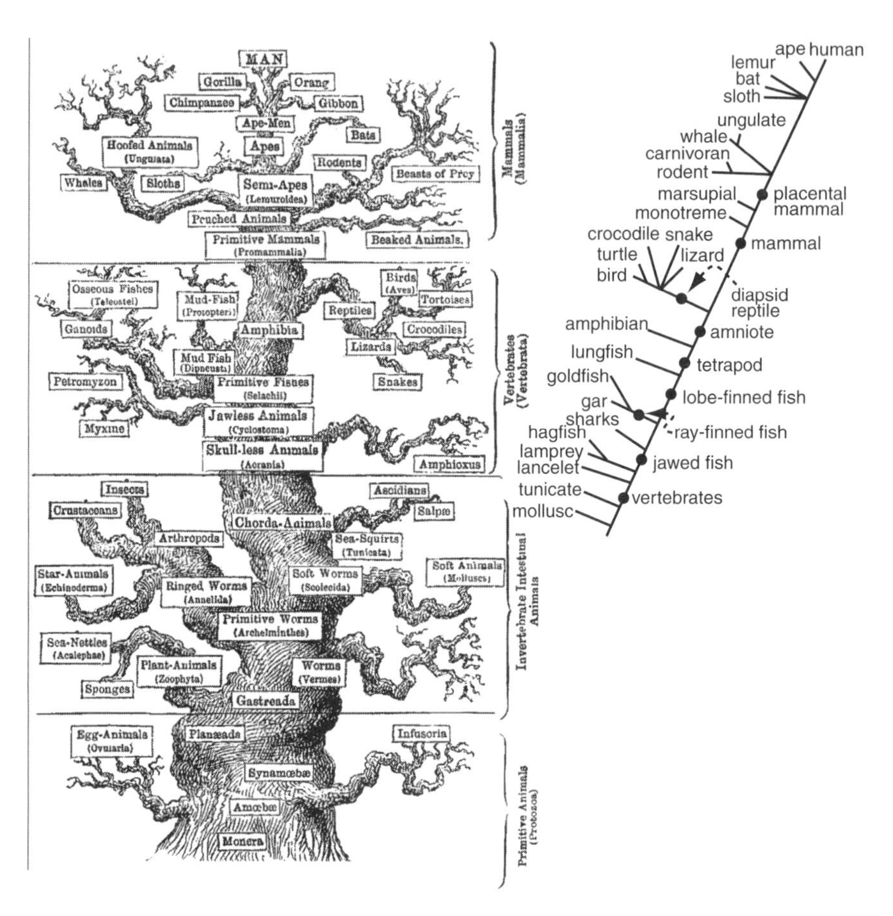

Figure 4.2. At left is a diagram published in the 1897 English edition of Ernst Haeckel's *Evolution of Man*, volume II, repeating his previously articulated views (e.g., Haeckel 1873: 513) on how different vertebrate groups were united by common ancestry in an age when neither the fossil record nor the molecular biology of vertebrates was well understood. At right is his tree redrawn using the modern convention of a "cladogram" or evolutionary genealogy (branch lengths are arbitrary). His arrangement differs from modern ones in only a few regards (e.g., tunicates, turtles, some groups within mammals).

> protection and nutrition, and finally enter the world by the help of the same mechanism? The reply is not doubtful for a moment. ... Without question, the mode of origin and the early stages of the development of man are identical with those of the animals immediately below him in the scale: without a doubt, in these respects, he is far nearer the Apes, than the Apes are to the Dog.[10]

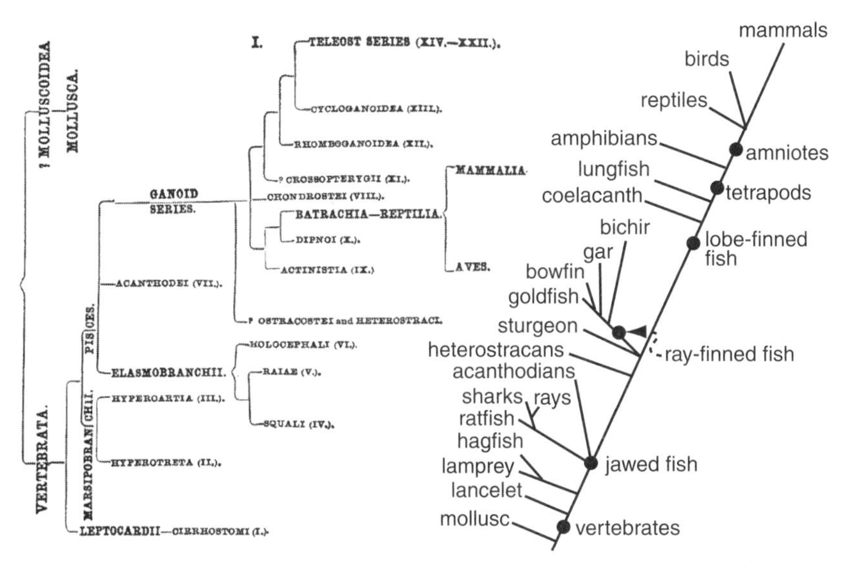

Figure 4.3. At left is a diagram published by Theodore Gill in 1872 (p. xliii), encapsulating his ideas on how different vertebrate groups were united by common ancestry in an age when neither the fossil record nor the molecular biology of vertebrates was well understood. At right is his tree redrawn using the modern convention of a "cladogram" or evolutionary genealogy (branch lengths are arbitrary). Our understanding of this genealogy differs from his in a few regards, such as placing birds closer to other "reptiles" than mammals, fossil jawless fish close to hagfish and lamprey, and sturgeons within ray-finned fish. In other regards, his tree is close to modern ones based on the fossil record and molecular biology. The correspondence of these distinct bodies of evidence toward the same pattern supports the role of natural selection as the process behind the pattern of vertebrate biodiversity.

In this book, Huxley continues to elaborate on the ways in which humans resemble other animals, particularly chimpanzees and gorillas, both in terms of their development and gross anatomy as adults. On p. 125, he concludes

> If man be separated by no greater structural barrier from the brutes than they are from one another—then it seems to follow that if any process of physical causation can be discovered by which the genera and families of ordinary animals have been produced, that process of causation is amply sufficient to account for the origin of Man. … At the present moment, but one process of physical causation has any evidence in its favour … that propounded by Mr. Darwin.

Huxley has just used Darwinian evolution to explain why animals, including humans, resemble each other in terms of their development and anatomy. Furthermore, he has implied that the level of similarity corresponds with the recency of common ancestry, and he further outlined an idea of common ancestry consistent with the evolutionary trees subsequently made by Haeckel (Figure 4.2) and Gill (Figure 4.3). As we'll see later on, this arrangement is supported by analyses of molecular data as well.[11] Haeckel, Gill, and Huxley constructed their classifications based on developmental and comparative anatomy, and without most of our current understanding of paleontology and no clue at all about molecular biology. They could have classified animals in a number of different ways, placing, for example, the shark in the same category as whale, lungfish with tuna, or raccoon with opossum. But they didn't, because they saw that such categories would not be compatible with the then-evident patterns of development and anatomy. One-and-a-half centuries later, we see a remarkable correspondence between their categories and the first appearances of each group in the fossil record (Figures 4.1–4.3).

At a finer scale, the story is of course more complex. Paleontologists are generally not under the illusion that we're out to identify the literal, direct ancestor of modern groups. Nor do modern paleontologists claim that geologically older fossils always represent ancestral organisms. In fact, many fierce debates exist about the extent to which the fossil record accurately records the first appearance of a given group,[12] and paleontologists realize that a first appearance in the rock record is an underestimate of the actual first appearance of that species on our planet.[13]

Indeed, a sure-fire way of making a splash in paleontology is to push back the record of a group by finding the oldest [insert favorite animal here]. Such discoveries happen quite frequently, for example, the recent description of a trackway of a land-walking vertebrate from rocks in southeastern Poland that date to the middle Devonian, about 397 million years ago.[14] These prints show that some animals at that time were capable of propelling themselves with muscular limbs that ended in digits. They predate the oldest skeletal remains of animals with muscular limbs and digits (e.g., *Tiktaalik*,[15] perhaps better known as "fishapod"[16]) by about 18 million years, and they predate other trackways by about 12 million years or perhaps even less, given uncertainty on the age of other trackway sites.[17] Relative to what had been thought to be a 385-million-year fossil history of tetrapods, the Polish discovery extends the record of this group by about 3%. This is an important find, but does not change at all the fact that, as shown in Figure 4.1, the first appearances of major groups correspond quite closely with the genealogical relationships inferred from anatomical and molecular evidence.

Hence, on a broad scale, the signal is unambiguous,[18] but at finer levels of resolution there is definitely some noise, and both current and nineteenth-century biologists have made mistakes. For example, Huxley entertained ideas about a long, independent history of lineages within mammals that are no longer tenable today. He thought that mammals evolved from multiple common ancestors from within reptiles, what he once called "Hypotheria," and believed that individual orders (e.g., whales, primates, rodents, etc.) independently progressed through stages in which they exhibited primitive features such as lizard-like jaw joints and absence of mammary glands.[19] These ideas are wrong, but they do not diminish the credit that he (among other nineteenth-century biologists) deserves for getting much of the Tree of Life right, including the observation that humans belong to the primates, one of several kinds of mammals, closer to one another genealogically than any is to a bird, lizard, frog, or lungfish (Figures 4.2, 4.3).

As it turns out, we now know that the fossil record of humans is far younger than that of all monkeys, which is in turn far younger than that of all mammals, which is in turn far younger than that of all tetrapods (Figure 4.1). No one in the mid-nineteenth century knew the extent to which the fossil record of, say, amniotes (i.e., lizards, snakes, crocodiles, birds, and mammals) predated that of mammals, or that their supposition that lungfish are closer to land animals than they are to other fish is now supported by a dense fossil record.[20] Neither Huxley, Haeckel, nor Gill could have known during their lifetimes that the fossil record we have in the twenty-first century would emphatically support their conclusions on life's branching pattern, but it does.

MICRO- AND MACROEVOLUTION

Natural selection as outlined in the Darwin–Wallace theory states that because species generally produce more offspring than can survive, and because those offspring differ in features that can be passed down to future generations, it follows that any heritable feature that helps certain individuals survive and reproduce will become more and more prevalent in breeding populations. This is an intuitive proposition that even some creationists accept, as it is easily observable within the course of a single human lifetime. The difficulty some people have is not with this form of so-called "microevolution," but with this pattern extrapolated over a very long expanse of geological time. This is "macroevolution,"[21] and it typically includes such major events as the evolution of land-dwelling animals from those restricted to an aquatic environment, or the evolution of birds from terrestrial dinosaurs. Many people think it also includes the evolution from a common ancestor of modern humans and chimpanzees, although the anatomical differences in our case are small compared to those of other species, e.g., between a sparrow and the common ancestor it shares with *Velociraptor*. Nevertheless, the fact that it's hard to draw the line between micro- and macroevolution is precisely the point; the two exist on a continuum and the distinction between them is for the most part arbitrary. (We will discuss at the end of this section the limited ways in which "macro"-evolution can genuinely differ from "micro"-evolution.)

An interesting application of the micro/macro dichotomy regards "baraminology," or the creationist reinterpretation of some natural phenomena to fit into their literal reading of Genesis. You may recall our discussion in Chapter 2 of Noah's ark. To help mediate the obvious difficulty of fitting representatives of all the Earth's species onto a "mid-size cargo ship," creationists now argue that they don't have to fit a pair of every sexually breeding organism onto this boat because they allow for "microevolution." The ark need not have contained every species of Earth-bound animal, they say, but only those that served as the forerunners of each "kind." Hence, the ark did not host pairs of both coyote and jackal (for which they accept evolutionary descent from a wolf-like canid), but it must have had at least one breeding pair each of cat and dog, with the variations on each "kind" following later, thanks to microevolution. Therefore, the authors of one creationist website[22] do in fact subscribe to Darwinian evolution, but draw a line between cats and dogs. Given what we know about "micro"-evolution as an empirical fact, easily observable in the course of a single human lifetime, and about the abundance of living and fossil intermediates between major animal groups alive today (among other lines of evidence), this line is very fuzzy when viewed in the wholeness of geological time.

Descent with modification is a major engine behind both micro- and macroevolution, and with only a few qualifications, the primary difference between them is one of perception. Within one human lifetime, we can easily see evolution at a certain level: animal and plant domestication, resistance to medication among infectious viruses and bacteria, as well as other changes in natural populations of many sorts of organisms.[23] Many of these cases of evolution-in-action involve dramatic changes: bacteria that have evolved the capacity to digest a synthetic chemical unknown to nature before the 1930s,[24] the ability to thrive on a substance that is useless to natural populations,[25] or the appearance of a new species of amoeba that contains a bacterial symbiont.[26] Other, specific cases include great differences in age at maturity and body size in some vertebrates within only a couple dozen generations,[27] the evolution of body armor in small, freshwater fish from the Pacific northwest[28] (discussed further in Chapter 11), the rapid disappearance of migratory activity through selection in both natural and experimental observations of birds,[29] or the substantial changes in the skeleton of certain rodents tracking human urban populations.[30]

Darwin's own observations on animal domestication amply showed just how anatomically flexible a species can be over a short period of time. Despite the fact that Charles Darwin had access to an exceedingly bad fossil record (by today's standards), he ultimately made the connection between species' lability across a few generations and the mechanism by which species themselves arise. He made the rational, but challenging, step from a world limited by a personal sense of time to one that went far beyond the scale of a single human life. Within a couple decades of 1859, he had convinced the scientific community to do the same.

PUNCTUATED EQUILIBRIUM

While the mechanism of natural selection is still vital to macroevolution, one of the reasons for the prefix "macro" relates to how the process of evolution usually shows up in the fossil record. Should we expect to see the results of macroevolution in the geological record as clearly as we see microevolution in, say, dog breeds over the last two centuries? Not necessarily, and here is where much misinterpretation has occurred in both the public and scientific arenas. "Punctuated equilibrium" is the phenomenon by which, in 1972, paleontologists Niles Eldredge and Stephen Jay Gould applied a certain pattern of speciation to the fossil record.[31] They concluded that long periods of stasis, or "equilibrium," within a lineage "punctuated" by what appear to be sudden episodes of change, is often what we should expect to find in the fossil record given what we know about how species actually diverge from one another.

An analogy with language can help to explain Eldredge and Gould's concept. My friend Tim was born into an English-speaking family. Over the years, during particular bouts of his formative and professional life, he's lived in both Spain and Germany (not to mention densely Latino areas of New York), and has managed to learn the dominant languages of both countries. He's still got an accent, and now that he's back in a country where he doesn't use either Spanish or German every day, his mastery of them is not improving. Nevertheless, at this moment, he's pretty good at using both.

If you had access to recordings of his spoken voice constantly over the past 20 years, you'd most frequently hear him use English. Furthermore, the rate at which he's learned foreign languages during that time has not been at all steady. He took some Spanish in high school in the mid-1980s, but not really enough to be able to carry on a decent conversation. He only really started learning the language when he arrived in Spain in 1989. His proficiency shot up those first few months, and by December his Spanish was pretty good. Another visit to Spain in 1991, followed by a close circle of Latino friends (several of whom were his housemates) during his early PhD career, led to additional bursts of linguistic improvement.

With German it's a similar story. Tim had no speaking ability in that language until the summer of 1998, when he arrived at the Goethe Institut in Freiburg. His German improved dramatically during the eight weeks spent there and in the following year as a research student at the University of Tuebingen. He returned to the United States in late 1999 with a decent command of German. This started to improve again when he moved to Berlin in 2003, a city that he called home for the next few years.

Listening to those hypothetical recordings of Tim over the past 20 years, you'd notice a sharp rise in vocabulary and grammar in the respective language while he was most exposed to it, and little improvement during those times when he was surrounded only by English speakers. This reflects the normal way in which we learn. You take classes, go shopping, listen and talk to people, and even develop relationships with some of them. By virtue of such daily exposure, a foreign language develops to the point where you don't have to think too much about making yourself understood. Without exposure to these languages, you could say that Tim's foreign-language ability stagnated, or at best entered a kind of stasis during which his ability didn't improve.

So would you expect that those recordings over the past 20 years would exhibit a perfectly graded chain of "no–little–some–moderate–decent–good–fluent" speaking ability in each language, with each stage taking the same amount of time as the previous one? Of course not. Tim's Spanish and German have undergone periods of "equilibrium," representing little change, "punctuated" by bouts of learning when he was frequently exposed to the language. This understanding of how people learn languages leads to a realistic expectation of how Tim's personal linguistic history would sound to anyone listening over the past couple decades. No one would expect to hear a gradual, daily improvement of his Spanish and German over the past two decades; no one would reasonably conclude from any single period of stasis that at some other time he learned nothing.

Punctuated equilibrium applied to biological evolution is similar. A realistic appreciation of how species can arise involves the concept of "allopatric" (sometimes more precisely referred to as "peripatric") speciation. This means that a population of one species becomes reproductively isolated from its parent group, but is just large enough to survive and continue breeding. Imagine a storm blowing a flock of birds out toward an island on the continental shelf, or a mudslide changing the course of a river that ends up separating a group of gazelles from their larger community. Because these groups are small, genetic variations have a better chance of spreading throughout the group over fewer generations, compared to a large group

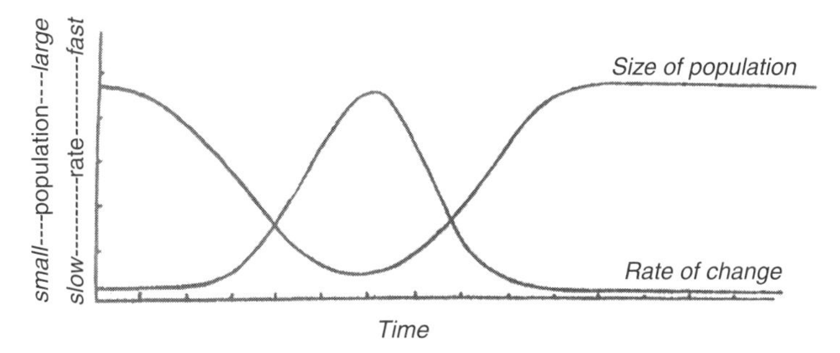

Figure 4.4. Figure 17 reproduced from p.118 of George Simpson's 1944 book, *Tempo and Mode in Evolution*. Simpson made the observation that small populations will be the ones that exhibit the most evolutionary change. At the same time, by virtue of their rarity, they will also be much less likely to be recovered in the fossil record than representatives of large populations. This idea foreshadows the concept of "punctuated equilibrium" (Eldredge and Gould 1972), and helps to explain why "gaps" in the fossil record do not contradict evolution by natural selection.

in which such novelties take longer to become fixed. At the same time, the more numerous you are as a population or species, the greater your chances are of ending up in the fossil record. Fossilization is not a likely event, and if you're part of a very small population, the chances that a member of your group will become immortalized as a fossil are much more remote than for groups with many individuals. In other words, one of the primary models of species' origin (allopatry) coincides with a major drop in the chances of ending up in the fossil record. This concept was illustrated by the famous paleontologist George Simpson in 1944 (Figure 4.4), long before anyone was misquoting scientific articles on "punctuated equilibrium."

This is not to say that all speciation events are allopatric, i.e., that they result from reproductive isolation of small founder populations. Nor does it mean that anything with "intermediate" morphology has no chance of becoming a fossil (as we'll discuss at length later on). However, it does lead to an understanding of how an important model of speciation should typically manifest itself in the fossil record: a given species should typically show stasis, or "equilibrium," during those times when it is well-adapted to its particular environment and sufficiently numerous to leave behind evidence of its existence in the fossil record. An isolated, small daughter population—one in which an evolutionary change is prone to become widespread—will be less likely to leave behind fossil representatives simply because it is small. After many generations, and possibly eons of geological time, if and when that population becomes big enough to leave behind fossil representatives, the paleontological "equilibrium" of the parent species will be accompanied (and not necessarily replaced) by the apparently sudden appearance in the fossil record, or "punctuation," of the daughter species.

Hopefully the analogy with language is clear. No one expects foreign-language fluency to come about in precise, hourly increments starting with the first class. That's not the way humans learn. In the same way, no one should expect fossils to appear as an unbroken chain of perfectly graded intermediates connecting their living descendents. That's not the way animals become part of the fossil record.

In his writings about punctuated equilibrium, Stephen Jay Gould's appreciation of natural selection as an important mechanism by which evolutionary change occurs was sometimes hard to recognize for all the rhetoric. Gould himself deserves some blame for this due to his flamboyant tone, courtship of ambiguous mechanisms,[32] and understatement of genuine examples in the fossil record that do not exhibit punctuated equilibrium.[33] Some of his peers therefore regarded punctuated equilibrium as having little substance. For example, in chapter nine of his 1986 book, *The Blind Watchmaker*, Richard Dawkins took issue with Gould's portrayal of paleontologists' expectation of change in the fossil record, an expectation known as "gradualism." Dawkins argued that Gould made an unfair caricature of "gradualists" who expected every subtle change in shape and size associated with speciation to show up in the fossil record. Such a caricature would be similar, said Dawkins, to the claim that biblical historians believed that the Israelites traversed the 200 miles from Egypt to Israel in their 40-year journey in daily increments of about 24 yards, always in the same direction. This claim is clearly absurd, just as no one would believe that language proficiency accumulates at a constant, hourly pace from the first uttered word toward fluency. Dawkins' point was that Gould set up a straw man to misrepresent the paleontological thinking of the time. Not only was "gradualism" an inaccurate representation of how most twentieth-century paleontologists viewed evolutionary rate, noted Dawkins, but some of the so-called "gradualists" had explicitly acknowledged the substance of punctuated equilibrium decades before Gould and Eldredge coined their now-famous phrase (see Figure 4.4).

I think Richard Dawkins (among other biologists[34]) has made a valid point. Nevertheless, the extent to which mid-twentieth-century paleontologists had a realistic view of how speciation shows up in the fossil record is debatable; some did, others did not. Gould and Eldredge emphasized the latter; Dawkins and many others thought their emphasis was not representative of the field as a whole. Whatever their disagreements were about how speciation is paleontologically apparent, all of these authors were aware of the importance of natural selection as a mechanism behind evolutionary change. Moreover, and as we'll discuss below, Stephen Jay Gould was an unabashed and talented advocate for the basic mechanism proposed by Darwin for evolutionary change throughout time.[35]

CREATIONISTS AND QUOTE-MINING

If you've ever listened to a debate or lecture involving creationists, you've heard citations of evolutionary biologists presented as evidence against evolution itself. These generally come in two flavors: absurd and crafty. In the "absurd" category are the naïve creationists who, due to the threat they perceive from evolutionary biology to their theistic views, and their eagerness to counter it, grasp at some apparent discord among genuine biologists by skimming websites and/or media reports. They then present the controversy as a problem of evolution itself. In the "crafty" category are those who seem to have a better understanding of biology, but because they are so deeply and publicly wedded to the anti-Darwinian lobby, an honest presentation of the issue does not serve their interests. In each case, and to varying degrees, genuine discussion among biologists regarding certain evolutionary topics has been taken out of context to portray evolution by natural selection as a theory "in crisis" or otherwise flawed.

Punctuated equilibrium is probably the mother lode for creationist quote-mining, most of which belongs in the "absurd" category. Rabbi Shmuley Boteach of the Lubovitcher brand of Hasidic Judaism provided a remarkably clear example of absurdity during a 2008 debate with the secular journalist Christopher Hitchens. Based on his statements in this debate, Boteach places himself alongside ill-informed anti-Darwinian fundamentalists of all denominations. Consider his portrayal of the late Stephen Jay Gould, transcribed here verbatim from a recording of his January 30, 2008 debate with Christopher Hitchens at the 92nd St. Y in New York City:[36]

> My friends, most people today accept evolutionary principles without being knowledgeable of them. I actually studied them and debated some of the world's greatest evolutionary biologists ... Richard Dawkins... John Maynard Smith. ... I'm not the only person who doesn't believe in evolution without guidance. Let me be clear about that. The greatest scientists ... he passed away a few years ago, but I met him at Oxford, from Harvard, Stephen Jay Gould. He didn't believe in evolution. He believed in punctuated equilibrium. I saw him debate Richard Dawkins. He destroyed evolution, and he's an atheist. He said that the fossil record does not reflect evolution. Sir Fred Hoyle said that because there were trillions of mutations that would have to take place, and none of those organisms would have survived, because the possibility of a genetic mutation is 99.9 negative, detrimental and lethal to the organism, so Hoyle said, you would have to find trillions of destroyed organisms

> that don't exist. In fact, it was Darwin himself who said that the greatest objection to his theory, 'cause let's remember, the *Origin of Species* published in 1863 [sic[37]] was not a tome about the origin of life. It was a theory to explain the fossil record. But the problem was that there were missing transitional links. So what he said was, I'd like to give you the exact quote, Darwin said that the imperfection of the fossil record comes entirely to the fact that we have not yet dug sufficiently. In 1863 that would make sense. But my friends, we've now dug up the whole world ... we've dug up everywhere; we've barely found them. So Stephen Jay Gould, as Christopher Hitchens well knows, doesn't believe in evolution, or didn't. He argued for punctuated equilibrium. Great great leaps, which explains the dearth of fossils. To still be an evolutionist today, you have to have faith. There is no evidence; the transitional links barely exist. That's why ... when Dawkins ... friend of mine... when he debated Gould, he had nothing to say. What evidence? Where are these trillions and trillions and trillions of transitional links?

Let us set aside for a moment his inaccurate portrayal of Gould, and his ignorance of paleontology (which includes his misrepresentation of the *Origin* as a "tome ... to explain the fossil record"). This is a great example of creationist absurdity because Rabbi Boteach repeats popular misunderstandings of evolutionary biology for which easily accessible explanations have been available for decades. First of all, natural selection is not at all comparable to entropy, to random associations of parts to build a complex system, the odds against which any single such association might be one to "1000 to the millionth power." Well-trodden examples of such randomness include a hurricane sweeping through a junkyard[38] or monkeys pounding away at typewriters. Boteach thinks that the evolution of biological complexity is akin to a hurricane assembling a functional jumbo jet, or of monkeys typing out the works of Shakespeare by accident. If he had read even one of many popular texts on evolutionary biology, or Wikipedia for that matter, he would know that such analogies are irrelevant to evolution. He would know that for the vast majority of life history, natural selection has operated upon an array of functional organisms that exhibit variety, that produce more offspring than can practically survive, and that undergo differential survival over many generations and across geological time. He would know that sources of novelty result from not only mistakes in DNA replication or repair, which may in fact occur in regions of the genome without destructive consequences for the organism (as discussed in Chapter 10 and in many other popular texts[39]), but also derive from changes in patterns of individual growth. By altering the rate, onset, or offset of growth in an individual organism, or the age at which sexual maturity occurs, substantial novelty may occur across generations.[40] Similarly, novelty can result from sexual recombination, that is, the mix of male and female chromosomes via the union of sperm and egg. By combining parental genotypes, sexually reproducing organisms exhibit yet more novelty from one generation to the next.

Most importantly, he would know that the objects of natural selection do not start from scratch every generation. To make the analogy of monkeys at a typewriter relevant to natural selection, one would have to modify it so that one iteration of monkey-generated text builds on its predecessor. After the initial "draft," further rounds of typing would have to preserve adjacent consonants and vowels. With a mechanism to maintain these connections (to make the procedure analogous to selection in nature), further iterations would retain groups of consonants and vowels that resemble words in the English language, and again until actual words are built, and so on until these words resemble phrases from English literature. Such a procedure has been empirically demonstrated to generate English text from this non-random process,[41] and is discussed in more detail in Chapter 11. It is true that the odds of generating Hamlet by flailing your primate digits upon a keyboard are essentially nil, as are the odds of a hurricane accidentally assembling a jumbo jet in a junkyard. But such examples have nothing to do with natural selection, and anyone who talks or writes about evolution today has had several decades to inform him or herself about this simple fact.[42]

Now consider Boteach's portrayal of Stephen Jay Gould, who on many occasions before his death in 2002 crafted prose precisely because of people like Shmuley Boteach. Take this 1987 example:

> The theory of punctuated equilibrium ... is not, as so often misunderstood, a radical claim for truly sudden change, but a recognition that ordinary processes of speciation, properly conceived as glacially slow by the standard of our own life-span, do not translate into geological time as long sequences of insensibly graded intermediates ... but as geologically "sudden" origins.[43]

Or consider this somewhat more precisely worded rejoinder, identified in the index of his 2002 book, *The Structure of Evolutionary Theory*, under "creationism, errors about punctuated equilibrium and":

> Since modern creationists ... can advance no conceivable argument in the domain of proper logic or accurate empirics, they have always relied ... upon the misquotation of scientific sources. They have shamelessly distorted all major evolutionists in their behalf. ... Since punctuated equilibrium provides an even easier target for this form of intellectual dishonesty (or crass stupidity if a charge of dishonesty grants them too much acumen), no one should be surprised that our views have become grist for their mills and skills of distortion. ... [Creationism] rarely goes beyond the continuous recycling of two false characterizations: the

> conflation of punctuated equilibrium with true saltationism ... and the claim that no intermediates exist for the largest morphological transitions between classes and phyla. ... I have written numerous essays in my popular series, spanning ten printed volumes, on the documentation of ... intermediacy in a variety of lineages, including the transition to terrestriality in vertebrates, the origin of birds, and the evolution of mammals, whales, and humans—the very cases that the usual creationist literature has proclaimed impossible.[44]

Stated differently, the now famous theory reconciling the fossil record with biological theories of how a species arises, what Eldredge and Gould called "punctuated equilibrium" in 1972,[45] is not an alternative to evolution, but a prediction of how evolution typically manifests itself in the geological record. Punctuated equilibrium is a way of describing evolution, and a Darwinian one at that. Stephen Jay Gould was among the most eloquent evolutionary biologists of the twentieth century. To argue that he "didn't believe in evolution" is no less absurd than saying Albert Einstein believed in a flat Earth, or that Carl Sagan didn't travel to its four corners because he was afraid of falling off.

STEPHEN JAY GOULD AND NEO-DARWINISM

Because he was so prolific, Stephen Jay Gould provided many potential subjects for misquotation besides his writings on punctuated equilibrium. One of the most famous is from a 1980 paper:

> I well remember how [neo-Darwinism] beguiled me with its unifying power when I was a graduate student in the mid-1960s. Since then I have been watching it slowly unravel as a universal description of evolution. ... I have been reluctant to admit it ... but if [neo-Darwinism] is accurate, then that theory, as a general proposition, is effectively dead, despite its persistence as textbook orthodoxy.[46]

Why did Gould say this if he really was such an ardent evolutionary biologist? The answer has to do with a style of framing biological questions during the mid-to-late twentieth century that Gould helped to change, and to some extent to his penchant for hyperbole in his writing.[47] This somewhat harsh critique concerns "neo-Darwinism," not "Darwinism" per se, and is related to naïve practices of some early/mid-twentieth-century evolutionary biologists. It is not a denial of evolution itself. A fuller explanation requires a bit of background about what it means to place "neo" in front of the term "Darwinism."

When the first edition of Darwin's *Origin* was published in 1859, no one understood how attributes of an organism were passed on to the next generation. They knew it happened, but the actual carriers of inheritance, i.e., strands of DNA tightly coiled inside a cell's nucleus, remained a complete mystery until Theodor Boveri (1862–1915) and Walter Sutton (1877–1916) independently observed in 1902 that chromosomes played an important role in this function. At that time, Boveri, Sutton, and others were ignorant about the molecular make-up of chromosomes, that is DNA itself, until the work of James Watson (1928–) and Francis Crick (1916–2004). In 1953, they proposed the currently accepted model of DNA structure, consisting of four building blocks, or nucleotides, pairs of which form bonds with each other to make a ladder-like chain in the shape of a double helix.

Around the time when chromosomes were identified as the part of the cell that contained heritable information, geneticists Hugo deVries (1848–1935) and Thomas Morgan (1866–1945) helped to codify how inheritance operates. DeVries in particular was among the first to rediscover the work of Gregor Mendel (1822–84), who recognized that features of parents are typically passed on in a particulate fashion to the next generation. Even if rare, some traits (like blue eyes) may persist across generations, rather than being diluted to the more common state (e.g., brown eyes), as would be the case if the units of inheritance were blended with other such units over time. The Cambridge biologist and statistician R.A. Fisher (1890–1962) provided a basis for quantifying such patterns of inheritance in his 1930 book, *Genetical Theory of Natural Selection*.[48] Shortly thereafter, a student of Morgan, Theodosius Dobzhansky (1900–75), also helped to clarify the relationship between genetics and natural selection in a 1937 book, *Genetics and the Origin of Species*.[49]

Nineteenth- and early-twentieth-century paleontologists subscribed to all sorts of strange notions about how certain fossils they observed exhibited "trends" over time, apparently with some kind of capacity for "progress." With the publication of *Tempo and Mode in Evolution*[50] in 1944 by George Simpson (1902–84), the hard, empirical practice of population biology was applied to the fossil record. Simpson demonstrated how the mechanism of natural selection can explain many of the observations of scientists working on fossils. He showed that the "trends" or "innate drives" discussed at the time by many paleontologists were fictitious. For example, some early paleontologists implied in their writings that early relatives of animals, such as the horse, were intrinsically destined to become what they are today: large, speedy, single-toed, long-toothed grass-eaters. On the contrary, Simpson showed in his 1944 book and in later writings[51] that an early member of the horse family was not ineluctably destined to become any of these things independently of the actual selective pressures acting upon its progeny through the course of geological time.

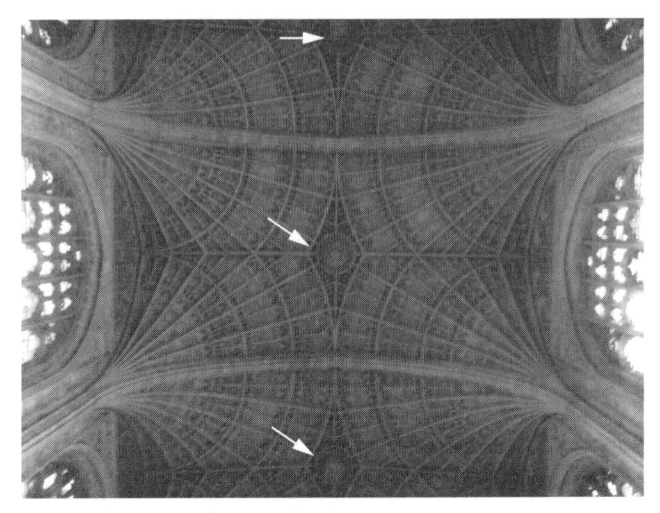

Figure 4.5. King's College Chapel, Cambridge. Ceiling (above) and chapel hall (below). White arrows indicate the diamond-shaped spandrels, formed by the "fan-vaulted"

Another key publication around that time was *Systematics and the Origin of Species*,[52] by Ernst Mayr (1904–2005). Mayr was an ornithologist who integrated his extensive field observations among the birds of New Guinea (among other places) to help define the very elusive category of the species, and theorized how a species relates to other such zoological categories both in and across time. This helped to form the biological discipline now known as systematics, which he presented in the context of Darwinian evolution by natural selection.

Ultimately, the coalescence of books by Fisher, Dobzhansky, Simpson, Mayr, and a few others formed the basis of what is today understood as "neo-Darwinism." At face value, "neo-Darwinists" differ from Charles Darwin in having a more sophisticated understanding of genetics, systematics, and paleontology. They know what the units of inheritance are, to what extent populations can interact with one another to delineate a species, and how these facts can help make sense of the fossil record. Darwin knew quite a bit about many of these subjects, but 80 years after the *Origin* was first published, much progress in evolutionary biology had been made, so much so that Julian Huxley epitomized this progress by giving it a name: "The Modern Synthesis,"[53] a phrase now synonymous with "neo-Darwinism." This represented the synergy of systematics, genetics, and paleontology into a coherent framework upon which most of twentieth-century biology was based, and continues in many forms to this day.

A problem with some interpretations of neo-Darwinism concerns the concept of "adaptationism." In 1979, Stephen Jay Gould and his colleague Richard Lewontin published a short paper entitled "The spandrels of San-Marco and the Panglossian paradigm."[54] Figure 2 in that paper (and Figure 4.5 in this one) shows the ceiling of King's College Chapel, Cambridge. It may have been only an accident of typesetting (plus the authors' preference for soft rather than hard alliteration) that St. Mark's chapel in Venice got Figure 1 and that this highly cited paper was not entitled "The spandrels of King's College Chapel, Cambridge." In any event, Gould and Lewontin made a very important analogy between architecture and biology.

Next time you're visiting King's College Chapel, look up. You'll see along the midline of the ceiling a series of diamond-shaped spaces, or spandrels,

Caption for Figure 4.5. continued

ceiling, i.e., the intersection of concentric circles radiating out from each of the supporting columns on either side. Within each spandrel are alternating images of the Tudor rose and portcullis, an architectural by-product of the intersecting circles radiating out from each supporting column. Designs within each spandrel are therefore contingent on this structure. Photograph used by kind permission of the Provost and Scholars of King's College, Cambridge.

within which are placed alternating images of the Tudor rose and portcullis, symbols of English royalty. These spandrels exist as a by-product of fan-vaulting, or the pattern of concentric circles radiating out onto the ceiling from each of the supporting columns. When a circle from one side of the ceiling intersects with its counterpart from the other, you automatically get a diamond-shaped space in the middle. While this provided a convenient space to illustrate symbols of British royalty, no one would argue that the carved rose or portcullis within that space came first, or was the "cause" of the columns supporting the chapel.

A century before, Mark Twain enumerated a similar point regarding the Eiffel tower.[55] No one would argue that its great height was designed to support the strip of paint on its uppermost spire. Rather, paint and spandrel-decorations exist for protective and/or aesthetic purposes; they're not the primary reasons for which towers or chapels are built. The "Panglossian paradigm" represents this error of reversed causality, and is based on a character in Voltaire's *Candide*, Dr. Pangloss, who believed that noses were for holding spectacles and legs for wearing trousers. Such an error is one of contingency: legs and noses (or towers and chapel ceilings) arose first, for reasons that have nothing to do with trousers or spectacles (or paint and spandrels). The latter are contingent upon the former, not vice versa.

Gould and Lewontin argued that the explanatory practices of neo-Darwinism during the mid-twentieth century committed errors of Panglossian proportions. Some authors atomized every bump on every bone and acted as if each had an independent, heritable basis. For example, humans differ from other apes and fossil hominins such as *Australopithecus* in having a chin. Hence, a protuberance at the midline of the lower jaws, according to a Panglossian paleoanthropologist, must have conferred some advantage to its owner, which would have enjoyed higher reproductive success as a result, leading to the evolution of the chin across hominine generations preceding anatomically modern humans.

But consider an alternative: the *lack* of a chin is a by-product of having a large facial skeleton (supporting your teeth, nose, and eyes) and a relatively small braincase. This happens to represent the proportions in a chimp or australopithecine (in which the brain occupies relatively less space than in a human), accentuated in adult members of these species as their face becomes relatively larger compared to the rest of their skull as they grow. Now consider that natural selection favored an increase in size of the braincase, perhaps via a slower rate of juvenile growth, which would maintain juvenile proportions longer and yield an adult with a slightly bigger braincase and smaller face. As the braincase increased in relative size over the course of many generations via natural selection in our lineage, the proportions of the facial skeleton and braincase changed, until the face was relatively small.

However, because we depend on our lower teeth to occlude precisely with our uppers, there is a limit on the decrease in size of the lower face, corresponding more or less to the position we know as the chin. Some of our close relatives, such as adult neanderthals, don't have much of a chin because their face is relatively large. Nevertheless, there's a strong case to be made that our chin "evolved" for reasons that have nothing to do with the immediate adaptive utility of its pointiness, whatever that might be. Rather, our chin is an architectural side-effect, or spandrel, resulting from differential growth in other regions of the skull, specifically, a small face and a large braincase.

Again, as we noted above in the case of punctuated equilibrium, many biologists criticized Gould and Lewontin for setting up a straw man.[56] Neo-Darwinists, they said, are not so naïve as to so readily ignore important natural influences as structural constraint. Indeed, if you look carefully at the writings of Mayr, Simpson,[57] and other architects of neo-Darwinism, and indeed at those of Darwin himself,[58] you will find substantial acknowledgment of factors in species divergence beyond the obvious fitness of the animals themselves, in addition to realistic descriptions of how we should expect to observe species-change in the fossil record (Figure 4.4). Neo-Darwinians, and Darwin himself, were aware of many circumstances under which forces in addition to natural selection could influence the evolutionary process, including structural constraint and contingency.

In the 1980 paper in which Gould eulogized "Mayr's characterization of the synthetic theory" (or neo-Darwinism, as paraphrased above), he wanted to make the case that awareness of these factors in the cutting-edge biology of the time[59] had displaced the naïve adaptationism present in some mid-twentieth-century biology publications. Even given his exaggerated prose (which he later acknowledged[60]), I think his basic point is correct, that natural selection working on atomized phenotypic traits, without acknowledging constraint or contingency, does not form an adequate basis upon which to understand evolution. However, the recognition of constraint in evolution does not diminish the major role of natural selection as the engine driving change over time. And no serious biologist, including Gould, ever claimed that natural selection was irrelevant as a mechanism behind evolutionary change. Hence, the "effective death" of neo-Darwinism about which Gould wrote in 1980[61] concerned his view of the increased appreciation of contingency and constraint in biology. It was not a denial of the importance of Darwinian natural selection to evolution.

"MACROEVOLUTION": SPECIES AS INDIVIDUALS

I've previously stated that the difference between "micro" and "macro" evolution is one of perception, with a few qualifications. These qualifications are worth mentioning now, as the awareness of them within the modern paleobiological community comprises one of Stephen Jay Gould's several positive legacies. In his last and largest book, *The Structure of Evolutionary Theory*, Gould summarizes a few reasons why we should consider at least some distinctions between micro- and macroevolution. Again, making this distinction does not deny or replace the basic Darwinian mechanism behind species' change. It adds to it.

In Gould's own words, he and several close colleagues who made the case for a distinct role for macroevolution (such as Niles Eldredge, Stephen Stanley, and Elizabeth Vrba) "were often accused of trying to scuttle Darwinism, and to invent an entirely new (and fatuously speculative) causal apparatus for evolutionary change. ... We made no such claim."[62] What they did claim was that the concept of a species, a group of morphologically uniform and interbreeding organisms, could take upon itself some attributes of an "individual," distinct from those of other such species. Such species-level individuals, they said, would have the potential to be subject to a kind of selection analogous to that exerted upon individual organisms, as argued in a traditional Darwinian context.

Consider, for example, the fact that on the plains of Africa today, there are over 80 species of bovid artiodactyls (i.e., herbivorous, even-toed ungulates such as gazelles, wildebeest, buffalo, kudu, oryxes, and goats). This contrasts greatly with other large mammals in Africa, such as the three or so species of zebra, the single species of aardvark, and the (arguably) two species of elephant. Furthermore, no one can confuse a gazelle with a wildebeest (both bovids), whereas the differences between non-bovid species (e.g., Burchell's and Grevy's zebras) generally must be explained to the novice safari tourist. Why are bovids so much more diverse and speciose than other large African mammals?

One possible explanation for this difference concerns macroevolution. Certain factors about the environment where an animal lives and its social dynamics may differ substantially across species. A group that makes its living in a forested habitat, one in which meandering streams divide patches of forest and over time are prone to changing course, would be more subject to geographic isolation than another group that lived in a more open habitat, or one which was less subject to division by geographic and environmental factors. Over the course of geological time, each group would potentially undergo very different rates of population isolation. The closed-habitat group would more frequently be divided into small populations, ones which by their nature as few in number are more prone to go extinct or exhibit evolutionary change. This difference is not immediately due to the anatomical or genetic make-up of any individual animal, but rather to the surrounding environment in which a certain species (but not others) makes its living. Most importantly, this difference in environment could be regarded as an emergent property of a given species, one which on a paleontological scale would endow it with attributes (increased susceptibility to geographic fragmentation) that are absent in other such species.

Gould also articulated a scenario in which the social system of certain large, herbivorous mammals (including some bovids) favored large body size in males toward the center of the species distribution. Smaller males are driven to the periphery of the species' range by female choice and the animals' territorial nature. Now consider the possibility that here, too, the periphery of a species range frequently exhibits geographic instability. Populations are more likely to break off from the parent group along the periphery than in the center. While successful matings for any single low-ranking male in such a scenario would be relatively few during his lifetime, they would still occur at a substantial pace over the course of geological time. High frequency of geographic isolation along the periphery of a species range, coupled with the small body size along that periphery due to the social structure of that species, could yield a long-term trend of repeated small body size in daughter species. Note that this example of species selection runs directly counter to the selection acting on most individuals in that population, one in which the most dominant and reproductively successful males (in the center of the range) are the largest.

Importantly, such group-defining features do not have to be environmental. Some animals are constrained by certain patterns of development that influence the kind of diversity they and all of their close relatives can exhibit. We noted in Chapter 3 how marsupials differ from placental mammals in their peculiar form of birth. Newborn marsupials are very under-developed compared to placentals; at birth, they must climb from the maternal reproductive tract into the pouch, where most of their growth into an independent animal takes place. This requirement for every known marsupial to climb toward the pouch at birth has important consequences, such as the fact that no marsupial, living or fossil, shows hooves, flippers, or wings. Their forelimbs have to be sufficiently developed at birth so as to enable this formidable trek into the pouch. Evidently this requirement has precluded a certain level of adaptive flexibility in the marsupial forelimb, a limitation not seen among placental mammals. Thus, consider a hypothetical answer to the question "why doesn't any marsupial have wings for powered flight?" One might be tempted to say "because such adaptations were never beneficial to marsupial populations during their evolutionary history." However, our understanding of marsupial development renders this answer too simplistic. Rather, developmental constraint present throughout marsupials has closed off this potential adaptive route for the group as a whole.[63] Due to the presence of this constraint throughout marsupials, this is less of an individual phenomenon than a macroevolutionary one.

In the above cases, features of a species derived from its development or environment operate above the level of the individual and influence that species' capacity to generate novelty. In this sense, "macroevolution" treats species in an analogous way as "microevolution" treats individuals. Nevertheless, the actual mechanism of natural selection, i.e., animals connected by common descent exhibiting differential survival, is implicit and necessary in all of the above examples; it comprises the means by which one generation of organisms contributes its genetic and anatomical information to its descendants throughout geological time. Rather than replacing this mechanism, the macroevolution endorsed by Stephen Jay Gould was an elaboration upon it, one which helps paleobiologists to better understand the larger patterns of biodiversity over time.

So yes, it is in theory possible to make a case for distinguishing "macro" versus "micro" evolution. Stephen Jay Gould did so,[64] even though others regard the distinction as lacking much difference.[65] Whatever your opinion on treating species as individuals, recognizing macroevolution in paleobiology does not diminish the importance of Darwin's basic mechanism of descent with modification as the engine behind the evolution of biodiversity.

FIVE

THE ROOTS OF MAMMALS

Let us return to the predictions made by the theory of evolution by natural selection outlined in Chapter 2. One prediction we made arising from the occurrence of natural selection over time is that certain animals should mix adaptations and morphologies seen in others, comprising what are popularly known as "missing links," or species that stretch the definition of exactly what constitutes a given category of organisms. In fact, as described for the coqui, platypus, bandicoot, and tarsier in Chapter 3, you don't need fossils to observe suites of transitional features that link major animal groups. But of course we have them—lots of them—and to the unbiased observer they document the previous existence on Earth of an extraordinary array of life, now extinct. These fossils are in many cases representative of the common ancestors shared by major groups recognized today. You've heard of at least some of the most impressive fossils that enjoy the status of "missing link." Starting with the letter "A," you might recognize the names *Acanthostega*, *Archaeopteryx*, or *Australopithecus*. The list goes on (see Table 8.1)—-*Adapis*, *Aetiocetus*, *Apateon*, *Apternodus*, *Ardipithecus*, *Asioryctes*—but most of these aren't quite as famous.

Archaeopteryx is probably the most famous fossil in the world. It's a bird-like animal from the late Jurassic (about 150 million years ago) of southern Germany. Modern birds show feathers, powered flight, beaks, and lack tails and teeth. *Archaeopteryx*, in contrast, blurs the definition of "bird" because it combines features we associate with very different groups in one animal.

It has the flight-adapted forelimbs and feathers of a bird, but the teeth, hip bones, and long tail of a lizard. For well over 100 years, up until a rash of discoveries in the 1990s–2000s,[1] *Archaeopteryx* was one of only a few fossils that challenged our understanding of what a "bird" really is, anatomically speaking.

In contrast, mammals have such an abundance of extinct relatives that one of the great challenges of twentieth-century paleontology has been to define what "mammal" actually means. As for birds, it is not at all difficult to come up with a list of features present only in living mammals (Figure 5.1). They generally have fur, different kinds of teeth (such as incisors, canines, and molars) which are replaced no more than once during an individual's lifetime, have upper and lower molars that interlock, a single jawbone, three small ear bones, cessation of growth at adulthood, a chest cavity divided into thoracic (ribbed) and lumbar (ribless) regions, few bones in their shoulder skeleton, limbs that extend underneath the body rather than to the side, and they drink mother's milk. Now what would you call an animal that showed incisors, canines, and molars, limbs that extend underneath the body, a chest with thoracic and lumbar regions, but with several bones in both its shoulder and jaw, just one in its ear, more than one generation of replacement teeth, and which didn't stop growing as an adult? No such animal is alive today, but hundreds of such species have been found in deposits around 200–300 million years old, from the Permian and Triassic of (for example) Canada, Russia, South Africa, and Texas. Many combinations of "mammal" features can be found among three kinds of early relatives of mammals, often referred to as "mammal-like reptiles": pelycosaurs, therapsids, and cynodonts. These organisms could with equal justification be known as "reptile-like mammals" but, again, this is the point I want to make: how we name them using references to living organisms is arbitrary. Scientifically, they're part of the very large and diverse group known as the Synapsida. This group is so large that we're part of it too. All mammals are, in fact, any organism that's closer to us on the Tree of Life than to a bird or lizard is a "synapsid." It so happens that because of extinction, mammals are the only synapsids alive today.

Very early signs of the synapsid fossil record occur in one of the Atlantic provinces of Canada, Nova Scotia, at what is now a UNESCO World Heritage site: the Joggins fossil cliffs. This region was publicized for its geologic importance early in the nineteenth century, and classified as representing a portion of the Pennsylvanian Epoch of the Carboniferous period (see Figure 4.1), now known to be about 315 million years old. Because it preserves the remains of large tree trunks, still many feet tall and sometimes in a near-upright position, Charles Darwin referred to these outcrops in the *Origin*[2] as an example of how thickness of a geological formation does

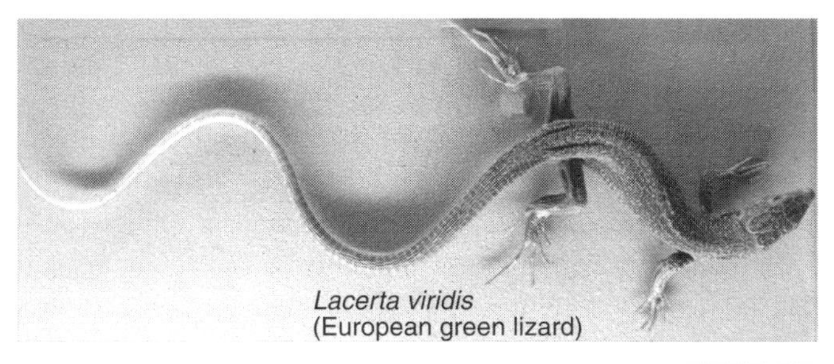

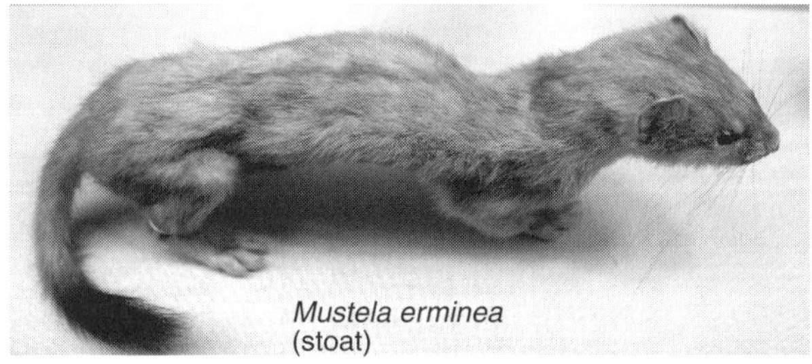

Figure 5.1. Living mammals and reptiles differ in many ways. Mammals have fur, drink mother's milk, have different kinds of teeth (such as incisors, canines, and molars) which are replaced no more than once during an individual's lifetime, have upper and lower molars that interlock, a single jawbone, three small ear bones, cessation of growth at adulthood, a chest cavity divided into thoracic (ribbed) and lumbar (ribless) regions, few bones in their shoulder skeleton, and limbs that extend underneath the body rather than to the side.

not necessarily correspond with the time required for its deposition. What Darwin and his contemporaries didn't fully appreciate was that among the fossils inside of those tree trunks were ancient remains of tetrapods more closely related to modern birds and reptiles (diapsids) and mammals (synapsids) than to amphibians such as frogs and salamanders. As we discussed in Chapter 3, diapsids and synapsids comprise living amniotes. They are united by features of their embryo that enable development independently of water, in particular outgrowths of the embryo that surround important structures inside the egg.

You're probably familiar with the fact that fossils do not typically preserve soft-tissue anatomy, and we don't know if the Joggins vertebrates, including small lizard-sized animals such as *Archaeothyris* and *Protoclepsydrops*, exhibited the same kinds of embryonic membranes (amnion, chorion, allantois, yolk sac) shown by modern mammals, reptiles, and birds. Perhaps they exhibited a chorion and yolk sac but not an allantois or amnion, like the direct-developing coqui frogs we discussed in Chapter 3. Although we don't have direct evidence regarding their soft-tissue anatomy, we do know a lot about their skeleton. For example, unlike other land-dwelling animals from the Carboniferous, Joggins amniotes show just one bone in their ankle (the astragalus) that connects with the bones of the lower leg, a distinct, tooth-bearing ridge formed by the pterygoid bone in the roof of their mouth, and a skull attached to the vertebral column via a single condyle. These features did not occur in other tetrapods, but are shared with later representatives of reptiles and mammals,[3] supporting the interpretation that some of the Joggins vertebrate fossils represent amniotes.[4]

The small, lizard-like Joggins fossils already show distinctive features with which we can align them specifically to mammalian (synapsid) or reptilian (diapsid) lineages, the two major divisions within amniotes.[5] Hence, the actual divergence between these two groups took place even earlier than the *ca.*315-million-year age assigned to these deposits. One of the fossils I've already mentioned, *Archaeothyris*, exhibits features of the posterior skull that occur in later synapsids (mammal ancestors), but not in later diapsids (reptile ancestors). This includes the presence of a single opening in the skull behind the eye socket and the make-up of the bones that surround this opening. *Archaeothyris* is thus related to the much bigger and younger sail-backed pelycosaurs, such as the famous *Dimetrodon* and *Edaphosaurus*. An attentive seven-year-old should be familiar with pelycosaurs, and will emphatically (and justifiably) correct you if you call her/his sail-backed models "dinosaurs." *Dimetrodon* is not a dinosaur;[6] like *Archaeothyris*, it's a synapsid (which includes pelycosaurs) and is closer to mammals than to extinct dinosaurs or any living diapsid. Pelycosaurs of the Carboniferous and Permian were very diverse and included many diverse forms besides those at Joggins. Some *Dimetrodon* individuals were over 3 m in length, and wouldn't find it too challenging to dispatch an *Archaeothyris* under their front toes. Of course, the two animals would never have come into contact, as the best-known sail-backed pelycosaurs were substantially younger than the fossils represented at Joggins. In the grand scheme of geological time, pelycosaurs existed for a brief spell starting in the late Carboniferous, and exhibited some diversity until the middle of the Permian, a duration of over 50 million years (Figure 4.1).

By the late Permian, pelycosaurs had greatly declined in diversity. Taking their place were other synapsids that had a similar skull: therapsids. For example, *Biarmosuchus* from the late Permian of Russia is like *Dimetrodon* in having a large canine, a big rostrum (part of the skull supporting the upper teeth and housing the nose), and a large hole behind the orbit defined by a similar complement of bones around its margin. Many other therapsids are known, comprising dozens of species from Permian localities in Africa, Eurasia, and the Americas. Three particularly remarkable groups, dicynodonts, therocephalians, and cynodonts, appear in the middle to late Permian and managed to survive the largest mass extinction event in Earth history at the Permo-Triassic boundary, around 250 million years ago.[7] In geological deposits immediately after this event, including some in South Africa, dicynodont fossils are abundant, therocephalians and cynodonts less so. As the Triassic drew to a close around 210 million years ago, just one of these groups remained: the cynodonts. Within this group of synapsids, we find the ancestors of all living mammals.

Pelycosaurs, therapsids, and cynodonts are all synapsids that belong on the lineage leading to modern mammals.[8] They comprise an exceedingly rich array of big and small animals, some carnivorous, some herbivorous, some good at burrowing, others at swimming, all adapted for life on an Earth that looked very different compared to the one we've got today.

The reason these fossils are so important to understand mammalian evolution has to do with several anatomical regions, in particular the teeth, jaw, ear, and the shoulder and hip skeletons. In all of these parts of the body, as we've previously observed, living mammals are very different from a lizard or crocodile. However, in the lineage joining pelycosaurs to therapsids to cynodonts to modern mammals, we find a series of animals that blurs this distinction. There are a lot of turns and detours along the way, and these fossils do not represent a linear "progression" from modern reptiles to modern mammals. Rather, these early ancestors of mammals fit quite well with the notion of "intermediate and transitional links" between the modern groups and their common ancestor, predicted to have existed by Charles Darwin in 1859.[9]

SYNAPSID JAW AND EAR BONES

Let us now use the fossil record to examine how the skull and jaw have changed anatomically during the course of synapsid evolution. We've previously observed that a mammal has just one jaw bone, the dentary, which simultaneously holds its teeth in place as well as contributes to the joint around which the jaw opens and closes (Figure 5.2). On the skull side, it is the squamosal bone which completes the joint. Hence, all modern mammals have what's called a "dentary–squamosal" jaw joint. In medical terminology, the squamosal may also be called the temporal bone, and the jaw the mandible, so you may also have heard of the "temporomandibular" joint, or TMJ.

In other jawed vertebrates, the bone holding the lower teeth is not the same one that forms the jaw joint. In a crocodile, for example, the dentary still holds all of the teeth, but behind the dentary sits the surangular, splenial, coronoid, angular, and articular (Figure 5.2). The articular bone sits

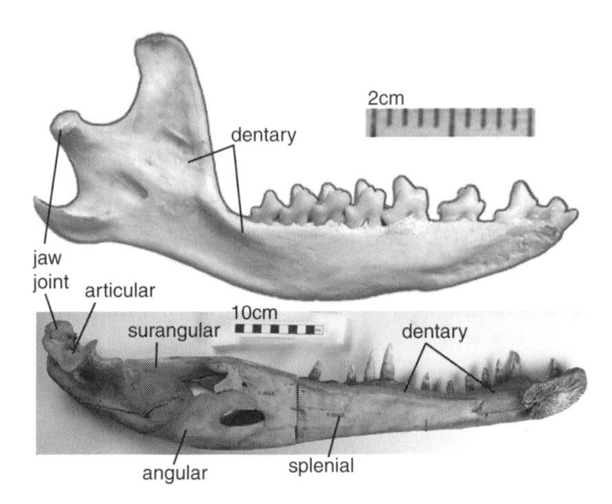

Figure 5.2. Lower jaws of a mammal (above, *Setifer setosus*) and a crocodile (below, *Crocodylus porosus*) from the collections of the Museum of Zoology, Cambridge. Note that the mammal jaw consists of just one bone, the dentary, which therefore comprises part of the mammalian jaw joint. The same bone exists in the crocodile and also serves to hold the teeth, but is accompanied by numerous other bones behind it, such as the articular, which contributes to the reptilian jaw joint.

at the back of the jaw and contacts the quadrate bone of the skull, forming an articular–quadrate joint. Although there is a lot of variation in terms of the size of individual bones, other land-dwelling animals (e.g., birds, lizards, snakes, and turtles) have a similar layout, including the quadrate–articular jaw joint. It is far from obvious by looking at a modern adult reptile or mammal skull, but these various bones close to the jaw joint played key roles in the evolution of mammalian hearing.

In order to hear, animals living on land such as frogs, lizards, turtles, birds, crocodiles, and mammals have to deal with the problem of converting relatively weak, airborne sound waves into something stronger that the brain can interpret. They do this by means of an eardrum that collects sound waves, transmits them via a narrow, delicate bone called the stapes (or sometimes the stirrup or columella), toward a very small opening in the skull called the oval window. Inside the oval window is the fluid-filled inner ear, within which is a cranial nerve that conveys to the brain electric impulses derived from vibrations of the stapes. Mammals and most other vertebrates have this basic setup, but mammals differ in that the bone framing our eardrum is the tympanic, as opposed to the quadrate (in reptiles and birds) or the squamosal (in frogs). We've also got two more small bones between the oval window and eardrum besides the stapes: the malleus (informally called the hammer) and incus (or anvil).

The Carboniferous and early Permian ancestors of mammals, reptiles, and birds showed no sign of an indentation or notch on the back of the skull where an eardrum could fit, and their stapes bones were rather large, built to support the skull, not to convey sound to the inner ear. Paleontologists agree that the sensitivity to high-frequency sound provided by an eardrum was not present among these animals at the base of the amniote tree.[10] However, although they were less sensitive to high-frequency sound, they were not totally deaf. As demonstrated by modern salamanders and snakes, low-frequency sound can be recognized quite well without an eardrum connected to a delicate stapes. Perhaps more importantly, the lack of a modern hearing apparatus in these early amniotes, including the ancestors of both reptiles and mammals, means that these groups acquired their hearing capacity independently.

Pelycosaurs, therapsids, and at least some cynodonts possessed a jaw joint like most other land-dwelling animals, consisting of the quadrate and articular bones. But look carefully at the photos of synapsid jaws in Figure 5.3, and the evolutionary tree that joins them. What you'll see is that compared to the synapsid fossils from the Carboniferous and Permian, those from the younger deposits of the late Permian and Triassic show a dentary bone that forms a much larger proportion of the jaw, receiving most muscle insertions, and have correspondingly reduced posterior jaw elements. Now this is not a perfectly graded transition[11] with every younger fossil showing a dentary of increasing size. Some branches of the therapsid family tree contained animals with very strange-looking jaws, such as the turtle-like, keratinous beaks of the early Triassic dicynodonts. In contrast, some late Triassic cynodonts show a jaw with the dentary so enlarged, and the posterior jaw bones so reduced, that they actually possess two jaw joints. These animals (such as *Brasilodon*, *Diarthrognathus*, and *Pachygenelus*) show contact between both dentary and squamosal and between articular and quadrate.[12]

The incremental nature of how the jaws and teeth changed at different points in the synapsid lineage is based on a large body of paleontological research,[13] and is partly illustrated in Figure 5.3. Early synapsids, such as the pelycosaur *Dimetrodon*, differed from other land-dwelling vertebrates of their time because they showed different kinds of teeth—for example, large canines and small incisors in a single individual. In addition, the presence of a gap (or anatomically speaking, a "fenestra") behind the eye socket served as a strong anchor-point by which jaw-closing muscles could attach to the skull, thereby increasing the force by which they could pull the jaw shut. Furthermore, the dorsal surface of the jaw, just behind the toothrow, formed a structure called the coronoid eminence (Figure 5.3), to which that same jaw-closing muscle could attach. This kind of jaw is not far off from that seen in modern crocodiles, but modern crocodile teeth are more homogeneous in size; their jaws lack a coronoid eminence, and they possess two fenestrae behind the orbit, not one.[14]

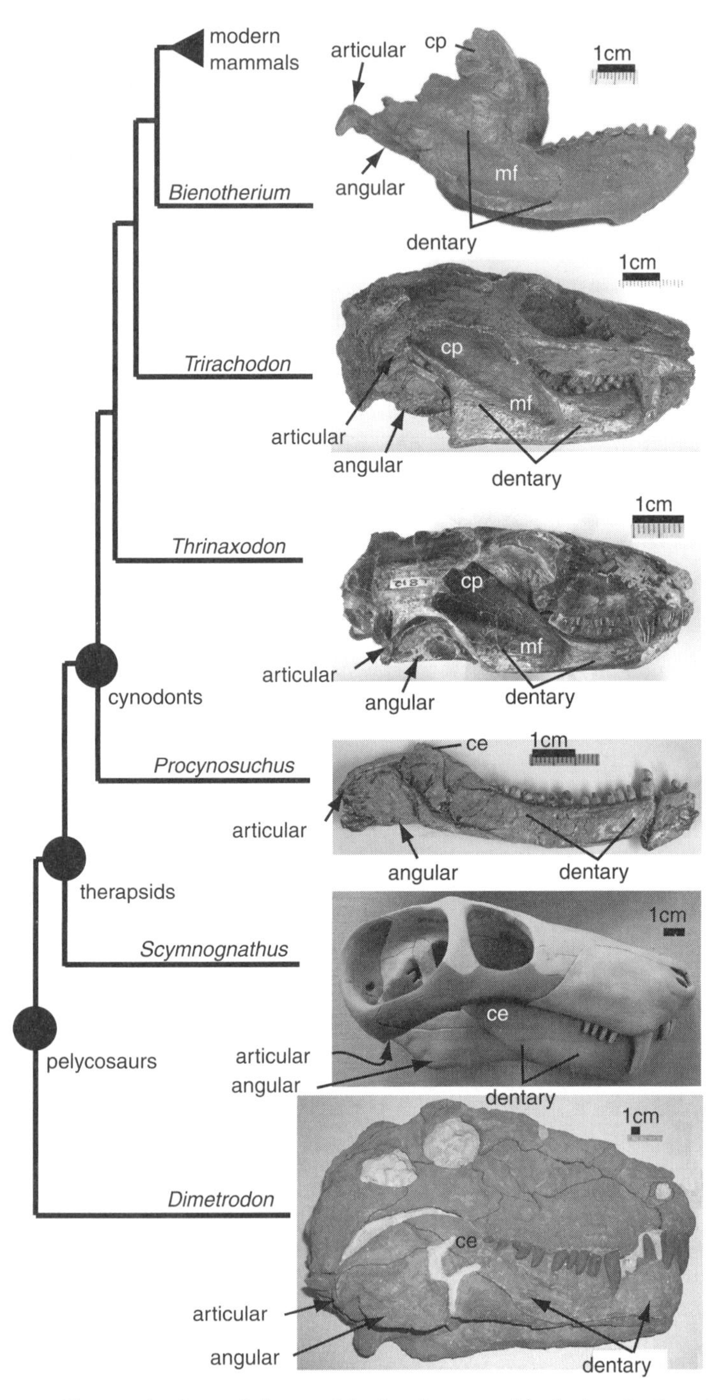

Figure 5.3. Changes in size and shape of the jaw in synapsids during the Permian and Triassic. Pelycosaurs (like *Dimetrodon*) and therapsids (like *Scymnognathus*) show

Compared to *Dimetrodon*, therapsids (for example *Scymnognathus* from the Permian of South Africa, Figure 5.3) possessed a larger postorbital fenestra into which jaw-closing muscles could attach. In addition, the canines and incisors differed more in size, implying that individual parts of the toothrow had specialized functions (e.g., grabbing incisors and stabbing canines). The coronoid eminence now had an even larger contribution from the dentary bone, resulting in a correspondingly smaller contribution from other jaw bones. This large coronoid eminence received a particularly well developed slip of jaw-closing musculature, called the temporalis muscle in modern mammals. Therapsids such as *Therioognathus* also show elaboration of another part of the jaw situated behind the dentary at its posterior and ventral aspect, consisting of a lamina of the angular bone. Muscles for jaw opening could attach here (originating from behind the jaw), in addition to a small expansion of temporalis jaw-closing muscles (originating from above the jaw). To withstand the slightly changed forces of this enlarged set of jaw-closing muscles, therapsids such as *Therioognathus* showed mobility not only between articular and quadrate (forming the jaw joint proper), but also between the quadrate and an adjacent bone on the skull, the squamosal. Flexibility at this articulation gave the jaw joint itself mobility and reduced stress in responding to the forces of opening and closing.[15]

The late Permian cynodont *Procynosuchus* exhibited several key traits that more closely resembled the chewing apparatus of modern mammals. The teeth were more variable in shape and size, depending on their position in the toothrow. This animal had not only small incisors and a stabbing canine, but also more posterior teeth (molars) with multiple cusplets on each tooth, with which it could masticate its food. *Procynosuchus* shows distinct muscle scars on both the internal and external surface of its coronoid process, and it also possessed a cheek bone (or zygomatic arch) to which another jaw-closing muscle (the masseter) could attach. By virtue of the muscle attachments on either side of the coronoid eminence (Figure 5.3), these muscles held the jaw in a sling, enabling the animal to tightly close its

Caption for Figure 5.3. continued

a jaw consisting of about half dentary and half postdentary elements, such as the angular and articular, and a coronoid eminence ("ce") demarcating the site at which jaw-closing muscles attach. Geologically younger cynodonts (e.g., *Thrinaxodon* [image reversed], *Trirachodon*, *Bienotherium* [image reversed]) show a much larger dentary bone, including an enlarged coronoid process ("cp") and a deep groove for jaw-closing muscles laterally on the jaw, called the masseteric fossa ("mf"). The late Permian cynodont *Procynosuchus* shows an intermediate condition with a small coronoid eminence and a very slight masseteric fossa. Note that branch lengths on the tree are arbitrary.

jaw while minimizing stress at the quadrate–articular jaw joint. Rather than pulling the jaw upwards and backwards with a muscle just on the internal aspect of the jaw (which would have entailed stress at the jaw joint to resist the backwards pull, as in a crocodile or pelycosaur), cynodonts began to pull the jaw more directly upwards with muscles on either side of the jaw. This configuration enabled the animal to not only close its mouth with great force, but also to more precisely control side-to-side movements of the jaw.

The dorsal part of the jaw in the Triassic cynodont *Thrinaxodon* is enormous and made up entirely of the dentary bone (Figure 5.3), comprising what is typically called the coronoid process of the dentary—basically, a big version of the coronoid eminence. Its cheekbone was broader than that of *Procynosuchus*, and supported a masseter that, as in *Procynosuchus*, inserted onto the external surface of the back of the jaw (you can see the scar left behind by this muscle, labeled "mf" in Figure 5.3, just as you can in the jaw of most living mammals). *Thrinaxodon* differed from *Procynosuchus* because the masseter inserted more deeply on the posterior dentary, closer to the base of the coronoid process, and because the angular and other postdentary bones were greatly reduced. In an animal with muscles attaching to the back of the jaw both medially and laterally, the very substantial closing forces are balanced such that the jaw joint did not have to withstand major stresses of the kind present in a modern crocodile or a pelycosaur. In those animals, jaw-closing muscles pull back from their insertion sites on the internal aspect of the jaw, pulling up and in; they are not balanced by an additional muscle pulling up and out. Hence, in a crocodile, the quadrate–articular joint must resist the resulting strain so that the jaw does not become disarticulated. Because of the balanced muscular force of jaw closure, this strain was much reduced in cynodonts such as *Thrinaxodon*.[16]

The dentary bone makes up nearly the entire jaw in other Triassic cynodonts, such as *Bienotherium* (Figure 5.3). Here, a portion of the dentary extends backwards to support (but not replace) the quadrate–articular jaw joint. As in *Thrinaxodon*, jaw-closing muscles inserted on the internal and external surfaces of the very large coronoid process, and the jaw joint itself therefore did not require as much of a strong, immobile suture with the tooth-bearing dentary to resist forces generated by jaw closure. In other Triassic cynodonts, including *Brasilodon*, *Diarthrognathus*, and *Pachygenelus*, the dentary extends sufficiently far backwards on the jaw so as to approximate the skull and buttress the small quadrate–articular jaw joint. In these animals, wear on the upper and lower teeth shows that they could control side-to-side movements of the lower jaw against the upper, resembling modern mammals in the way they chew. Again, posterior jaw bones such as the angular and articular were quite small. Importantly, this meant that the connections of these bones with the dentary could be more flexible. The posterior jaw bones were therefore more sensitive to airborne vibrations, which would have been conveyed via the articular and quadrate bones of the jaw joint toward the stapes bone.

This does not mean that synapsids were the only vertebrates to incorporate their jaw into their hearing system. Indeed, we know that modern salamanders and some snakes do this too (among others). Cynodonts such as *Pachygenelus* have the same series of bones in and near their jaw joint as a therapsid and even modern reptiles, such as dentary, angular, articular, quadrate, and stapes. But in cynodonts these bones are smaller, and this small size enabled cynodonts to use this chain of bones to perceive a wider range of frequencies than other vertebrates.

The oldest synapsid fossils to exhibit a real mammalian jaw joint, i.e., one between the dentary and squamosal bones, belonged to animals called morganucodonts. The first representatives of this group are from the Triassic–Jurassic boundary (roughly 200 million years ago) and were first described in the mid-twentieth century from localities in England and Wales. Morganucodonts ranged in size from shrew to small rat, and are now known from relatively complete skeletons and skulls from localities in Eurasia, Africa, and North America. Their jaw was composed of the dentary bone, which showed not just marginal contact with the squamosal of the skull, but had an enlarged condyle of the dentary with a corresponding joint surface on the squamosal, similar to the jaw joint you and I have got. Major differences remained, however. Specifically, in adults, the morganucodont angular, articular, and quadrate still had a connection with the back of the jaw (Figure 5.4). As in *Thrinaxodon*, *Pachygenelus*, and other cynodonts, the angular framed some kind of eardrum. Just inside this eardrum was the articular and quadrate which, as we've just observed, comprised the jaw joint in Carboniferous, Permian, and Triassic synapsids.

For the first time in the history of the known vertebrate fossil record, *Morganucodon* shows an exclusive function of these posterior jaw bones in hearing, not in supporting the jaw joint. That is, vibrations picked up by the membrane supported by the angular would vibrate the articular, which would in turn vibrate the quadrate and then stapes bones. As described above for most land-dwelling vertebrates, the stapes in *Morganucodon* sits in a very small opening in the skull, which corresponds to the opening of the fluid-filled inner ear called the oval window. The nerve of hearing would pick up subtle movements of the stapes as it displaced the inner-ear fluid via the oval window, and convert this fluid-borne energy into impulses that could be interpreted by the brain. *Morganucodon* was not the first synapsid to use its posterior jaw bones to convey sound,[17] but it was the first one in which those posterior jaw bones were not essential for maintaining the jaw joint.

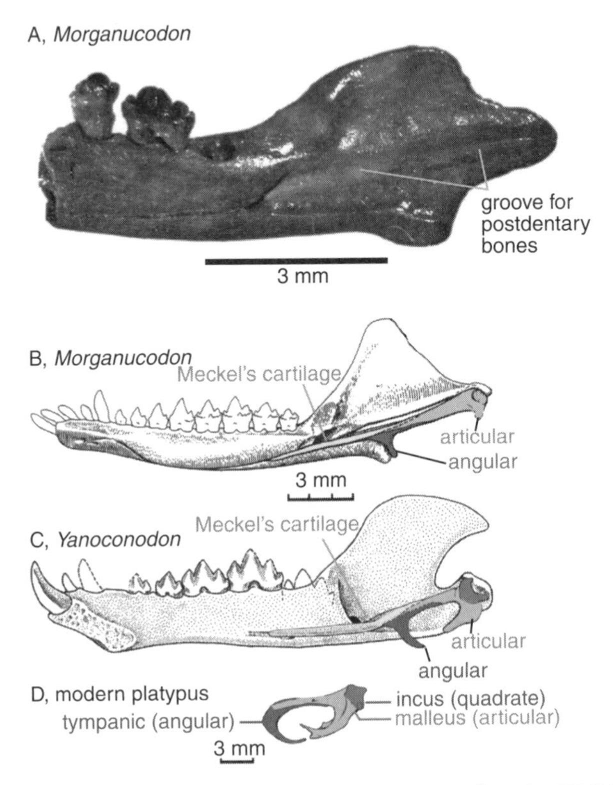

Figure 5.4. Internal views of the lower jaw in *Morganucodon* (A = UMZC specimen, image reversed) showing the groove left behind by the postdentary bones. B = the postdentary bones reconstructed in *Morganucodon*. These are preserved in the early Cretaceous fossil *Yanoconodon* (C), including an ossified Meckel's cartilage, angular, articular, and quadrate. D = the ear bones of a modern platypus. Note the similarity of each bone in the modern platypus and fossil jaws. Early in development, the ear bones of the platypus and every other modern mammal are connected to the jaw via Meckel's cartilage, just as they are in fossils *Morganucodon* and *Yanoconodon* (images B–D are from Luo *et al.* 2007, figure 3, reprinted with permission from Macmillan Publishers Ltd).

The angular, articular, and quadrate bones of a cynodont or early mammal such as *Morganucodon* are the same ones that convey sound in you and me, but because the connection to the jaw has disappeared entirely in adult humans and other modern mammals (but not in embryos!), they are known by a new suite of names: the tympanic (angular), malleus (articular), and incus (quadrate). The developmental and evolutionary continuity of these bones across animals is why they are the same—or homologous—in reptiles (angular, articular, quadrate) and mammals (tympanic, malleus, incus).

EAR BONES AND DEVELOPMENT

While the paleontological basis for inferring the homology of crocodile jaw bones and mammalian ear ossicles is strong, comparative anatomists had suspected some kind of connection between the two long before anyone recognized the significance of synapsid fossils. In fact, while Darwin was still a 20-something gentleman-naturalist aboard the HMS *Beagle*, a German anatomist by the name of Karl Reichert was collecting data for his doctoral dissertation at the University of Berlin. In 1837, more than 20 years before *On the Origin of Species*, Reichert published his observations on the development of the jaw in pig, bird, and frog embryos.[18] From these developmental observations alone, without any data from the fossil record, the theoretical framework of Darwinian evolution, or even a decent stereo microscope, he came to a clear, startling conclusion:

> Rarely does one find a part of the animal organism in which changes from its original formation are so clearly evident as the middle ear bones of mammals. One can hardly believe that their curious, diverse construction is comparable with simple, long-formed, cartilaginous visceral arches, or that from the latter any sort of complicated form may be derived; yet nevertheless it is in fact so. Namely, the first two visceral arches [of the embryo] comprise the structures out of which building blocks for the individual components of the middle ear bones derive.[19]

To appreciate the importance of Reichert's discovery, you should be familiar with a bit of embryonic anatomy. Visceral arches (Figure 5.5) were first recognized in the 1820s by another German anatomist, Martin Rathke. These are subtle, slit-like features present during early development in all vertebrates, such as lungfish, sharks, salmon, frogs, birds, and humans. Medical embryology textbooks generally refer to the presence of six arches on either side behind the head, although not all of these are necessarily visible in a given animal at any one time. Inside of each of these arches are primordial tissues from which adult structures develop, such as major blood vessels, glands, and certain bones of the skull, jaw, and neck. Importantly, in very different animals, the same adult structures derive from the same visceral arches.

For example, you'll find that in all animals with a backbone (e.g., humans, kangaroos, birds, frogs, goldfish, lampreys), tissue from the fourth embryonic arch contributes to the major arteries as they attach to the heart, carrying blood away from it. Among reptiles, birds, and mammals, vascular tissues near the heart and supplying the head and lungs arise only

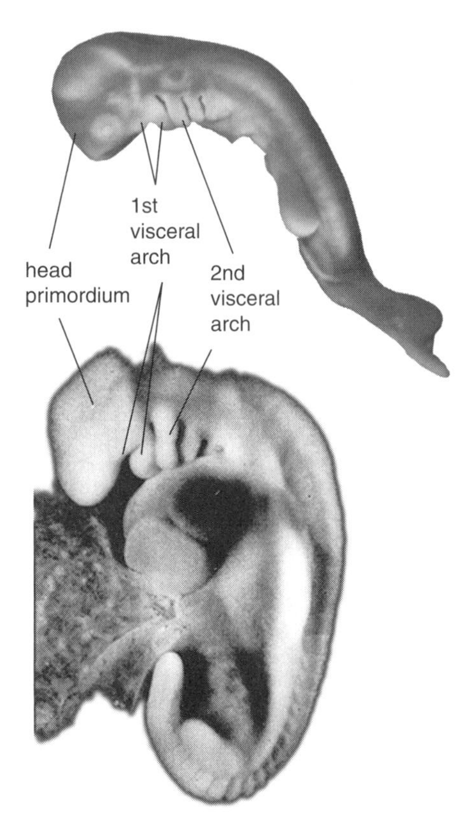

Figure 5.5. Photographs of early embryos of a sea turtle (*Lepidochelys olivacea*, above, courtesy of I. Werneberg and M. Sánchez-Villagra, Zürich) and human (below, courtesy of R. O'Rahilly and M. Richardson, Leiden). Ernst Haeckel and other nineteenth-century biologists correctly noted the greater similarities among embryos of different species than among adults. No serious contemporary biologist questions the profound similarities shown in early stages of many different kinds of animal embryos. This includes the presence of visceral arches from which adult structures of the head and thorax develop throughout all vertebrates.

from the third, fourth, and sixth arches, not from the first, second, or fifth. Such relationships between embryonic structures and their adult derivatives are consistent across very diverse animals. Because of this common pattern of development, embryos of human and chicken appear very similar to one another (Figure 5.5). You might have heard that the German biologist Ernst Haeckel took liberties with his drawings of early animal embryos, and indeed he has been criticized since the 1890s[20] for missing important details. However, much of his other scientific illustration was extraordinarily detailed and accurate,[21] and in two essential points he was absolutely correct: embryos of different organisms resemble each other far more than do adults, and embryology provides strong support for Darwin's theory of evolution.[22]

During the course of pig development, Reichert observed that mammalian middle-ear bones take shape, become relatively small, and eventually migrate behind the jaw joint, detaching from the jaw at a relatively advanced stage. He observed that in embryos and young fetuses some of these elements can be traced to the first visceral arch. When you were a developing embryo, your jaw surrounded an elongate rod called Meckel's cartilage, just as in the animals observed by Reichert. At the very back of this rod were attached the precursors of two of your ear bones (the malleus and incus), plus the bony frame of your eardrum (the tympanic). By the time you were about 20 weeks old, your Meckel's cartilage had partly disintegrated, but the posterior elements attached to this cartilage remained and formed two of the ear bones that are now essential for your hearing. Just as in any other mammal, the precursors of your tympanic, malleus, and incus were still attached to your jaw early in your development.

Figure 5.6 shows such an image of a fetal African elephant shrew. These small mammals are similar in some regards to rabbits, but related to other African species like hyraxes and aardvarks, and are found in places like Kenya, Tanzania, and South Africa. The development of their ear bones is pretty much the same as in humans. Importantly, the largest connection between the jaw and skull in embryonic stages of both elephant shrews and humans was via two of these ear-bone precursors, the malleus and incus (Figure 5.6). Stated differently, at an early stage during development, all mammals, including marsupials, elephant shrews, and even creationists, had a joint between two of their ear bones that looked very much like the jaw joint of an early synapsid, or even modern bird or crocodile, attached to the back of Meckel's cartilage derived from their first visceral arch.[23]

Reichert also traced the embryonic origins of the stapes bone, and found that it was continuous with the second visceral arch, along with the upper part of a small bone called the hyoid. In humans, the hyoid bone is located in the top of the neck, just above the voice-box, behind and below the tongue. All of us have a ligament that persists into adulthood called the stylohyoid ligament, connecting the top of the hyoid bone to a narrow projection at the base of the skull, which originally had a connection to the stapes bone. Just as the malleus and incus lost their connection with Meckel's cartilage, so the stapes lost its connection to stylohyoid ligament and hyoid bone during early development.

Reichert observed that elements of the jaw and ear also derive from the first two visceral arches in birds and frogs. They both have a Meckel's

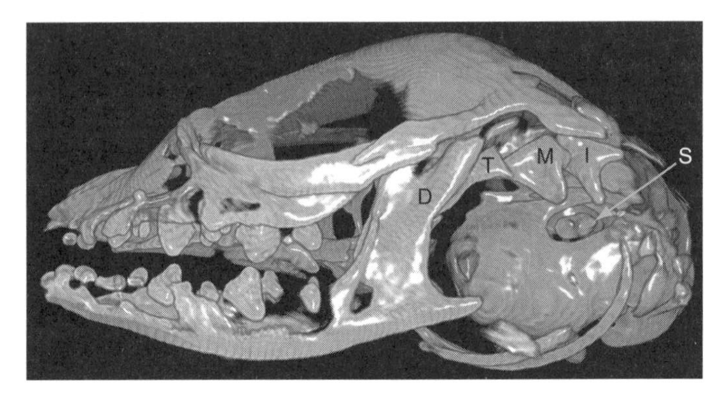

Figure 5.6. Computerized tomographic (CT) image of the developing skull in a ten-day premature newborn of the species *Macroscelides proboscideus*, also known as a sengi or elephant shrew. Note how the ear bones (T = tympanic, M = malleus, I = incus) are still situated adjacent to the back of the jaw (D = dentary). The stapes (S) frames the entrance into the inner ear. At this stage, the joint between the malleus and incus resembles the articular–quadrate jaw joint in non-mammalian vertebrates. During development, mammals show the migration of the ear bones into the middle-ear space and away from the back of the jaw.

cartilage that (to varying extents) is superseded by ossified jaw bones late in development. They both have a stapes which conducts sound into the inner ear and is embryonically contiguous with the top part of the hyoid bone. However, Reichert noticed a fascinating difference: in birds and frogs, the small elements at the back end of the first-arch Meckel's cartilage do not migrate toward the stapes bone or frame the eardrum. Instead, they remain close to the back of the jaw throughout adulthood, forming the jaw joint as articular and quadrate bones, and contributing to the posterior and ventral part of the jaw as the angular bone.

Hence, even though the jaw bones of a reptile are given different names than the ear bones of a mammal, Reichert established that they share an origin from the same embryonic structures. In addition, he and many other anatomists have noted that muscles attaching to them receive innervation from the same nerves. Muscles that attach to the avian quadrate and articular bones (to help close or adduct the jaw) are innervated by a cranial nerve called the trigeminal; those attaching to the stapes (the stapedius) are innervated by another cranial nerve called the facial. Mammals have a muscle that attaches to their malleus, the tensor tympani, which is innervated by the trigeminal, and they've got a stapedius muscle, innervated by the facial. Now the mammalian tensor tympani differs drastically in function from the avian jaw adductors: it is similar to the stapedius muscle in that it helps to keep the ear bones from vibrating too much when faced with a sudden, loud noise. In contrast, avian jaw adductors close the jaw. Nevertheless, both the mammalian tensor tympani and avian jaw adductors are developmentally continuous with the first visceral arch, along with the articular (malleus) and quadrate (incus) bones. This explains their common innervation from the trigeminal nerve.

Reichert's observations on the continuity of the ear bones with the embryonic first and second visceral arches in mammals have been repeatedly confirmed since 1837. Although there has been some debate regarding his conclusions since the nineteenth century, and although some eminent evolutionary biologists (including T.H. Huxley) were not entirely in agreement, later study of Reichert's theory[24] along with modern studies of genetics and development[25] have sealed the case: Reichert was right about mammalian ear bone homologies. His idea is widely accepted throughout biology, and indeed has been greatly supplemented by data from genetics and the synapsid fossil record.

Darwin (1809–82) does not seem to have been familiar with the work of Karl Reichert (1811–83) on the development of mammalian ear bones, although the two overlapped almost entirely during their lifetimes. By the time of publication of Darwin's *Origin*, Reichert occupied a fairly senior position in the German academy as Professor of Anatomy in Berlin. Throughout his career he was not known to be sympathetic to then-novel ideas in biology,[26] which included not only Darwinian evolution but also the Cell Theory.[27] As for his attitude towards Darwinian natural selection as a mechanism behind evolution, he wrote little specifically on the subject in his publications and letters.[28] One of his students noted that among some of the more senior comparative anatomists in Germany around 1860 (including Reichert), "almost generally at that time Darwin's teaching was regarded as entirely mistaken."[29] Reichert presumably exhibited the same skepticism regarding the Darwinian theory as did many other senior naturalists on the European continent immediately following its publication.

Whether or not Reichert viewed his own work on the development of the mammalian ear as compatible with evolution by natural selection is uncertain. Reichert was a diligent follower of his very empirically oriented mentor, Johannes Müller.[30] From their perspective, science was about data-collecting first, followed (and then only reluctantly, if at all) by theorizing on a broad scale. Reichert knew about French and German natural philosophy, in which anatomical similarity across animals was viewed as evidence of some kind of derivation from a basic, natural design. Such ideas were espoused by eighteenth- and nineteenth-century European naturalists such as Johann Wolfgang von Goethe, Jean Baptiste de Lamarck, Etienne Geoffroy St. Hilare, and Charles Darwin's grandfather Erasmus.[31] The mechanism by which evolution occurred was never specified in any detail by these authors, and certainly not by Reichert, but their observations about nature were in fact compatible with Darwinian evolution by natural selection.[32]

Karl Reichert made what is today regarded as a major evolutionary discovery: the homology of mammalian ear bones with reptilian jaw bones. The fact that he did so with essentially no knowledge of the fossil record or understanding of Darwinian evolution demonstrates how the pattern left behind by evolution is independent of the theories by which one understands that pattern. Reichert discovered an important part of the pattern of vertebrate biology, but it was up to later authors[33] to connect this pattern to a larger theory of evolution.

Broadly speaking, Darwin's theory leads to the understanding that variation during growth, including subtle changes in the timing of development from one generation to the next, forms a basis upon which natural selection can operate. Note how greatly this differs from the false characterization that evolution must "create information" from scratch. A major anatomical novelty, such as the incorporation of small ear bones into the hearing mechanism of mammals, does indeed seem to present an unbridgeable gap when we ignore development and fossils and compare the hearing mechanisms of modern, adult reptiles and mammals. However, both groups show sufficient anatomical variation during their development so as to make the differences between them derivable from differences in the growth and timing of certain parts of the skeleton. Combined with data from the fossil record, we can connect the processes of development evident to us in the lifetime of any living mammal or reptile with the adult appearance of real animals that existed in the past.

Consider the resemblance between the dentary, angular, and articular in *Bienotherium* (Figure 5.3) and the dentary, tympanic, and malleus of a fetal elephant shrew (Figure 5.6). The fact is, as for the elephant shrew, you yourself had a jaw articulation like that of a crocodile or *Bienotherium* at an early stage during your own development. Fossil synapsids from the Permian, Triassic, and Jurassic, part of the ancestry of modern mammals, would have shown slight variation in the timing of jaw development, such as when the posterior jaw elements derived from the first visceral arch ossified.[34] When acted upon by natural selection over many generations, such variation could have major influence on the size and shape of those jaw elements in the adult, including the dentary, angular, articular, and quadrate.

Changes in the size of the dentary—for example, contributing to a larger coronoid eminence—were entirely gradual when viewed from the perspective of each generation of early synapsid. Furthermore, selection for a larger coronoid eminence would have had little impact on the animal's ability to hear. However, by virtue of their large dentary and correspondingly small postdentary bones, those postdentary bones in later synapsids (e.g., *Thrinaxodon*) would have been "preadapted" for relatively high-frequency hearing, amenable to further natural selection for increased acoustic sensitivity. Subtle, gradual changes from one generation to the next, leading to smaller and more flexible connections among angular, articular, and quadrate, would have increased the sensitivity of such an animal to auditory stimulation. Stated differently, and to draw upon the example of the architectural "spandrel" discussed in the previous chapter, one could say that the decrease in size in synapsid postdentary bones was, at least initially, a side-effect of the selection for increased power during jaw closure via enlargement of the dentary bone. Only later during synapsid evolution did acoustic sensitivity comprise a major selective force driving the evolution of what would become our very distinctive mammalian ear bones. (This touches on the phenomenon of "exaptation" which we will examine in more detail in Chapter 11.)

Development in modern animals comprises a concrete mechanism which anyone can observe in a scientific laboratory, and it has profound implications for how we understand evolutionary change over time. The fact that the developmental changes first uncovered by Reichert (Figure 5.6) so closely mirror paleontological changes found later in the fossil record (Figure 5.3) provides major support for the idea that natural selection is behind evolution, and that Darwin was correct.

MAMMALIAN LACTATION

The mammalian fossil record is good enough to inform us not only about the evolution of our jaws and ear bones, but also about a feature that does not normally evoke images of paleontology or fossils: lactation. Mammals are the only living vertebrates that provision their young with a fat- and nutrient-rich liquid administered via a modified gland. Unfortunately, paleontologists cannot depend on the fossilization of milk or the suckling behavior that is a hallmark of being a mammal. However, there are very tangible consequences of lactation for hard-tissue anatomy which we can observe. First, a suckling mammal has to be able to breathe through its nose while it is connected to the maternal teat. It can do this through a bony separation of the nasal passage from the mouth, a structure called a secondary palate. In addition, because an infant mammal doesn't have to find its own food but is provisioned by its parents, mammals grow very quickly. They also stop or greatly slow growth once adulthood is reached. In terms of skeletal anatomy, this means that if you were to examine 1000 domestic cat skulls from one population, from newborns to adults, there would be an upper limit in skull size beyond which adults do not grow, and fewer skulls at small sizes, since cats (and mammals generally) speed through development until they reach adult size.

The mammalian pattern of dental replacement is another correlate of this uniquely mammalian feeding strategy, facilitated by the precise muscular control on either side of the jaw, as described above for several early synapsids such as the cynodont *Thrinaxodon*. Remember that jaw-closing muscles of cynodonts could move the jaw from side-to-side; this is correlated with the development of cusps on the upper and lower teeth that could interlock with each other. You can vouch for the precision of this occlusion yourself by trying to move your jaw from side to side while gritting your teeth. Any luck? Neither you, nor opossums, dogs, lemurs, hyraxes, or most other mammals can do this because the cusps and basins of the upper teeth fit neatly into their counterparts in the lowers. This kind of interlocking occlusion is a remarkably effective means of breaking down and exploiting the nutrient content of food.

In cynodonts such as *Thrinaxodon* or *Pachygenelus*, interlocking occlusion was effective but not as precise as in modern mammals. Animals such as these were frequently replacing their teeth. This made maintenance of precise interlocking occlusion difficult, since when a lower tooth falls out and a new one erupts below it, any occlusal fit with the corresponding upper tooth is lost. For most vertebrates—for example, crocodiles and lizards—teeth are similar in shape and size, serving not to crush but to stab and puncture prey items. Hence, as long as whole stretches of adjacent teeth do not simultaneously fall out (which is why most animals typically replace teeth in a staggered fashion throughout the toothrow), continuing waves of replacement at any given tooth position does not negatively impact the ability of a reptile or most early synapsids to ingest food. This is important, as most non-mammals continue to grow throughout adulthood, and therefore require the addition of more and/or bigger teeth into the growing jaws.

Because mammals benefit from maternal provisioning of milk early in their development, they are initially less dependent on teeth to eat. Of course, this changes once a young mammal has been weaned and must forage for itself. Hence, mammals have two generations of teeth: a deciduous or "milk" dentition that generally erupts between birth and weaning, and a permanent dentition that replaces the milk teeth after weaning, usually by the time the individual reaches adult body size. As mammals, once we reach adult body size, we've got relatively few teeth, which have to last for the rest of our lives. This is a major difference between modern mammals on the one hand, and Permian/Triassic synapsids and reptiles on the other. Unlike mammals, early synapsids did not stop growing and possessed multiple generations of tooth replacement in both upper and lower jaws.

Hence, a combination of anatomical features that are not hard to spot in the fossil record can be associated with the presence of lactation, a phenomenon that at first glance you might think could only be inferred with soft-tissue or behavioral information. Lactating animals are associated with (1) a secondary palate, (2) fast juvenile growth, and (3) determinate adult growth. The fast growth and reduced dependence on teeth for food acquisition during early development enabled by lactation, capped by the cessation of growth in adulthood, further correlates with the precise interlocking occlusion of upper and lower teeth and the absence of more than one generation of replacement teeth. The combination of these hard-tissue features makes sense only in an animal that exhibits lactation. If you know what to look for, all of these features are visible in the fossil record, enabling the reconstruction of this important behavior in animals over 200 million years old.

Evidence for lactation is exactly what paleontologist Zhe-Xi Luo and colleagues[35] found by examining abundant remains of two extinct synapsids at the very base of the mammalian family tree. Both *Sinoconodon* and *Morganucodon* show a true dentary–squamosal jaw joint, not quadrate–articular. For this reason they are classified as true mammals. They are also known from hundreds of specimens, some of which are fragments of lower and upper jaws, within which replacement teeth can be observed adjacent to their deciduous precursors. In a 2004 paper,[36] Luo and colleagues observed first that there was a fairly large size-range represented by their array of *Sinoconodon* fossils, and second, that no matter how big individuals of *Sinoconodon* got, you could always find the occasional specimen showing a replacement tooth poking out from below its precursor. Relatedly, at a single tooth position more than one replacement could take place. For example, as an individual *Sinoconodon* grew from *ca.*10 to 500g in weight, its canine tooth would have been replaced about five times. In terms of its indeterminate growth and continuous dental replacement, *Sinoconodon* was similar to non-mammalian cynodonts mentioned above, including *Pachygenelus* and *Thrinaxodon*, and differed from modern mammals.

Morganucodon, in contrast, exhibited a pattern very much like modern mammals. The fossils of this animal from the locality in China scrutinized by Luo and colleagues show a fairly limited upper size, estimated at about 90g. Furthermore, Luo and colleagues observed that these largest specimens never showed replacement teeth, but only ones that were fully erupted. Instances of tooth replacement were observed only in relatively small specimens. Finally, just the incisors, canines, and premolars were replaced; the posterior-most chewing teeth (the molars) erupted without any precursors, just as in modern mammals. Luo and colleagues thus interpreted *Morganucodon* as the oldest and most basal known mammal to have exhibited both determinate growth and dental replacement limited to two generations, both hallmarks of modern mammals that provision their young with milk.

ROOTS OF MAMMALS: SUMMARY

At certain times and places in Earth's history, the pelycosaurs, therapsids, and cynodonts that show intermediate morphologies of the jaws and ear bones were very common land-dwelling animals. They represent "missing links" that succeeded to the point where their populations were quite large—large enough so that their members contributed to the fossil record in great numbers. Recalling our discussion about punctuated equilibrium earlier in this chapter, here is a very well-documented reason why generalizations about "stasis" in the fossil record have been viewed skeptically by many paleomammalogists.[37] The mammalian record has an abundance of fossils that exhibit "transitional" features, perhaps to a greater extent than the fossil record of invertebrates highlighted by Eldredge and Gould.[38] Pelycosaurs, therapsids, and cynodonts thrived in their Permian and Triassic environments; they did not only exist in small populations with little chance of entering the fossil record. There are many other such examples.[39]

Consider again a few of the gross anatomical differences between the skull of a typical modern reptile and that of a mammal. The mammal has just one bone in its jaw and three in its ear; it shows determinate growth and only two generations of teeth, as opposed to the constant growth and ever-replaced teeth in the jaws of a reptile. Consider also Darwin's prediction that "the number of intermediate and transitional links, between all living and extinct species, must have been inconceivably great."[40] In a previous chapter, we distilled this observation into the prediction that, in the fossil record, features now regarded as characteristic for certain groups—such as feathers and milk—occurred independently of the animals they now serve to define—such as birds and mammals. In the preceding pages we have gone through a small but important number of such features related to mammalian jaws, teeth, ear bones, and lactation. Far from occurring simultaneously in an animal that shows up suddenly in the fossil record, and is obviously a "mammal," these features of the skull (and we've not even mentioned the similarly mosaic nature of the synapsid limb skeleton[41]) appear piecemeal over the course of about 110 million years among animals that don't really fit into the categories we recognize among modern species.

In fact, it is only an arbitrary act of consensus that scientists have chosen the squamosal–dentary joint that defines the connection between jaw and skull as the key feature that defines a mammal. Note that by this definition, given what we've reviewed above concerning the pattern of growth and dental replacement in *Sinoconodon*, one of the very first "mammals" did not nurse its young with mother's milk. In other regards, it was much more similar to a mammal than a lizard: it had a tympanic–malleus–incus series conveying sound via its stapes bone into the inner ear; the same bone (dentary) held its teeth and formed the main jaw joint; and it possessed different kinds of teeth in its mouth with different functions (slicing, stabbing, chewing).

The fossil record of synapsids that we've reviewed above came to light largely after publication of the *Origin* in 1859, and was fairly good by the early twentieth century. In the mid-nineteenth century, Charles Darwin had no way of knowing the extent to which the fossil history of mammals would bear out his predictions on life's past. We now know that the occurrence of features associated with mammals at various points in the Tree of Life, spread over the course of 100,000 millennia, corresponds closely to what he predicted would be found.

SIX

A BRIEF HISTORY OF ELEPHANTS

There are many cases in the fossil record of extinct species bridging the apparent anatomical gap between living groups, such as what we observed in the previous chapter between living mammals and the common ancestor we share with crocodiles, birds, lizards, and snakes. One of the other cases that we'll look at in some detail concerns modern elephants, a group zoologically classified in the Order Proboscidea. At present, there are at least two and probably three species. African elephants belong in the genus *Loxodonta* and are most commonly grouped in the species *L. africana*. A smaller species of "forest elephant" has also been proposed, *Loxodonta cyclotis*. Both are distinct from the Asian elephant, *Elephas maximus*. These species comprise a minute fraction of the diversity of proboscideans that have existed over the past 60 million years.

The fox-sized, 55-million-year-old fossil mammal *Phosphatherium* (Figure 6.1) does not look like a modern elephant, to say the least. It's got teeth with just a few cusps on its large crown, with crests connecting the cusps on either side, not much of a forehead, and a pretty small braincase. The fossil is small and eminently squishable by the footfall of any modern elephant, which has bizarre, high-crowned chewing teeth composed of plates of enamel, sometimes called lamellae, closely squeezed together and an exceedingly large skull. The modern elephant skull not only has a big braincase and lamellar teeth, but also large air sinuses (to make it lighter), gigantic tusks (developmentally the same as the second upper incisors of

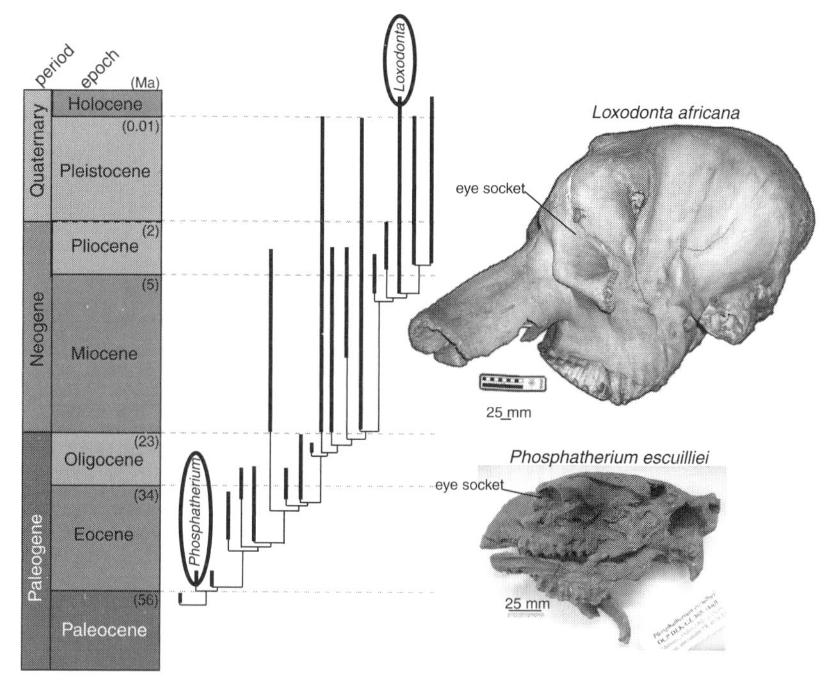

Figure 6.1. Images of two proboscideans, the modern African elephant (*Loxodonta africana*) and the extinct *Phosphatherium escuilliei*, known from the early Eocene of North Africa. The difference in size between the two is larger than depicted (note scale bars). *Phosphatherium* is fox-sized, a fraction of the bulk of a modern elephant. While the two animals differ greatly in overall size and shape, several shared features (for example, the anteriorly situated eye socket), plus the existence of many intermediate fossil taxa (see Figure 6.2), show that they are descended from a common ancestor to the exclusion of other, non-proboscidean mammals.

other mammals[1]), a large depression in its face for muscles that support its long trunk, and a nasal aperture which is located toward the top of the skull, not its anterior end (Figure 6.1).

Phosphatherium shows none of these things. However, despite these differences, features of *Phosphatherium* provide strong evidence that modern elephants are more closely related to this 55-million-year-old animal from North Africa than they are to any mammal alive today. How do we know? In the case of the Proboscidea, it is not only because we see certain key anatomical features shared by modern elephants and *Phosphatherium* (Figure 6.1), but also due to the fact that we have a long string of fossils that show aspects of their bones and teeth intermediate between those of ancient

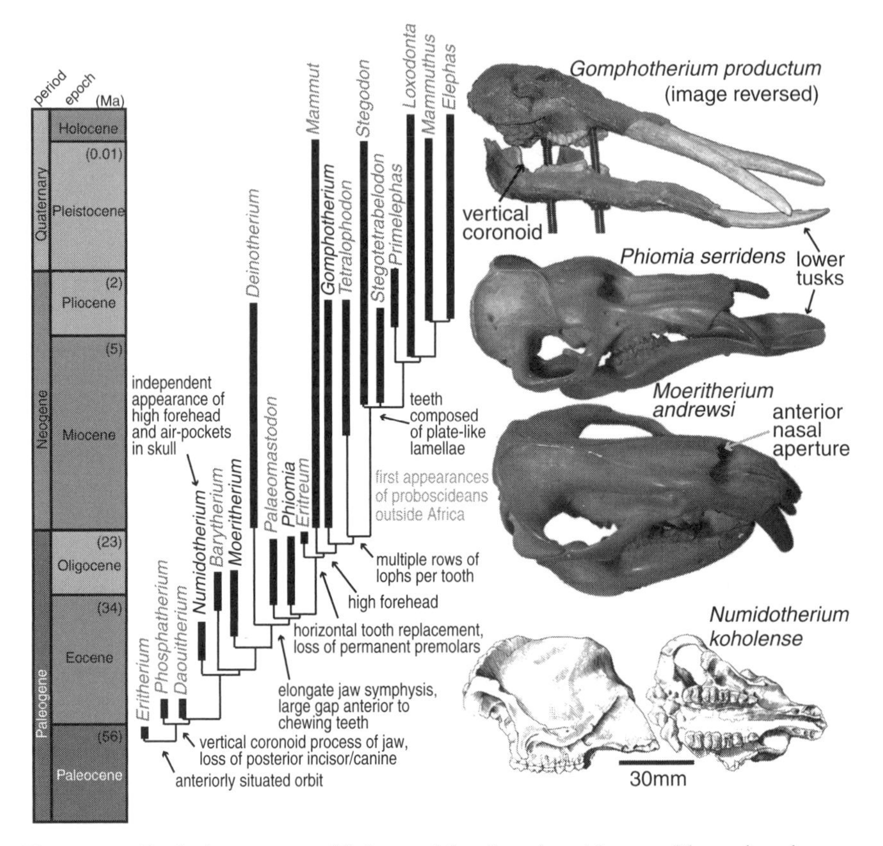

Figure 6.2. Evolutionary tree of living and fossil proboscideans calibrated to the geological timescale. The independent appearance of key morphological features are indicated along the nodes of the tree, demonstrating their mosaic or piecemeal evolution. Known proboscidean fossils, including the images shown at right, do not represent a direct ancestor–descendent series, but rather show some ways in which fossil proboscideans exhibit a variety of features, some in common with modern elephants (e.g., vertical coronoid of jaw), others with mammals generally (anterior nasal aperture in *Moeritherium*), and yet others that are specialized (adult lower tusks in *Phiomia* and *Gomphotherium*). The tree of genealogical relationships (or cladogram) is derived from Tassy (1996: 39–48) and Shoshani (1996: 149–77). Nodes in the tree basal to deinotheres follow Gheerbrant and Tassy (2009). Scale bar corresponds to *Numidotherium* only.

and modern proboscideans.[2] These intermediates document with some precision how features that we know today in elephants have appeared mosaically, or piecemeal, in many extinct animals over the past 60 million years, including fossils of *Phosphatherium* (Figure 6.2).

The fossil record of the modern African genus, *Loxodonta*, is quite good, and includes numerous skulls and skeletons over the last several million years throughout Africa. One of these fossil species, *Loxodonta adaurora*, is known from many localities in Kenya, Ethiopia, and Tanzania. In terms of the characters shown in Figure 6.2, *L. adaurora* is very close to the modern African elephant. One of its major differences concerns the shape of its lower jaw. In modern elephants, the two sides of the jaw join in front via a fairly short, spout-shaped connection, or symphysis. This region is longer in *L. adaurora*, and paleontologists used to believe that its length resulted from the root of an enlarged lower tusk in geologically older elephants, such as *Stegotetrabelodon*.[3] It now seems that this interpretation is not quite accurate,[4] and that the chamber in *L. adaurora* contained a plexus of nerves and blood vessels, not a tooth. However, modern elephants do occasionally develop lower tusks, as do some of their extinct relatives such as mastodons and an extinct dwarf (*Elephas celebensis*) from the island of Sulawesi in southeast Asia. Moreover, the presence of an elongate symphysis in recently extinct elephants, such as *L. adaurora*, is consistent with the presence of lower tusks in their more distant relatives, such as *Stegotetrabelodon*.

Looking at a different part of the proboscidean family tree (Figure 6.2), again very close to the modern species, it is not difficult to see why a woolly mammoth (e.g., *Mammuthus primigenius*) is understood to be a close relative of modern elephants. It's got the same "lamellar" chewing teeth consisting of several vertically oriented, narrow plates of enamel (which are a bit narrower in the mammoth and Asian elephant than in the African elephant); it has similar tusks rooted in the upper jaw with none in the lower, the same dorsally situated nasal aperture around which is a large depression for trunk musculature, and the same big skull with a large braincase and air sinuses.

Now consider a mastodon (*Mammut americanum*), or a South American gomphothere (*Cuvieronius platensis*; Figure 6.3). They too are quite similar to modern elephants: the tusks, large braincase and sinuses, the eye sockets situated anteriorly in the skull, and the spout-like anterior process of its lower jaw. However, unlike modern elephants, juvenile mastodons usually possessed a lower tusk which fell out, unreplaced, as the animal matured. An even greater difference is evident in the shape of their more posterior teeth (Figure 6.3). Mastodon and gomphothere chewing teeth look very different from those of mammoths and modern elephants. They're not composed of narrow enamel plates pressed against one another, but of much more primitive-looking molars with distinct cusps and wide roots (Figure 6.3). These teeth possess strong crests that join the cusps on either side of the tooth, and provide the imagery responsible for the term "mastodon." This was coined by the grandfather of comparative anatomy, Georges Cuvier, in the early nineteenth century, and may be roughly translated as

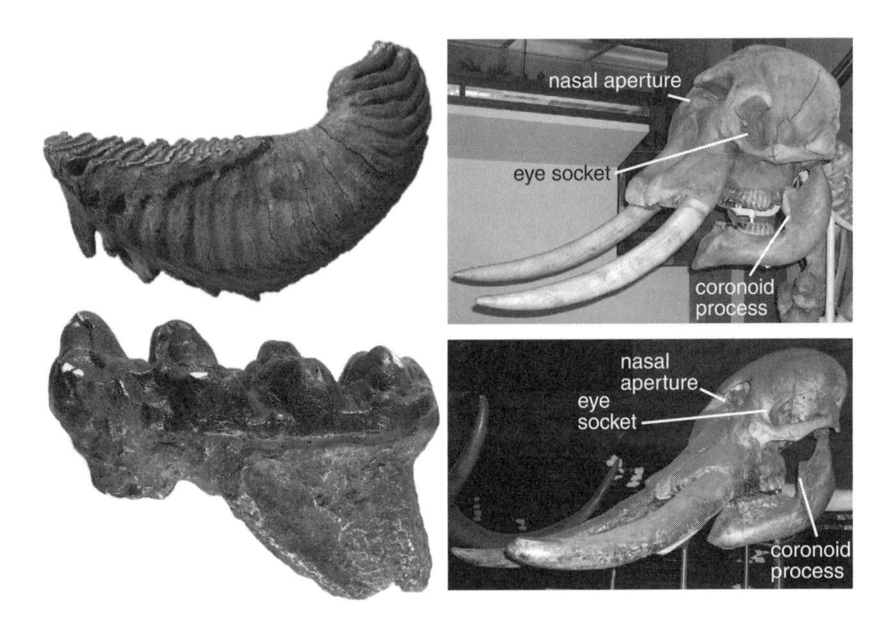

Figure 6.3. Modern elephants (*Loxodonta africana*, above right; lower posterior molar of *Elephas Maximus*, above left) and extinct proboscideans such as mastodons (*Cuvieronius platensis*, below right) and some gomphotheres (lower posterior molar of *Gomphotherium angustidens* [image reversed], below left), have very similar skulls and share such features as an anteriorly placed eye socket, a dorsally situated nasal aperture, a vertically directed coronoid process of the jaw, large upper tusks, and absent lower ones. However, note the very distinctive teeth. Chewing teeth of modern elephants (top left) show multiple enamel plates closely fused together, whereas mastodons, gomphotheres, and other fossil proboscideans had more primitive-looking teeth with more conspicuous cusps and lophs (bottom left). These fossils demonstrate some of the many ways in which the features we observe in modern elephants have appeared independently of one another, or mosaically, over the course of proboscidean evolution.

"nipple tooth" in reference to the apparent similarity of each cusp to a human breast.[5]

Cuvier had intended "mastodon" to be this animal's formal, zoological name. However, since *Mammut* was proposed some 18 years earlier for what turned out to be the same animal, that name has priority and remains the one in use today. So we're left with a slightly confusing combination of common and scientific names: *Mammut* is a mastodon, *Mammuthus* is a mammoth, and the two are very different kinds of proboscideans. Just remember that the nipple-toothed mastodon (*Mammut*) has a very elephant-like head, but very un-elephant-like teeth. The mammoth (*Mammuthus*) is much more closely related to modern elephants, and is even closer to the Asian elephant (*Elephas*) than the latter is to either species of modern African elephant (*Loxodonta*), as evident by its chewing teeth.

While the external shapes of mastodon and modern elephant teeth are quite different, we can infer that their patterns of dental eruption are similar based on an abundance of mastodon fossils of varying ages, some juvenile, others adult. As a child, you may have earned some cash when your baby incisors, canines, and premolars (also known as milk or deciduous teeth) fell out and were replaced by their permanent successors, each of which emerged from underneath its corresponding milk tooth (known as vertical replacement). In contrast to humans, mastodon and modern elephant anterior chewing teeth, or premolars, are not replaced by permanent premolars from below, but by larger molars from behind. A worn milk tooth in living elephants, mammoths, and mastodons is pushed out as it reaches the anterior end of the toothrow and is replaced in the jaw by the one behind it in a conveyor belt fashion. It is not replaced by a successor that grows in the jaw from underneath the milk tooth, as is (or was) the case in you, me, and most other mammals.

Other proboscideans, such as Miocene–Pliocene *Stegotetrabelodon* and *Stegodon*, Miocene *Gomphotherium*, Oligocene *Phiomia* and *Palaeomastodon*, Eocene/Oligocene *Moeritherium*, and early Eocene *Phosphatherium* exhibit a variety of differences from modern elephants (Figure 6.2). *Stegodon* resembled modern elephants and mastodons in the shape of its head, the dorsal position of its nose, the muscle scars that would have supported a long trunk, the conveyor belt pattern of dental eruption, and the tusks only in its upper jaw, but had molars with lower crowns and a different pattern of cusps. *Stegotetrabelodon* is very similar to modern elephants in terms of its cranial shape, but retained low-crowned molars and large tusks emerging from both its upper and lower jaws throughout its life. *Gomphotherium* also had large tusks in its lower jaw, showing four elongate front teeth jutting out of its mouth.

Phiomia (Figure 6.2), a smaller animal known from its fossilized remains in northern Egypt, shares these peculiar upper and lower tusks with *Gomphotherium*, along with non-lamellar molars with distinct cusps. However, *Phiomia* has a nasal opening in front of its orbits (rather than above them), its tusks are smaller, and it appears to have lacked the conveyor belt pattern of tooth eruption of geologically younger proboscideans. Instead, it has two generations of premolars, the second of which erupts from underneath its milk precursor (as in humans and most other mammals). Palaeomastodontids also lack the horizontal tooth displacement of modern elephants, mastodons, and gomphotheres. Furthermore, they have much smaller tusks and less pronounced cranial features such as muscle scars in the nasal area, indicating that they did not have a large trunk. Neither did the Egyptian Oligocene fossil *Moeritherium* (Figure 6.2), which was substantially smaller than a modern elephant. Yet their teeth, anteriorly situated orbits, and jaw shape unambiguously place both *Phiomia* and *Moeritherium* close to other proboscideans on the Tree of Life.

Despite the substantial differences between *Phosphatherium* and modern elephants on either extreme of our proboscidean lineage, it is not hard to recognize where this strange North African fossil belongs in the scheme of mammalian diversity, thanks to the abundant record of other fossil proboscideans over the past 60 million years (Figure 6.2).

ELEPHANTS AND THE BIG PICTURE

In the same way that modern astronomers do not spend time proving that the Earth revolves around the sun, paleobiologists modern generally do not spend their time proving that evolution via descent with modification is a major explanatory mechanism behind modern biodiversity. It's already been done. Of course, when the need arises, it is very easy (particularly with mammals) to step back and evaluate competing ideas on the causes of biodiversity, as we are doing now. Note, therefore, how two of the predictions of evolution by natural selection we've outlined previously are amply supported by the fossil record of proboscideans: (1) the features we observe in modern elephants occur in varying combinations in many kinds of extinct mammals, and (2) fossils that most closely resemble modern forms are, by and large, geologically younger. Thus, Eocene proboscideans have less in common with modern elephants than do, say, those from the Miocene or Pliocene (Figure 6.2).

A "designer" who is sufficiently talented so as not to need any of the natural phenomena studied by biologists would have no restrictions on the order in which she/he/it created elephant-like mammals. For example, we could have found mastodons (*Mammut americanum*) in the Eocene fossil record, showing large tusks in the upper jaw, no posterior incisors or canines, no horizontal succession of chewing teeth, and gargantuan size. We could have found proboscideans with compressed, elephantine dental lamellae in geological deposits below those of the late Miocene, with the smaller *Moeritherium* and its nasal aperture in the front of its skull, and its nearly full complement of incisors, canines, premolars, and molars, appearing higher up in the Pliocene. Even ignoring for a moment the mosaic acquisition of elephant anatomical features throughout the fossil record, we could have found the series of fossils shown in Figure 6.2 in virtually any order. In fact, had they all been deposited simultaneously in a single, catastrophic flood, we should find not only representatives of the various proboscidean species but also individual skeletons within a species dispersed throughout much of the stratigraphic column. But we don't. What we do find is the relatively more primitive species lower in the record, and those that most resemble living species toward the top. This is what we would expect if descent with modification is responsible for the evolution of proboscidean mammals.

Now there is some noise, and the relationship between stratigraphic order and proximity to modern species is not perfect. Take the Eocene form *Numidotherium* (Figure 6.2), for example. This animal would not for a microsecond ever be confused with a modern elephant. With a skull about 70 cm long, it's not much bigger than a horse. Furthermore, its nasal passage is at the front of its skull, not on top as in modern elephants, and it retains most of its anterior teeth, all of which are far smaller than elephant tusks. Not enough juvenile material is yet known to be certain about how those teeth were replaced (from below as in most mammals versus from behind as in modern elephants), but the overall shape of *Numidotherium*'s molar teeth are fairly primitive. However, in other aspects it shows some surprisingly elephant-like features, despite its very great geological age at over 50 million years. For example, it's got a short but tall skull, and the bone underneath the braincase and between the ears (the basicranium) is situated well above the palate (Figure 6.2). Near its nasal aperture are a number of depressions and other bony indications that large muscles attached to the front of the skull, suggestive of the possibility that this animal possessed an elongate nose, perhaps even a trunk. Furthermore, the skull shows numerous internal air cavities, or pneumatizations, which in modern elephants serve to decrease weight in a skull that still needs to protect sensitive organs such as the brain.

Thus, in a few regards, *Numidotherium* shows more similarity to modern elephants than do geologically younger specimens of, say, *Moeritherium* (Figure 6.2), which possessed no evidence of a trunk, no skull pneumatization, and had a more primitive, flat skull. This observation does not change the big picture that by far the closest similarities to modern elephants occur in fossil proboscideans that are much younger (Figure 6.2), but it does show that evolution is not an inexorable march toward animals that are alive today.[6] Rather, it is analogous to a bush, growing in whatever directions it can given local conditions and adaptive opportunities, directions which can change substantially over geological time.

Furthermore, paleontologists do not have the luxury of always sampling the oldest environments in which a given lineage represented today first evolved. Some of the oldest records of proboscideans, in the earliest Eocene of Morocco, already contain a diverse fauna of elephant relatives. I've already mentioned *Phosphatherium*; at the same locality is a much larger proboscidean called *Daouitherium*, which at over 100 kg may have been ten times bigger.[7] Clearly, therefore, proboscideans existed earlier than the age represented by the localities where *Phosphatherium* and *Daouitherium* have been found. In fact, paleontologists have already begun to make progress uncovering yet earlier fossils relevant to this group.[8]

Even so, we cannot be 100% certain that slightly older deposits somewhere else will not yield another relevant fossil. As we observed in our previous discussion of punctuated equilibrium, there are solid biological reasons (e.g., allopatric speciation) to suspect that the very earliest populations that gave rise to a group, including proboscideans, were unlikely to have left behind an immediate fossil record. It is only once a new species becomes sufficiently widespread and numerous that it is likely to appear in the fossil record, and such conditions may occur much later than the actual speciation event. Thus, it is neither problematic nor surprising that, on occasion, fossil animals that appear primitive relative to their modern relatives (e.g., *Moeritherium*) appear in the record simultaneously with or even after fossils that show some specializations (e.g., *Numidotherium*).

My focus on proboscideans results in part from the fact that this group has a good fossil record, but as we saw in the beginning of this chapter, the point I'm trying to make is broadly applicable. Even with the qualifications such as the ancient-but-derived *Numidotherium* just described, the link between ascending stratigraphic order and increased relatedness to modern species has been the subject of a lot of formal scientific investigation among mammals[9] and many other kinds of organisms.[10] It is safe to say that, while the system does show noise, a proportional relationship between the two comprises a thoroughly tested and generally accepted proposition in paleobiology. In fact, it is so broadly agreed upon that the fossil record reflects a genuine biological signal, i.e., older species tend to be ancestral to younger ones, that a few scientists have proposed using such data to help build evolutionary trees.

This has been dubbed "stratocladistics,"[11] the idea being that a tree of interrelationships (like that shown in Figure 6.2) should minimize the gaps in the fossil record between the first appearances of closely related animals. According to that tree, which was constructed using anatomical information only,[12] the proboscidean *Deinotherium* occupies a more basal evolutionary branch than either *Palaeomastodon* or *Phiomia*, despite the fact that the latter two are known from the Oligocene, and *Deinotherium* is younger, unknown before the Miocene. Such a tree therefore incurs "stratigraphic debt." This amounts to the added supposition, not supported by direct evidence, that pre-Miocene fossils related to *Deinotherium* will eventually be found to correspond with its divergence with known Eocene species.

Now the anatomical basis for placing *Deinotherium* basal to *Phiomia* and *Palaeomastodon* in Figure 6.2 is fairly robust,[13] and the existence of "stratigraphic debt" in this case does not mean that this idea about common ancestry is wrong. It only makes clear that, given what has been recognized about the proportional relationship between stratigraphic order and the proximity of a common ancestor, a very specific part of the fossil record (in this case the Eocene and Oligocene records of proboscideans related to *Deinotherium*) is assumed to be incomplete. Furthermore, if for some reason the anatomical data became weaker (e.g., due to additions or changes of characters used to build the tree), input of stratigraphic data might push the Miocene *Deinotherium* toward other species that do not differ so much in terms of the stratigraphic range, such as gomphotheres or mastodons.

Interestingly, we see in *Deinotherium* and animals closely related to it the first proboscideans outside of the African continent. Neither *Phiomia* nor *Palaeomastodon* are known outside of Africa (as is the case for all of the other relatives of elephants prior to about 30 million years ago), but deinotheres, mammutids, gomphotheres, and elephantids (known after 25 million years ago) are found in Europe and Asia. This information could be considered as a kind of "geographic debt," i.e., like the stratigraphic signal, an indication that *Deinotherium* might well be better placed farther up the tree relative to *Phiomia* and *Palaeomastodon* due to its geographic similarities to proboscideans younger than 25 million years.

Importantly, the tree shown in Figure 6.2 does not include any such stratigraphic or geographic input. It represents a synthesis of anatomical studies carried out by Jeheskel Shoshani,[14] Pascal Tassy,[15] and Emmanuel Gheerbrant,[16] three widely respected paleontologists who have spent much of their careers on the fossil record of proboscideans. Part of this task includes constructing evolutionary trees for this group. As for most other mammalian paleobiologists, they use biological features of the animals themselves (such as morphology and DNA) to reconstruct those trees. In the papers upon which Figure 6.2 is based, they don't use stratigraphic or geographic data. If they had, and stratigraphy had been "front-loaded" into the process of building trees, it would clearly be more problematic to interpret the correspondence between stratigraphic succession and common ancestry as a genuine biological signal. The use of non-intrinsic, non-biological data such as stratigraphy or geography to build evolutionary trees among animals is an intriguing, but still controversial practice. Nevertheless, just as engineers design bridges under the legitimate assumption that gravity will act upon them, evolutionary biologists often consider methods, such as stratocladistics, that assume evolution has taken place. Stratocladistics is still a rare form of phylogeny reconstruction, but it is one that builds upon a broad consensus among scientists that the "theory" of evolution is, in practice, a fact.

ELEPHANTS AND ENGLISH

As we observed for punctuated equilibrium, certain ideas in paleobiology are analogous to certain ideas in human linguistics. For example, the histories of both elephants and English contain similar kinds of information that can be traced over a long period of time, during which each has had close evolutionary connections to (respectively) certain other animals or languages.[17] Take English and German, for example. Along with Swedish, Danish, Dutch, Icelandic, Norwegian, and Afrikaans (plus dialects like Platt, Faroese, Geordie, and Schwäbisch), these languages are Germanic. Even though all have been heavily influenced by other sources, ultimately these languages share a common ancestor, one that was actually spoken by humans in northern Europe at some point during the last two millennia. German and English share many similar words (e.g., *Man*, *Haus*, hand, sight) and, more tellingly, they share many aspects of grammar, such as verbs modified by prepositions: "I look the word *up*," or "I'll watch *out* for you." Uttering these phrases without the "up" or "out" can change the meaning of those verbs quite a bit (give it a try!). Romance languages derived from Latin, such as French and Spanish, do not use prepositions to modify verbs in this way.

Now because of the peculiar history of the British Isles, where for hundreds of years during the middle ages French was spoken in court and at home by the Norman aristocracy, English vocabulary has a great deal in common with French—sometimes it seems more than with German. Relying on vocabulary alone, words such as "history" (French *histoire*, German *Geschichte*), "utilize" (French *utiliser*, German *verwenden*), and "serve" (French *servir*, German *dienen*) might lead you to think that English was also a Romance language, rather than a Germanic one. Based on sentence structure, you might also reach this conclusion. To the ears of native English speakers, German grammar is notoriously strange: "When a phrase the supporting clause is, must I the verb at the end of the clause write."[18] German was apparently the inspiration of the famous muppeteer Frank Oz when he provided Yoda with his pidgin English in the *Star Wars* series: "Powerful is he, but to the dark side could he drawn be!"[19] In contrast, word order in modern French or Spanish is usually fairly close to that of English, with the notable exception of most adjectives.

But English is not a Romance language. We know this based on the structure of its modern form and as demonstrated by the literature left behind in previous centuries, showing the incremental nature of increasing French influence. We know, for example, that the "English" (or more accurately, Anglo-Saxon) written by Oxfordshire Kings of the ninth century (prior to the Norman invasion) exhibited grammar much closer to that of modern

German than to French, including lots of verbs placed at the end of supporting clauses and/or entire sentences.[20] The story of the early Christian scholar *Orosius*, written by the Saxon King Alfred (849–99),[21] contains sentences like "but to him the Carthaginians had the road blocked,"[22] using verbs in a similar fashion as they are used in modern German. Anglo-Saxon of ninth-century England represents a linguistic fossil, or missing link, between English and German, which have diverged substantially from one another since their common roots over 1500 years ago.

The proboscidean fossils discussed above can be compared to a series of Anglo-Saxon, Old-English, and Middle-English texts documenting the evolution of the English language. The modern form, with its abundance of French vocabulary, doesn't sound much like its ancient Germanic relatives—just like big-tusked, five-ton modern elephants do not closely resemble small, tuskless *Phosphatherium*. However, modern English shares with Germanic languages key features (such as prepositional verbs), just as modern elephants share with *Phosphatherium* some features such as the anteriorly situated orbit or the vertical coronoid process of the jaw. Furthermore, "fossil" English, as written, for example, in Alfred's *Orosius*, or in younger documents showing increased French influence in later centuries,[23] shows how "English" used to have a more distinct Germanic grammar, just as gomphotheres and phiomiids show increasing resemblance to *Phosphatherium* (e.g., chewing teeth with large cusps, vertical tooth replacement).

Modern English has in fact become so frenchified that you might even be misled as to its true heritage, just as some of the lineages in the "bushy" parts of the proboscidean tree have acquired odd specializations, such as the large, curved lower tusks in deinotheres. Note also that German did not "evolve into" English, but rather that German and English share a common ancestor. *Phosphatherium* is no more the literal great $\times 10^6$ grandfather of some modern elephant than contemporary German is the unchanged, medieval precursor of English. Both derived from a common ancestor; both accumulated changes at varying rates along the way.

Relatedly, keep in mind that the ninth-century Old-English spoken and written by King Alfred of Wessex was able to get the job done. He could communicate in it reasonably well, and he had no idea that an offshoot of his language was destined to become the first truly global *lingua franca*, heavily influenced by French. In the same way, the various elements of the proboscidean lineage that I've described were all perfectly capable of making a living, of surviving to reproduce, without any trend present within them necessitating that their descendents become what we now know as elephants. In both cases, what we rationally perceive as an unpredictable course taken by history has led to the present circumstances, both biologically and linguistically. Ultimately, English and German are connected to one another by the process of linguistic evolution, just as *Phosphatherium* and modern elephants are connected by the process of biological evolution. It's not a perfect analogy (*Phosphatherium* is extinct, German is not), but hopefully you get the point.

SEVEN

WHALES ARE NO FLUKE

Of all the groups of tetrapod vertebrates, none has done better at recolonizing the sea than mammals. And among the many mammals that make a living in water, none has done better than cetaceans, or whales. In this group are fully aquatic forms such as dolphins, porpoises, orcas, sperm whales, minke whales, blue whales, and humpbacks. The nature of their transition from terrestrial, to semiaquatic, to fully marine animals is very well documented in the fossil record.[1] In the following pages, I wish to add only slightly to previous accounts of their origins[2] from terrestrial, even-toed ungulates (including such animals as camels, pigs, deer, and hippos) by discussing the relevance to this origin of an extinct group of mammals called mesonychids. I will also discuss another cetacean evolutionary transition that has received somewhat less public attention than whale origins: the paleontological, genetic, and developmental links between toothed whales, such as dolphins and orcas, and baleen whales, such as minkes and humpbacks.[3]

CETACEAN ORIGINS: MOLECULES AND MESONYCHIDS

From birth to death, modern whales do just about everything in the water. A marine habitat has profound implications for the anatomy of any animal, and whales show numerous specializations throughout their skeleton and soft tissues. To name a few, all living whales possess an unusually shaped, thick tympanic bone which (as discussed above in the section on early synapsids) forms the outer edge of the bony housing of the middle ear. Their sense of hearing is specialized to pick up sound via water rather than air, and they are good at using echolocation to navigate under water. In toothed whales, such as dolphins and orcas, this is helped by an organ called the melon, composed of fatty deposits contained in a bowl-like depression near the forehead. All living whales show bones of the face and rostrum that overlap one another (as opposed to touching end-to-end, as in most other mammals), a nasal aperture on top of the skull rather than in the front, an elongate vertebral column without a functional pelvic girdle or hindlimbs (although modern whales frequently retain the remnants of hip bones), and paddle-like forelimbs with many more finger bones than other mammals.

Beginning in the nineteenth century, fossil finds from the southern United States and North Africa have documented the existence of large mammals with a specialized ear, an elongate head and vertebral column, and reduced hindlimbs. Fossils of *Basilosaurus* from many localities along the US gulf coast were regarded in the early nineteenth century as remains of "sea monsters," as suggested by the animal's reptilian-sounding name. It wasn't until the Victorian naturalist Richard Owen published his observations on the skull of this animal in 1841 that these fossils were recognized as those of mammals, and specifically whales.[4] About 150 years later, the unique association of living and extinct whales with even-toed ungulates, or artiodactyls, became decisively clear based on multiple lines of evidence.

Because of their single jaw bone, ear anatomy, vertebral column, and ribcage, Owen recognized that *Basilosaurus*, and whales generally, were mammals. It was clear to nineteenth-century naturalists that living whales nursed their young, had a long gestation, and a four-chambered heart with the main systemic artery on the left, like other mammals. However, it was not clear at that time where specifically among mammals whales should be classified. The first signs that whales might have a special evolutionary relationship to even-toed ungulate mammals (that is, artiodactyls or mammals that walk on two toes such as hippos, pigs, camels, cows, and deer) came from studies of placentation[5] and immunology[6] undertaken in the early to mid-twentieth century.

These ideas were noted in a 1966 study by paleontologist Leigh Van Valen,[7] who emphasized similarities among hoofed mammals generally, fossil whales, and an extinct group of mammals called mesonychids. Fossil skulls and skeletons of mesonychids from roughly 60–30-million-year-old deposits of Eurasia and North America enabled comparisons of their anatomy with living ungulates and whales. Van Valen was particularly impressed with the dental similarities between mesonychids and fossil whales (Figure 7.1). Both have lower teeth consisting of a large central cusp flanked anteriorly and

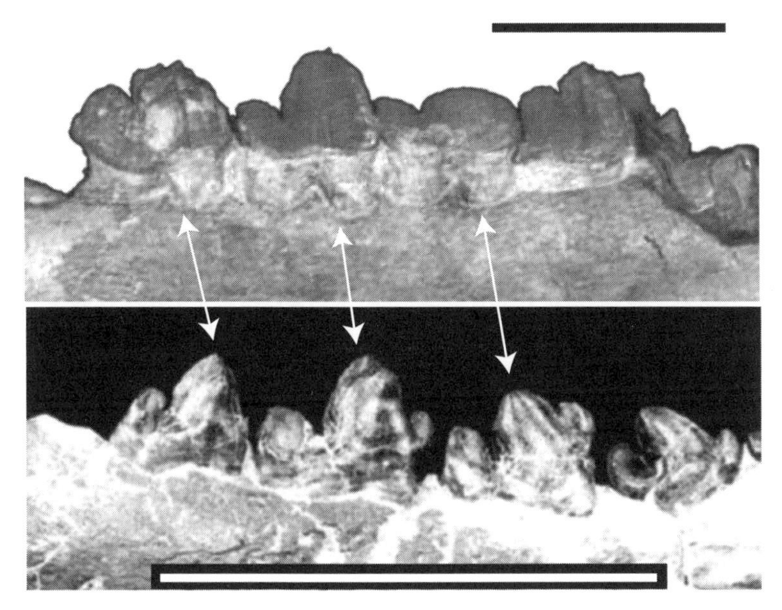

Figure 7.1. Dental similarities in the jaws of a fossil archaeocete whale (*Georgiacetus vogtlensis*, top, from Hulbert *et al.* 1998, figure 7, reproduced by permission of The Paleontological Society) and a mesonychid (*Ankalagon saurognathus*, from O'Leary *et al.* 2000, figure 3, reprinted by permission of the publisher—Taylor & Francis Ltd, http://www.informaworld.com). *Ankalagon* is from the early Paleocene of New Mexico (about 64 million years ago); *Georgiacetus* is from the Eocene of eastern Georgia (about 40 million years ago). Note the similar shape of the cusps on each of the posterior-most teeth (arrows). Other archaeocete whales and mesonychids also exhibit a high degree of dental similarity. Scale bars = 5 cm.

posteriorly by two smaller cusps, lined up from front-to-back, which show similar patterns of wear as the uppers and lowers occlude.

Starting in the early 1990s, discoveries of fossil whales[8] showed that, like modern artiodactyls, ancient whales such as *Basilosaurus* had hindfeet in which the third and fourth digits were the largest (Figure 7.2). Animals that support their weight between these two central digits are known as "paraxonic," as opposed to the "mesaxonic" pattern seen in odd-toed ungulates (such as horses, rhinos, and tapirs) in which the weight of the foot is balanced primarily on its central digit. Several extinct groups of fossil mammals, including mesonychids and *Basilosaurus*, also possessed a paraxonic foot. Combined with the dental similarities to mesonychids, this led an increasing number of paleobiologists to suggest that the origins of whales

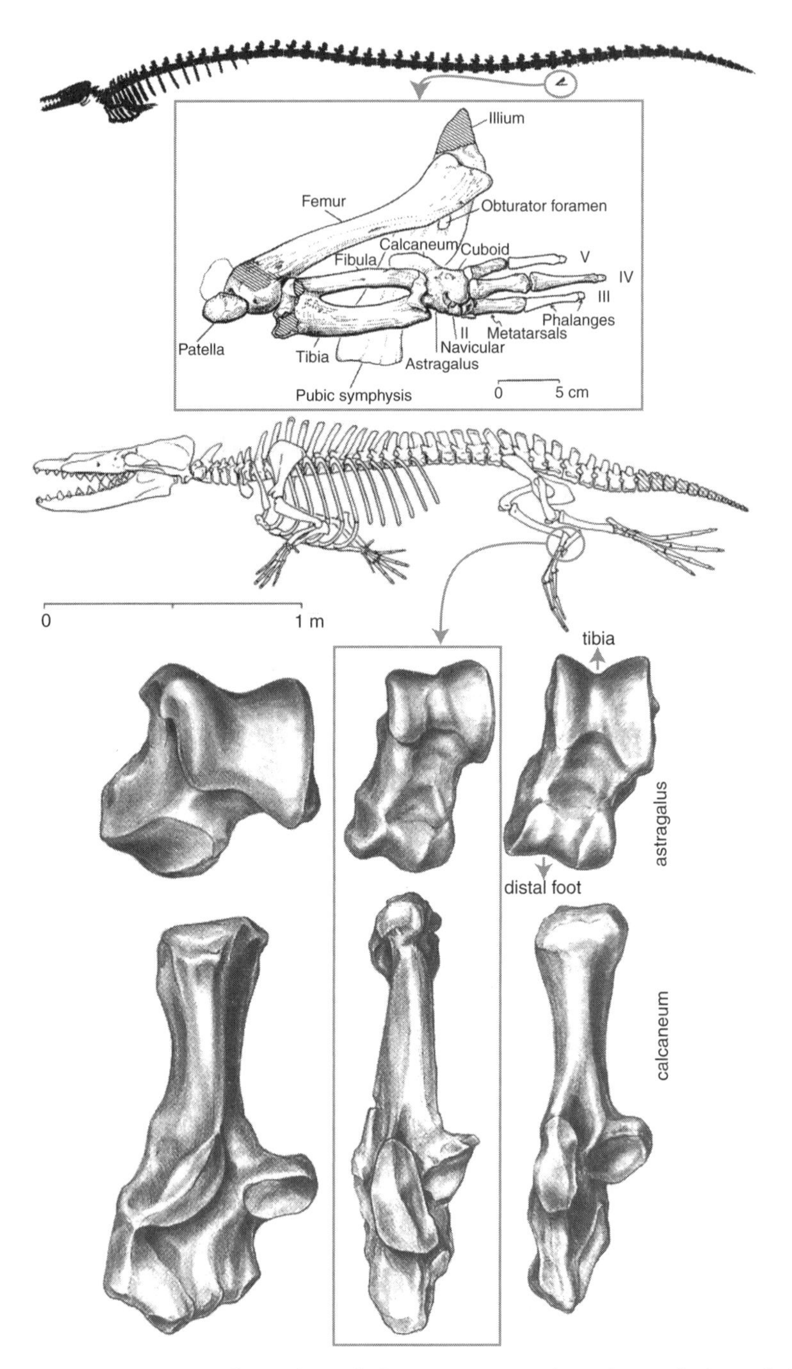

Figure 7.2. Important paleontological discoveries regarding the evolution of the whale hindlimb. The fully aquatic *Basilosaurus* (top) possessed a reduced hindlimb skeleton that was not capable of supporting the animal's weight, but still one which

were to be found specifically among fossil mesonychids, which in turn were likely to be closely related to living artiodactyls.[9]

The 1990s saw the molecular revolution applied to the Tree of Life, which meant that it became relatively easy to compare sequences of DNA across organisms. We will explore this subject in some detail in Chapter 9; for now, I'd like to briefly note the profound effect this revolution had on our understanding of whales. On the one hand, the close affinities suspected by anatomists between whales and artiodactyls were emphatically confirmed. However, the extent of the similarity as measured by DNA comparison was far more profound than any paleontologist had suspected. Studies of molecular data indicated that whales were not only close to artiodactyls, but that whales *were* artiodactyls. Specifically, molecular biologists[10] interpreted from their increasingly large datasets that whales share a common ancestor with the living hippopotamus to the exclusion of any other organism alive today, including other terrestrial artiodactyls, such as pigs, camels, and deer.

One of the features that all living artiodactyls share to the exclusion of any other mammal is their unique ankle joint. The ankle of all tetrapods consists of a number of small bones, the most important of which for this comparison are the astragalus and calcaneum. In particular, every land-dwelling artiodactyl shows what's called the "double-pulley" astragalus (Figure 7.2). It has a spool-shaped trochlea both proximally, to connect to the lower leg bone (the tibia) and distally, to anchor the other small ankle and toe bones. Furthermore, the calcaneum shows a large facet that connects with the astragalus laterally; it is not restricted to the underside of the astragalus as it is in you, me, and most other modern mammals.

Living whales, of course, have neither ankles nor functional hindlimbs. Despite the description of hindlimbs in *Basilosaurus*,[11] this particular fossil whale still had very odd feet, probably functional only as a copulatory guide

Caption for Figure 7.2. continued

showed digits III and IV as the largest, as in living even-toed ungulates such as hippos, cows, and deer. (Images reproduced from Gingerich *et al.* 1990 and Gingerich *et al.* 2001. Reprinted with permission from AAAS.) The reduced hindlimb of *Basilosaurus* may have functioned as a copulatory organ in males. In contrast, the hindlimb of *Rodhocetus* (middle) and *Artiocetus* (bottom central) were solidly connected to the vertebral column via a robust pelvis, could clearly carry the animal's weight, had feet that could deliver paddle-like thrusts under water, and had ankle bones extremely similar to those of hippos, cows, deer, pigs, camels, and the Eocene fossil artiodactyl *Bunophorus* (bottom right). Note the proximal and distal symmetry in the whale and artiodactyl astragali (leading to the double-pulley shape), and their lateral facet on the calcaneum. Fossil mesonychids (*Pachyaena*, bottom left) lack the double-pulley astragalus and lateral calcaneal facet, but resemble fossil whales in their dentition.

for males. Most of the ankle joint was fused into a single mass and left little basis for inferring the structure of individual ankle bones (Figure 7.2). No one knew, until 2001, what isolated ankle bones of fossil whales looked like. But we did know what mesonychid ankles looked like; they lacked the double pulley and were not particularly similar to those of artiodactyls (Figure 7.2). Since mesonychids have been extinct for about 30 million years, no molecular data are available to compare them with living whales or even-toed ungulates.

Thus, paleontologists of the 1990s faced a dilemma concerning the issue of whale origins. We knew that ungulate-grade mammals, particularly artiodactyls, were close living relatives of whales (Figure 7.2). We were impressed with the dental similarities between whales and mesonychids (Figure 7.1). But were the molecular data accurate that supported the placement of whales *inside* artiodactyls, next to the hippopotamus? If so, this would mean either that the dental similarities between fossil whales and mesonychids were misleading, or that one of the most unique and characteristic features of any mammal skeleton, the double-pulley astragalus, evolved and/or disappeared more than once. As we've noted, mesonychids didn't have this structure, but hippos did, and by keeping mesonychids and whales together, the implication was that ancient whales had a mesonychid-like ankle, without the double pulley.

The week of September 17, 2001 was an unfortunate time to announce one of the most important discoveries ever in mammalian paleontology. Due to their actions the week before, fundamentalist religious nuts deprived the world of what might have been a much more substantial level of media coverage of two very important finds, among other far more tragic crimes that need not be summarized here. Nevertheless, that week in the journals *Science*[12] and *Nature*,[13] two independent teams of paleontologists described the first relatively complete, associated fossil remains of whales with an unreduced hindlimb. Among these was a new species of *Rodhocetus*, and another was appropriately dubbed *Artiocetus clavis*, in homage to its "key" role in documenting a close relationship between whales and artiodactyls. These animals demonstrated without a shred of ambiguity that a double-pulley astragalus was present in fossil whales, just as you would expect if living whales were closer to artiodactyls, such as the hippopotamus, than to fossil mesonychids.

Since 1990, many key cetacean fossils have been described, documenting for example how early whales were capable of swimming by combining hindlimb paddling as seen in modern semiaquatic mammals (such as the Russian desman, a relative of moles)[14] and up-and-down movement of the vertebral column, similar to that of modern cursorial mammals that run on land.[15] We also have well-represented skeletons of the aquatic-adapted, but still largely land-dwelling animals such as *Indohyus*.[16] Besides a distinctive double-pulley astragalus indicative of its identity as a terrestrial artiodactyl, this animal showed one of the key features previously thought to be unique to whales: a thick tympanic bone with an emargination along its outer edge, called the involucrum. Animals such as *Indohyus* (collectively known as raoellid artiodactyls) have been shown to be the mammals closest to whales yet known, living or fossil, to walk on four limbs.[17] Yet they have unusually thick bones, with very narrow medullary cavities within them. In living semiaquatic animals such as the hippopotamus, this morphology helps the animal to maintain neutral buoyancy while it forages on the bottom of shallow lakes and rivers. Thus, many of the adaptations present in fully aquatic whales (tympanic shape, increased bone thickness) are exhibited in some of their partially aquatic, four-limbed fossil ancestors.

It is now agreed among biologists that whales and artiodactyls are each other's closest living relatives. Furthermore, recent studies support the placement of *Indohyus* adjacent to whales,[18] and all that sample a decent body of data support the living hippos as closer to whales than to any other artiodactyl, such as pigs, camels, or deer.[19] However, some uncertainty persists about the placement of mesonychids. One recent, well-sampled study suggests that they may indeed be closer to modern whales than whales are to hippos or raoellids,[20] despite mesonychids' very different-looking ankle skeleton. Even with this qualification about mesonychids, the association of whales with hippos and raoellids to the exclusion of other mammals is well supported in recent studies that broadly sample molecular and morphological data.[21] One of the persistent problems of whale origins is not really about whales, but hippos. Their earliest unquestionable fossils are Miocene, much younger than early Eocene fossil whales. There is also a lack of consensus among paleontologists as to which group of fossil mammals might conceivably comprise the closest terrestrial relatives of modern hippos. Since 1990, many of the "gaps" in the evolutionary history of whales have been dramatically filled in, and we may reasonably expect that at some point in the near future, the same will happen for the hippopotamus. An increased understanding of raoellid artiodactyls (the group to which the middle-Eocene *Indohyus* belongs) might indicate that these animals actually represent early cousins of the hippos, or we might find clear traces of hippo ancestry in parts of the world that have not yet been well sampled paleontologically, such as the Eocene and Oligocene of sub-Saharan Africa.

Indeed, I began writing the previous paragraph on Monday, June 28, 2010. In the June 29, 2010 issue of the *Proceedings of the National Academy of Sciences*, French paleontologist Maeva Orliac and colleagues[22] published a study of previously enigmatic fossils from East Africa, and in one stroke increased the age of the hippo fossil record by about one-third, from about 16 million to 21 million years ago to the early Miocene. In addition, the

French group supported an older idea that a group known as anthracotheriines, including some remains from Africa known to be over 35 million years old, have a closer evolutionary relationship to the modern hippo than to any other mammal.

The connection between anthracotheriines and modern hippos is not as obvious as that between, say, ancient toothed whales and modern ones. However, Orliac and colleagues have made a strong case for it. In so doing, they halved the duration of the missing record, or "ghost lineage," leading to fossil hippos of the Miocene. In other words, prior to their discovery, paleontologists had to assume that as-yet-unknown fossils would be found to document the existence of the hippopotamus lineage between the age of the oldest undisputed whales (*ca.* 53 million years ago) and the oldest undisputed hippopotamus relatives (*ca.* 16 million years ago). With their discovery, this gap decreased from 53–16 to 53–35, halving the previous "gap" from 37 million to 18 million years.

Despite some ongoing ambiguities about the artiodactyl and cetacean fossil record, no serious paleontologist or molecular biologist now disputes that the two groups are each other's closest living relatives. Nor does any modern evolutionary biologist doubt that the cetacean fossil record offers considerable resolution of the step-by-step adaptations toward fully marine life exhibited by numerous fossil whales during the Eocene.[23] Again, these examples comprise some of the many "intermediate and transitional links" predicted to exist by Charles Darwin in 1859.[24]

TOOTHED AND BALEEN WHALES

With the many important discoveries in cetacean paleontology and molecular biology made since 1990, whale origins have justifiably received a lot of media attention. However, there is another well-documented evolutionary transition within Cetacea: that between toothed and baleen whales. If you look into the mouth of a toothed whale (a member of the group Odontoceti) you'll find, unsurprisingly, teeth. These differ from the teeth of most other mammals in that they are conically shaped and good at stabbing, without much in the way of variation along the toothrow or interlocking occlusion between uppers and lowers. This is pretty much what a dolphin or orca needs to live on a diet of other sea-creatures.

In contrast, baleen whales, known by the taxonomic designation Mysticeti, do not have teeth at all, but rows of baleen—a keratinous substance similar to your fingernails which hangs down from the roof of the mouth. The largest mysticetes, such as humpbacks and blue whales, use baleen to filter out vast quantities of small, marine invertebrates from the "clouds" in which such plankton live. Feeding strategies among baleen whales are diverse,[25]

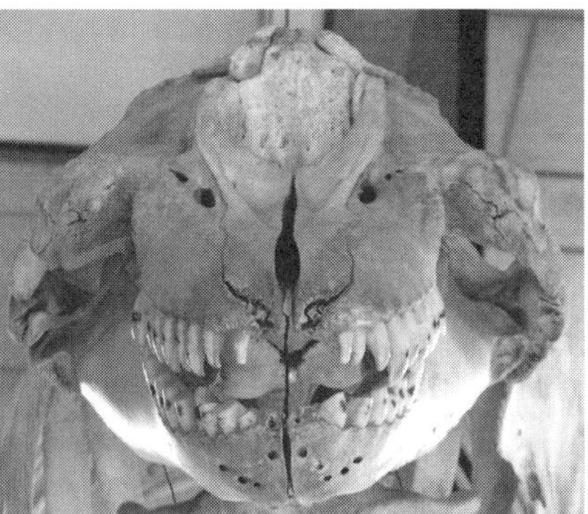

Figure 7.3. Differences in the skull of modern baleen (left) and toothed (right) whales. Most obviously, the orca (*Orcinus orca*, right) has a mouth full of teeth, whereas the fin whale (*Balaenoptera physalis*, left) has rows of baleen hanging down from its maxilla. Also note how the maxilla (the bone holding the upper teeth in the orca and the baleen in the fin whale) thins along its lateral margin in the fin whale (conveniently for the clips that anchor the skull to the metal supporting frame). In addition, note how the lower jaws of the fin whale are bowed laterally, and that they do not meet to form a joint, or symphysis, in the middle. In comparison, the orca has relatively straight jaws and a solid joint connecting right and left sides. The fossil record demonstrates that not all of these features distinguished the earliest members of the baleen whale lineage—for example, *Mammalodon* which possessed a laterally thin maxilla with teeth and a persistent connection between right and left jaws, or *Aetiocetus* which simultaneously possessed both teeth and baleen.

and may take the form of skimming prey from the ocean surface (e.g., right whales), filtering prey from the sea floor (e.g., gray whales), or engulfing and filtering prey at mid or shallow depths (e.g., rorquals). If you were to further examine a baleen whale skull, you would see a number of other differences from toothed whales: the left and right jaw bones are not solidly connected to each other in front, each jaw shows a strong outward curvature or "bowing", and the lateral margins of the upper jaw bones that anchor the baleen (i.e., the maxilla) are rather thin (Figure 7.3).

This suite of characteristics unites all living baleen whales, from right whales to minkes, and contrasts with the morphology seen in toothed whales, from dolphins to sperm whales. However, both groups are represented in the fossil record, and the history of baleen whales in particular shows that their ancestors did not appear with all of these characteristics simultaneously present. Rather, and as we've seen repeatedly throughout this book, features we now observe to co-occur in a modern group can occur piecemeal over the course of many millions of years.

During the middle to late Eocene, some 40–34 million years ago, fossil whales such as *Basilosaurus* and *Dorudon* were completely aquatic, dedicated to life in the sea and with no capacity to support their own weight with limbs on land. As noted above, *Basilosaurus* had a reduced hindlimb (Figure 7.2) that may have functioned in males as a copulatory guide, not unlike the hindlimb of some modern snakes. Based on the anatomy of its tail vertebrae and the clearly non-propulsory function of its hindlimb, *Basilosaurus* and its relatives depended in large part on a tail fluke to get around in the water.[26] Other fossil whales, including *Artiocetus* and *Rodhocetus* (Figure 7.2) dating to *ca.* 47 million years ago, and even older whales like *Pakicetus*,[27] were equipped to live in water but simultaneously had a robust hindlimb that would have given them the capacity to support their own weight in shallow waters or on land. None of these whales belonged to the modern groups of toothed and baleen whales alive today, but represent cousins related to the modern group's common ancestor, and are formally known as archaeocetes. The earliest relatives of baleen whales do not appear in the fossil record until the late Eocene and Oligocene, roughly 34–24 million years ago.

Remember our comparison between 55-million-year-old *Phosphatherium* and the modern elephant (Figure 6.1)? The fossil was the size of a fox, and to the untrained eye had little in common with a five-ton African elephant. But *Phosphatherium* did have a lot in common with other Eocene elephant ancestors, which in turn showed similarities to Oligocene phiomiids, which in turn showed similarities to Miocene gomphotheres and mastodons, which in turn showed similarities to Pliocene–Pleistocene mammoths and modern elephants (Figure 6.2). Baleen whales have a similar history. *Mammalodon* and *Janjucetus* are two early baleen whale cousins, known by relatively complete skulls and at least some skeletal material, recovered from *ca.* 30-million-year-old deposits from the southeastern coastline of Australia.[28] A partial skull of an animal called *Llanocetus* from the late Eocene of Seymour Island, Antarctica is even older, at about 34 million years.[29] Unlike modern baleen whales, these animals have a mouth full of teeth (Figure 7.4), but they do show a strikingly similar maxilla as modern baleen whales: this bone of their upper jaw becomes rather thin along its lateral margin.

Mammalodon, *Janjucetus*, and *Llanocetus* share a peculiar, flattened shape to the snout region of their skull (referred to as "platyrostral" by cetacean paleontologists[30]) with another group of Oligocene whales known as aetiocetids. These animals, including *Aetiocetus* and *Chonecetus*, are known from coastal regions of the Pacific coast in the USA and Canada. In addition to the laterally thin maxilla shared with both Oligocene and modern baleen whales, aetiocetids show a few additional similarities to modern baleen whales, and differences from archaic whales such as *Basilosaurus*: their mandibles are slightly curved or "bowed" laterally, and they exhibit a

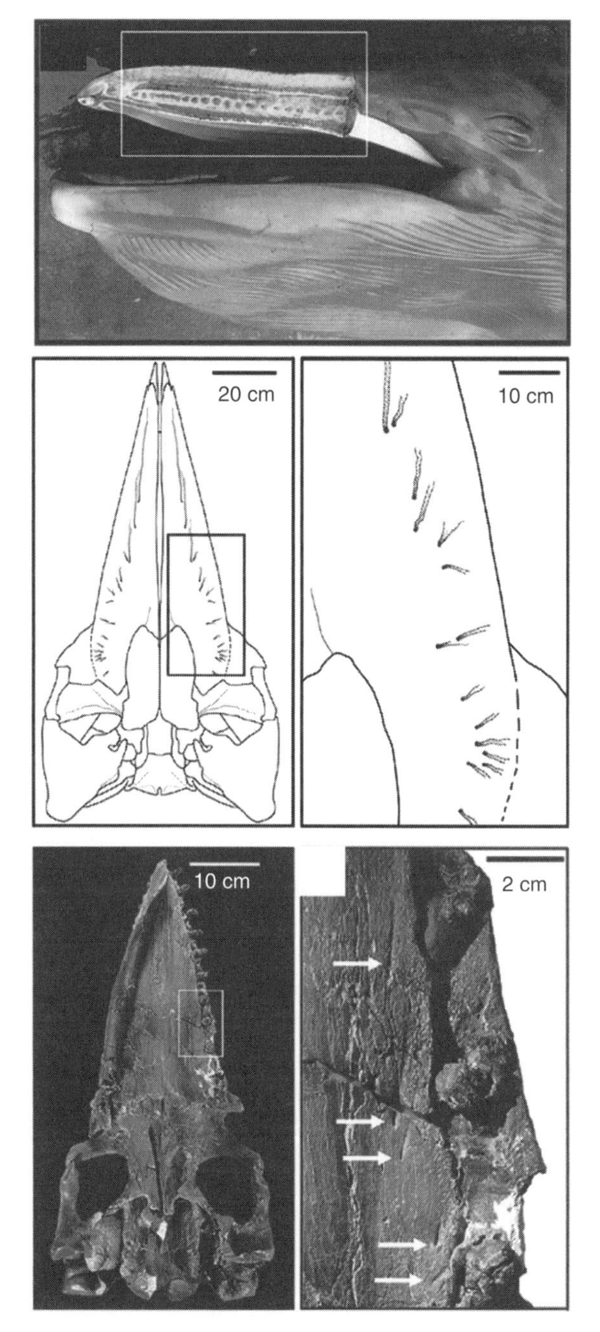

Figure 7.4. Adult baleen whales lack mineralized teeth and use baleen to filter-feed. However, fetal baleen whales, such as the image shown of a minke (top), show the embryological rudiments of teeth (boxed). The bony roof of the mouth in a modern

pattern of holes and grooves in the roof of their mouth, made by the passage of blood vessels in life, that among living mammals are closest to the pattern seen in modern baleen whales (Figure 7.4). This pattern has been interpreted as indicative of the presence of baleen.[31] Not all paleontologists are completely convinced about the presence of baleen in *Aetiocetus*,[32] but I think the case is strong.

Despite its likely possession of baleen and its laterally thin maxilla, *Aetiocetus* differed from modern baleen whales in several features, the most conspicuous of which you can clearly see in Figure 7.4. This animal had teeth. In addition, there was a weak but persistent connection between right and left jaws, one which was substantially weaker than that seen in modern toothed whales and archaic whales such as *Basilosaurus*. This connection, or symphysis, was not sutured and allowed for some flexibility, resembling the broad separation between the jaws seen in modern baleen whales (Figure 7.3). Most remarkably, *Aetiocetus* represents a fully aquatic, cetacean mammal that likely possessed some form of baleen along with hard, mineralized teeth.[33]

The most evolutionarily distant representatives of the baleen whale lineage to lack teeth were two animals from the middle Oligocene of the southeastern United States: *Eomysticetus* and *Micromysticetus*, the latter also known from fossils recovered in Germany. These animals unquestionably had some capacity for baleen filter-feeding, with laterally thin maxillae and lateral curvature to their lower jaws, which were detached in front. They occupied a similar ecological niche to modern filter-feeding whales such as minkes, but were preceded in the baleen whale lineage by several other forms (e.g., *Llanocetus*, *Aetiocetus*) with substantial differences in anatomy and ecology, differences that make sense given their evolutionary proximity to the common ancestor they share with living toothed whales such as dolphins and orcas.

BALEEN WHALE DEVELOPMENT AND GENETICS

Unlike some of the Oligocene ancestors of baleen whales, no living adult whale simultaneously has functional teeth and baleen. However, because

Caption for Figure 7.4. continued

baleen whale shows a very characteristic pattern of holes and grooves that supply nutrients to the baleen (middle). Some fossil cousins of baleen whales, like *Aetiocetus weltoni* (bottom), show the same pattern of holes and grooves (arrows, lower right), implying that they too possessed baleen. Note the presence of these grooves alongside functional teeth in this fossil relative of modern baleen whales. (Figure redrawn from Deméré *et al.* 2008. *Systematic Biology* 57(1): 15–37, by permission of Oxford University Press.)

of what we know about modern whale development and genetics, there are good reasons not to be shocked at the existence of such a combination in the fossil record. For example, modern baleen whales begin the process of tooth formation prior to birth. "Teeth" in a minke whale never fully form or break the gums, but they do at least begin to develop and their rudiments can be seen in fetal specimens (Figure 7.4).

The 2008 study of Tom Deméré and colleagues[34] incorporated data not only from the fossil record and development, but also genetics; they examined several genes known to be important for tooth formation. Mammal teeth consist primarily of three substances: an outer covering of enamel (the hardest substance produced by the body), filled with nearly-as-tough dentine, anchored to nearby bone with cementum. Many genes are known to pattern the formation of these substances, and Deméré *et al.* focused on three: *DMP1* (dentin matrix acidic phosphoprotein), *AMBN* (ameloblastin), and *ENAM* (enamelin). *DMP1* is known to contribute to the development of not only dentine but also other tissues such as bone and cartilage. The AMBN and ENAM proteins appear to express most strongly in the process of enamel formation in developing teeth.

So here's a question. Modern baleen whales completely lack mineralized teeth as adults. If evolution by natural selection were the process by which baleen whales descended from an ancestor that had functional, enamel-covered teeth (perhaps like those in *Basilosaurus*), and if genes serve as the instruction set by which various body parts are assembled during development, what would you predict about the existence of genes like *DMP1*, *AMBN*, and *ENAM* in modern baleen whales?

To answer this question, consider a very speculative thought experiment. What would a hyper-intelligent, human-like agent do if it were rolling up its sleeves with the explicit task of assembling a baleen whale, made in the same way a human engineer would build a device for transport? Such an engineer might borrow a few parts from one model to the next, but if you've got a fundamentally different mechanism, flying jib of a sailboat versus the heat and pressure of a jet engine, for example, you've just got to make your decision, use one technology and throw out the other. Jet propulsion depends on pressure and heat, not on viscous air pressing into a sail. What's more, no sane engineer begins with a different kind of propulsive force early in the building process—wind, for example—and then replaces it with internal combustion later on. You don't find a mainsail stuck inside a stealth bomber somewhere because such a mechanism has no relevance to the end product in the mind of a human engineer.

If a human-like designer were behind the construction of a baleen whale, what Deméré and colleagues found was that, for some inexplicable reason, he/she/it left a partially degraded blueprint for, and abortive development of, something the adult animal doesn't have: enameled teeth. They found that all three genes (*DMP1*, *AMBN*, *ENAM*) are present in baleen whales, but the two enamel-specific ones, *AMBN* and *ENAM*, have lost their enamel-producing function. Unlike the sequences in toothed whales (dolphin), even-toed ungulates (hippo, cow, pig, camel), and other mammals (human, mouse, rat, dog), their samples of these genes in modern baleen whales exhibited what are called frameshift mutations. That is, the basic sequences of *AMBN* and *ENAM* are present, but are missing critical elements that keep them from finishing what they do in other mammals, namely, synthesize proteins relevant to the formation of tooth enamel. Interestingly, such mutations were not present in the third protein, *DMP1*, which is demonstrably involved in processes besides tooth formation, such as bone and cartilage development.[35]

Applied to our speculative thought experiment, this discovery is analogous to an engineer hiding a degraded pair of oars inside a stealth bomber. Of course, no engineer would do such a thing, and human-like design is not the mechanism by which a baleen whale was "created." If you still believe that "design" is the actual mechanism behind the existence of a baleen whale, then you've got to somehow reconcile the absurdity of these animals not only showing non-functional teeth early in their development, but also the degraded blueprints for dental enamel still present in their genetic code. Note that the implication of "bad design" in this example is by no means the biggest problem for ID advocates. It is not simply that retaining the genetic instructions for, and abortive development of, structures that serve no function in an adult animal is a dumb thing for a human-like designer to do. Rather, these rudiments in a baleen whale are exactly what Darwinian evolutionary theory would predict, given its postulate that baleen whales descended from ancestors with teeth.

The discovery of Deméré *et al.*[36] is entirely consistent with the mechanism of natural selection. As teeth became less relevant to the feeding strategy of the lineage leading to modern baleen whales, mutations to enamel-regulating/producing genes crept in over many generations that gradually led to the loss of functional, enamel-covered teeth in the adult. These adult teeth are long gone, but their genetic rudiments are retained by living baleen whales, in part because these parts of the animal's genome, degraded though they may be, have no immediately negative influence on the animal's ability to contribute genetically to future generations.

We know that fetuses of modern baleen whales have the capacity to initiate the development of both teeth and baleen (Figure 7.4). The *AMBN* and *ENAM* genes present in baleen whales provide an example of what are known as genetic fossils, discoveries of which are becoming increasingly common elsewhere in the biological world.[37] Again, such genetic fossils are not simply cases of suboptimal design. They demonstrate not only that animals retain genetic rudiments of structures that are now of little or no use to them; they also represent signposts indicative of the specific part of the Tree of Life from which the animal in question evolved. Baleen whales descended from other animals with teeth, and their enamel pseudogenes (among other lines of evidence) prove it.

EIGHT

CREATIONISM: THE FOSSILS STILL SAY NO!

To wrap up the previous chapters on how fossils match the predictions of natural selection, I'd like to look more in detail at the skepticism about paleontology expressed by many in the anti-evolution crowd. A recent example is the 2007 textbook *Explore Evolution*,[1] co-authored in part by fellows of the Seattle-based Discovery Institute, a political think-tank that is home to a well-known anti-Darwin lobby. This book has a chapter called "fossil succession" which discusses if and how the fossil record provides evidence for evolution by natural selection. As in most of its other chapters, the book presents information by discussing the pros and cons, and it starts with a rendition of what Darwin believed, followed by a riposte from "the critics." It states that one of the cons facing those who subscribe to Darwinian evolutionary biology is the "stasis" we described earlier in the section on punctuated equilibrium:

> The sudden appearance of major new forms of life, and the stability of those forms over time, have led some scientists to doubt that the fossil record supports the case for common descent. ... Critics point out that discontinuity (abrupt appearance, followed by stasis) is the prevailing pattern of the fossil record [citation from R.L. Carroll 1997]. The transitional forms are the rare exceptions [quote from T.S. Kemp 1982]. ... Critics maintain that transitional sequences are rare, at best. For this reason, critics argue that Darwin's theory has failed an important test. ... In the overwhelming majority of cases, Common Descent

> does not match the evidence of the fossil record. ... Critics say that a scientific theory that only rarely matches the evidence fails the test of experience. ... Given the millions of different fossil forms in the fossil record, critics argue that we would expect to find, if only by pure chance, at least a few fossil forms that could be arranged in plausible evolutionary sequences.[2]

This passage is nonsense for several reasons. First of all, it dishonestly implies that genuine paleobiologists like Tom Kemp and Bob Carroll are among the "critics" who believe "Darwin's theory has failed an important test." Second, virtually every fossil found in rocks of about 500 million years and younger (comprising the vast majority of all known fossils) is manifestly constrained by the same general processes of structure and development seen in modern groups. Why is it that fossils share any anatomy with living groups? Why don't they exhibit something completely different, like wheels for locomotion or steel claws? The answer, of course, is that the animals we find in the fossil record are highly constrained by the developmental and genetic processes they share with living animals, a direct result of their common ancestry. Evolutionary biologists' understanding of these processes explains, among other things, why we don't find vertebrate fossils with wheels or steel.[3]

Furthermore, this passage's characterization of evolutionary stasis as implying the general lack of animals with transitional features throughout biological history is pretty similar to Rabbi Boteach's misinterpretation of punctuated equilibrium discussed in Chapter 4. The previously noted index entry about creationists and punctuated equilibrium in Stephen Jay Gould's 2002 book, *The Structure of Evolutionary Theory*, is just as relevant to *Explore Evolution* as it is to Rabbi Boteach: "modern creationists ... have shamelessly distorted all major evolutionists in their behalf. ... [Creationism] rarely goes beyond the continuous recycling of two false characterizations: the conflation of punctuated equilibrium with true saltationism ... and the claim that no intermediates exist for the largest morphological transitions between classes and phyla."[4]

If "pure chance" were the driving force behind paleontological interpretation, then why do we find in the fossil record animals with backbones, jaws, muscular limbs with digits, short rib cages, small ear bones, absent epipubic bones, a petrosal middle ear, closed orbits, anatomical bipedalism, and big brains, in precisely this temporal order, which happens to reflect the family tree of humans in their evolutionary context, as reconstructed by anatomy and DNA? *Explore Evolution* takes issue with the claim, made frequently in this book (e.g., Figure 4.1), that the correspondence between stratigraphy and ancestry supports a Darwinian interpretation of evolution:

> Another problem is that fossils don't always appear in the order they're predicted to by evolution.... Many "older" groups of animals (as depicted in cladograms) appear above, not below, the supposedly "younger" ones in the fossil record. Norell and Novacek, curators of paleontology at the American Museum of Natural History, say the primate fossil record "poorly reflects" the predicted evolutionary sequence.[5]

If you haven't been paying close attention to the paleontological literature over the last 50 years (and most regular folks have not), you would probably appreciate further explanation of how my very positive interpretation of Darwinian evolutionary biology as applied to paleontology, e.g., the early synapsids, elephants, and whales discussed above, can co-exist in a world with the authors of the above quote, who claim to interpret the same fossil record as I do. Have I just cherry-picked a series of random, fossilized forms and pasted them together to fit my own preconceptions about natural selection? Have I deluded myself in thinking that any animal group has a fossil record that shows anatomical continuity with other such groups, as well as a generally increasing level of similarity to modern forms over the entirety of geological time?

No. One of the easiest means of seeing past the vacuous claims of *Explore Evolution* is to read the sources it cites. In the above passages, the authors cite paleontologists Bob Carroll, Tom Kemp, Mark Norell, and Mike Novacek, all of whom I've met. In 2002, I was fortunate to have done fieldwork in the Gobi desert with Mark Norell; Mike Novacek was my postdoctoral advisor and we've co-authored several scientific papers together; and for the past several years Tom Kemp has contributed many lectures to a course I organize on mammalian evolution at my university. All of these individuals have spent their professional lives documenting something that this passage claims does not exist: an abundance of extinct animals that exhibit a mosaic of anatomical features found in different living groups, including early amniotes,[6] basal synapsids,[7] dinosaurs and birds,[8] and early cousins of placental mammals.[9] All of these scientists disagree emphatically with the creationist overtones of *Explore Evolution*, yet somehow their names appear spliced alongside phrases like "critics argue that Darwin's theory has failed an important test."

Particularly interesting is how *Explore Evolution* represents Mark Norell and Mike Novacek's interpretation of primate evolution, and how it "poorly reflects the predicted evolutionary sequence." Yes, there is a phrase to that effect in their 1992 paper in the journal *Science*—regarding primates. However, their actual conclusion in that paper for vertebrates in general was the opposite: "there is a noteworthy correspondence between the fossil record and the independently constructed phylogeny for many vertebrate groups. Statistically significant correlations ($P<0.05$) were found in 18 of the 24 cases examined."[10] Subsequent studies asking the same question but on different combinations of species (including primates![11]) and with different datasets have found similar results.[12] Mark Norell, Mike Novacek, and many other paleontologists have concluded that genealogical trees built with anatomical information typically match data from stratigraphy, despite the fact that *Explore Evolution* falsely attributes to them the opposite conclusion.

Table 8.1 summarizes some of the more conspicuous fossil animals known to mix anatomical features present in living groups, along with citations documenting current ideas on their place in the vertebrate Tree of Life. This table is by no means complete, but it's a good starting point for anyone who wants to consider creationist claims that "transitional sequences are rare, at best" or that "in the overwhelming majority of cases, Common Descent does not match the evidence of the fossil record." In contrast to these absurd statements from *Explore Evolution*, the reality is that the fossil record really does match the predictions of evolution by natural selection and goes far beyond the early synapsids, elephants, and whales that I've summarized in this book.

All of these fossil species exhibit similarities to one or more living groups, enabling in most cases a high level of confidence regarding their placement on the Tree of Life. Nevertheless, professional opinions on the affinities of some may vary precisely because none of these animals is identical to anything alive today. In the same way as the basal synapsids, elephants, and whales discussed above, these fossils show mosaics of anatomical features that are not found in any single living animal. What's more, their differences are not random relative to living animals, but rather reflect different combinations of anatomical features that correspond to their placement on the Tree of Life. I don't have the space in this book to go through each one in detail. However, my hope is that if you're still sympathetic to the claims of *Explore Evolution*, you've got access to a decent library and that you'll use the references in Table 8.1 to peruse the relevant literature and find out more for yourself.

You would learn, for example, that lagomorphs (the modern group of rabbits and pikas) are represented about 55 million years ago by Mongolian fossils of *Gomphos elkema*,[13] which exhibit squirrel-like teeth and tails alongside rabbit-like incisors and ankles. Lipotyphlans (shrews, talpid moles, hedgehogs, and *Solenodon*) are represented during the late Eocene of North America by animals such as *Oligoryctes*,[14] which like modern shrews have a bony "pocket" in the back of their jaw for chewing muscles, but with teeth that are very similar to those of the modern Caribbean *Solenodon*. Chrysochlorids (South African golden moles) are represented by the fossil species *Chrysochloris arenosa*[15] from deposits about five million years

Table 8.1. Age, classification, and references for some of the many fossils known to be related to living vertebrates, but that also exhibit mosaics of anatomical features present in distinct groups of their modern relatives. Note that this list is far from comprehensive. Summaries of these and other such fossil species are available in dozens of popular books published over the last century. Some of the more recent ones include Gee (2001), Clack (2002), Benton (2003), Kielan-Jaworowska *et al.* (2004), Kemp (2005), Rose (2006), and Prothero (2007).

GENUS	AGE	RELATED TO	PUBLICATION
Adapis	Eocene	Strepsirhine primates	Fleagle 1999
Algeripithecus	Eocene	Strepsirhine primates	Tabuce *et al.* 2009
Amphistium	Early Eocene	Percomorph flatfish	Friedman 2008
Ankotarinja	Early Miocene	Dasyuromorph marsupials	Long *et al.* 2002
Apateon	Late Carboniferous	Early amphibians	Froebisch *et al.* 2007
Apheliscus	Paleocene	Macroscelidid afrotherians	Zack *et al.* 2005
Apternodus	Eocene–Oligocene boundary	Soricid insectivorans	Asher *et al.* 2002
Archaeopteryx	Late Jurassic	Early birds	Ostrom 1976
Archaeothyris	Late Carboniferous	Early synapsids	Carroll 1988
Ardipithecus	Pliocene	Hominine primates	White *et al.* 2009
Artiocetus	Early Eocene	Archaeocete whales	Gingerich *et al.* 2001
Asiatherium	Late Cretaceous	Metatherian mammals	Szalay and Trofimov 1996
Asioryctes	Late Cretaceous	Eutherian mammals	Wible *et al.* 2009
Australopithecus	Pliocene	Hominine primates	Lockwood 2007
Balanerpeton	Early Carboniferous	Early amphibians	Milner and Sequeira 1994
Balbaroo	Early Miocene	Diprotodont marsupials	Long *et al.* 2002
Basilosaurus	Late Eocene	Archaeocete whales	Prothero 2007
Beipiaosaurus	Early Cretaceous	Therizinosaur theropod dinosaurs	Xu *et al.* 2009
Biarmosuchus	Late Permian	Therapsid synapsids	Kemp 2005
Biretia	Late Eocene	Anthropoid primates	Seiffert *et al.* 2005
Brachyrhinodon	Late Triassic	Early squamate reptiles	Fraser and Benton 1989

Table 8.1. (*cont.*)

GENUS	AGE	RELATED TO	PUBLICATION
Carpolestes	Paleocene	Early primates	Bloch and Boyer 2002
Catopithecus	Late Eocene	Anthropoid primates	Simons 1995
Centetodon	Late Eocene	Insectivoran mammals	Asher *et al.* 2005
Chambius	Eocene	Macroscelidid afrotherians	Tabuce *et al.* 2007
Chororapithecus	Miocene	Hominoid primates	Suwa *et al.* 2007
Cricetops	Oligocene	Muroid rodents	Carrasco and Wahlert 1999
Daphocoenus	Middle Eocene	Caniform carnivorans	Wesley-Hunt and Flynn 2005
Darwinius	Eocene	Strepsirhine primates	Seiffert *et al.* 2009
Deltatheridium	Late Cretaceous	Metatherian mammals	Rougier *et al.* 1998
Diacodexis	Early Eocene	Artiodactyl mammals	Prothero 2007
Dialipina	Early Devonian	Ray-finned fish	Schultze and Cumbaa 2001
Diatomys	Miocene	Hystricomorph rodents	Dawson *et al.* 2006
Dimetrodon	Early Permian	Pelycosaurian synapsids	Carroll 1988
Dinohippus	Pliocene	Equid perissodactyls	MacFadden 1992
Djarthia	Early Eocene	Australidelphian marsupials	Beck *et al.* 2008
Domnina	Late Eocene	Soricid insectivorans	Asher 2005
Dorudon	Late Eocene	Archaeocete whales	Prothero 2007
Eochrysochloris	Eocene	Afrotherian mammals	Seiffert *et al.* 2007
Eomaia	Early Cretaceous	Eutherian mammals	Ji *et al.* 2002
Eomanis	Eocene	Pangolins	Rose 2006
Eosimias	Eocene	Anthropoid primates	Beard *et al.* 1996
Eritherium	Late Paleocene	Proboscideans	Gheerbrant 2009
Escavadodon	Late Paleocene	Pholidote mammals	Rose 2006
Fallomus	Oligocene	Hystricomorph rodents	Dawson *et al.* 2006
Gerobatrachus	Early Permian	Amphibians	Anderson *et al.* 2008

Gomphos	Early Eocene	Lagomorph mammals	Asher *et al.* 2005
Gomphotherium	Miocene	Proboscidean afrotherians	Shoshani and Tassy 1996
Guiyu	Silurian	Lobe-finned fish	Zhu *et al.* 2009
Haikouichthus	Early Cambrian	Basal chordates	Shu *et al.* 1999
Haplomylus	Early Eocene	Macroscelidid afrotherians	Zack *et al.* 2005
Heomys	Paleocene	Rodents	Meng *et al.* 2003
Heptodon	Early Eocene	Tapiroid perissodactyl	Prothero 2007
Herpetotherium	Late Eocene	Metatherian mammals	Horovitz *et al.* 2009
Heteronectes	Early Eocene	Percomorph flatfish	Friedman 2008
Homogalax	Early Eocene	Perissodactyl mammals	Prothero 2007
Hyracodon	Early Eocene	Rhinoceratoid perissodactyls	Prothero 2007
Hyracotherium	Early Eocene	Equid perissodactyls	MacFadden 1992
Inkayacu	Late Eocene	Early penguins	Clarke *et al.* 2010
Indohyus	Early Eocene	Artiodactyls and whales	Thewissen *et al.* 2007
Janjucetus	Late Oligocene	Mysticete whales	Fitzgerald 2006
Jeholodens	Early Cretaceous	Primitive Therian mammals	Ji *et al.* 1999
Karanisia	Late Eocene	Strepsirhine primates	Seiffert *et al.* 2003
Kelba	Miocene	Afrotherian mammals	Cote *et al.* 2007
Kokopellia	Cretaceous	Metatherian mammals	Kielan-Jaworowska *et al.* 2004
Leptictis	Eocene	Insectivoran mammals	Novacek 1986
Madokoala	Early Miocene	Diprotodont marsupials	Long *et al.* 2002
Maelestes	Late Cretaceous	Eutherian mammals	Wible *et al.* 2009
Mammalodon	Oligocene	Mysticete whales	Fitzgerald 2010
Maotherium	Early Cretaceous	Therian mammals	Ji *et al.* 2009
Mayulestes	Paleocene	Metatherian mammals	DeMuizon 1998
Megalibgwilia	Pliocene	Echidnas (monotremes)	Long *et al.* 2002

Table 8.1. (*cont.*)

GENUS	AGE	RELATED TO	PUBLICATION
Merychippus	Miocene	Equid perissodactyls	MacFadden 1992
Mesohippus	Late Eocene	Equid perissodactyls	Prothero 2007
Metoldobotes	Oligocene	Macroscelidid afrotherians	Simons *et al.* 1991
Microbiotherium	Miocene	Australidelphian marsupial	Beck *et al.* 2008
Mimoperadectes	Paleocene	Didelphimorph marsupial	Horovitz *et al.* 2009
Mimotona	Paleocene	Lagomorph mammals	Rose 2006
Moeritherium	Oligocene	Proboscidean mammals	Shoshani and Tassy 1996
Monotrematum	Paleocene	Platypus (monotreme)	Rose 2006
Ngapakaldia	Early Miocene	Diprotodont marsupials	Long *et al.* 2002
Notharctus	Early Eocene	Strepsirhine primates	Rose 2006
Numidotherium	Eocene	Proboscidean afrotherian	Shoshani and Tassy 1996
Obdurodon	Late Oligocene	Platypus (monotreme)	Long *et al.* 2002
Odontochelys	Late Triassic	Turtles	Li *et al.* 2008
Omomys	Eocene	Haplorhine primates	Rose 2006
Onychonycteris	Early Eocene	Chiropteran mammals	Simmons *et al.* 2008
Oodectes	Early Eocene	Carnivoran mammals	Wesley-Hunt and Flynn 2005
Pachygenelus	Early Jurassic	Early mammals	Luo and Crompton 1994
Palaeolagus	Late Eocene	Lagomorphs	Rose 2006
Palaeoparadoxia	Miocene	Tethythere afrotherians	Rose 2006
Paljara	Early Miocene	Diprotodont marsupials	Long *et al.* 2002
Paradjidaumo	Eocene	Sciuromorph rodents	Korth 1980
Parahippus	Miocene	Equid perissodactyls	MacFadden 1992
Paramys	Eocene	Rodents	Rose 2006
Paranthropus	Pliocene	Hominine primates	Lockwood 2007

Patriomanis	Late Eocene	Pangolins	Emry 1970
Petrolacosaurus	Late Carboniferous	Diapsids	Carroll 1988
Pezosiren	Early Eocene	Sirenians	Domning 2001
Phiomia	Oligocene	Proboscideans	Shoshani and Tassy 1996
Phoberomys	Miocene	Caviid rodents	Sánchez-Villagra *et al.* 2003
Phosphatherium	Early Eocene	Proboscideans	Gheerbrant *et al.* 2005
Proconsul	Miocene	Hominoid primates	Fleagle 1999
Prolagus	Miocene	Ochotonid lagomorphs	Rose 2006
Prorastomus	Eocene	Sirenian afrotherians	Rose 2006
Prosalirus	Triassic	Frogs	Shubin and Jenkins 1995
Protictis	Paleocene	Carnivorans	Wesley-Hunt and Flynn 2005
Protosiren	Eocene	Sirenians	Rose 2006
Pucadelphys	Paleocene	Metatherian mammals	Macrini *et al.* 2007
Puijila	Miocene	Pinniped carnivorans	Rybczynski *et al.* 2009
Qatrania	Oligocene	Anthropoid primates	Simons and Kay 1988
Repenomamus	Early Cretaceous	Eutriconodont mammals	Hu *et al.* 2005
Rodhocetus	Early Eocene	Archaeocete whales	Gingerich *et al.* 2001
Rhombomylus	Eocene	Rodents	Meng *et al.* 2003
Saghatherium	Oligocene	Hyraxes	Gheerbrant *et al.* 2007
Saharagalago	Late Eocene	Strepsirhine primates	Seiffert *et al.* 2003
Sahelanthropus	Miocene	Hominine primates	Brunet *et al.* 2002
Seggeurius	Eocene	Early hyrax	Seiffert *et al.* 2007
Silurolepis	Silurian	Early jawed vertebrate	Zhang *et al.* 2010
Sinodelphys	Early Cretaceous	Stem metatherian mammal	Luo *et al.* 2003
Stegotetrabelodon	Miocene	Proboscidean	Sanders *et al.* 2010
Steropodon	Mid-Cretaceous	Monotremes	Long *et al.* 2002
Teilhardina	Eocene	Haplorhine primates	Ni *et al.* 2004

Table 8.1. (*cont.*)

GENUS	AGE	RELATED TO	PUBLICATION
Teinolophos	Mid-Cretaceous	Monotremes	Phillips *et al.* 2009
Tikiguania	Late Triassic	Squamate reptiles	Datta and Ray 2006
Tiktaalik	Late Devonian	Early tetrapods	Daeschler *et al.* 2006
Triadobatrachus	Triassic	Frogs	Rage and Rocek 1989
Trimylus	Late Eocene	Soricid insectivorans	Asher 2005
Tsaganomys	Oligocene	Hystricomorph rodents	Wang 2001
Ukhaatherium	Late Cretaceous	Eutherian mammals	Novacek *et al.* 1997
Utaetus	Eocene	Armadillos	Rose 2006
Victoriapithecus	Miocene	Catarrhine primates	Fleagle 1999
Widanelfarasia	Eocene	Afrotherian mammals	Seiffert and Simons 2000
Xenothrix	Quaternary	Pitheciid New World monkeys	MacPhee and Horovitz 2004
Yanoconodon	Early Cretaceous	Eutriconodont mammals	Luo 2007
Yarala	Early Miocene	Peramelian marsupials	Long *et al.* 2002
Zalambdalestes	Late Cretaceous	Eutherian mammals	Wible *et al.* 2004
Zhangeotherium	Early Cretaceous	Primitive therian mammals	Hu *et al.* 1997

ago in the western Cape of South Africa. This animal has teeth and skull bones that are exceedingly similar to those of the modern cape golden mole (*Chrysochloris asiatica*), but a very different elbow joint, one which closely resembles another living golden mole known from northwestern South Africa and southern Namibia: *Eremitalpa granti*. Some nine million years ago, giraffes, horses, sabre-toothed cats, mastodons, birds, giant tortoises, frogs, lizards, and many other animals roamed what is now central Spain,[16] yet they exhibited conspicuous differences compared to their modern relatives, such as the giraffe's short neck and the horse's three toes (Figure 8.1). These differences are easily reconciled with the developmental anatomy shown by modern groups. You would also learn that there is uncertainty

Figure 8.1. Paleontologists Adriana Oliver (left), Gema Alcalde (middle) and Juan L. Cantalapiedra (right) hard at work at the late Miocene locality Batallones (about nine million years old), just south of Madrid, Spain during the summer of 2008. In their midst are the neck vertebrae of a Miocene giraffe, related to an extinct south Asian genus called *Sivatherium*. "C2" refers to the second cervical vertebra, or the axis. This animal's skull was similar to that of a modern giraffe, but its neck was much shorter, beginning just one vertebra in front of C2 (toward Juan) and terminating at about the level of Adriana's left forearm. It was probably shorter than the neck of a modern okapi (*Okapia johnstoni*), known from the forests of equatorial central Africa. Adriana is carefully avoiding the skull of a three-toed horse (*Hipparion*) next to her right hand. Nine million years ago, Batallones consisted of a series of sinkholes that trapped a wide variety of ungulates, carnivorans, small mammals, turtles, lizards, and birds, closely related to those that exist today, but with notable differences such as the giraffe's short neck and the horse's three toes.

about the affinities of so-called "condylarths" such as *Haplomylus*, but that evidence exists to place at least some of them near or even within an endemic African clade of mammals called Afrotheria[17] (a group discussed in the next chapter).

I do not claim that the evolutionary affinities of all of the fossil species listed in Table 8.1 are now established beyond question. However, I do claim (and give the relevant citations to back this up) that every one of them exhibits anatomy that demonstrates that they are not just vertebrates, but ones representative of the many "transitional links" between vertebrate groups predicted by Charles Darwin to have existed. They are therefore strongly supportive of his theory of evolution. Thus, if you're sick of hearing about fossils of early synapsids, elephants, and whales, there is absolutely no shortage of other vertebrate lineages that exhibit a mosaic of features spread throughout other groups. Far from providing grist for the anti-Darwinian mill, paleontologists such as Bob Carroll,[18] Tom Kemp,[19] Mark Norell,[20] Mike Novacek,[21] and many others (e.g., Table 8.1), have demonstrated the opposite of what *Explore Evolution* is presenting to its target audience of high-school- and college-aged students. The fossil record does indeed support a view of evolution consistent with the mechanism proposed by Charles Darwin.

AGENCY AND CAUSE REVISITED

The passages quoted above from *Explore Evolution*, written not for any community of practicing scientists but for a younger and relatively uncritical audience, comprise the primary reason why the authors of *Explore Evolution* and others at the Discovery Institute elicit such anger and strong words from many scientists.[22] Even reasonable arguments made by ID advocates on certain, specific philosophical issues involving the origin of life or a possible agency behind the cosmos (see below) are received with hostility because most evolutionary biologists associate them with the blatant misquotation summarized above (and in Chapter 9).[23]

The reason creationist, anti-Darwin claims continue to be made, despite decades of rebuttal dating to the 1860s, probably has to do with the threat perceived in Darwinian evolution to morality and ethics by certain groups in society that tend to be religious. I believe this perception dissolves with careful consideration of evidence and the limits of science, just as it's possible to convince a friend and colleague that proof of gravity does not morally legitimize dropping bombs. I have stated above that if "intelligent design" means a mechanism analogous to a human building a jet aircraft, it is not the way in which a baleen whale was "created." Instead, I have made the case for descent with modification as the primary mechanism, or cause, behind the evolution and diversification of whales, and I hope I have convinced you that there is ample evidence to support this view.

However, I have made no claim to deny all forms of potential agency behind this cause. While I think there is no evidence supporting a role for a human-like, intelligent agent in this process, I cannot leap from this rational conclusion to the atheistic view that, therefore, no agency of any kind exists behind the evolution of elephants, whales, other mammals, or life in general. The concession that reasonable participants in this debate are obliged to make, I would argue, is that some of our scientist colleagues have made this leap, to the detriment of a sound public understanding of evolutionary biology. In so doing, atheistic evolutionists (i.e., those who espouse "philosophical naturalism" as discussed in Chapter 1) fuel the membership rolls of various anti-Darwin movements that perpetually feed off of this misunderstanding of how evolution relates to, and is limited by, human scientific inquiry.

Even though I think they are wrong in trying to admit supernaturalism into science, there are some ID advocates who would agree with me not only in rejecting philosophical naturalism, but also in recognizing Darwinian natural selection as a valid mechanism behind the evolution of not just vertebrates, but probably also multicellular life.[24] I only wish they emphasized this point of view for the non-scientific audiences to whom the ID movement most often preaches, such as the target audience of *Explore Evolution*.

According to a few unfortunate pronouncements of the well-known academic Richard Dawkins, he is an atheist because he accepts Darwinian evolution.[25] In my view, this is a total non-sequitur. As discussed in Chapter 1, he may as well claim that Thomas Edison is a myth because he knows an electric current in the vacuum of a glass bulb causes light. Understanding the mechanism by which a phenomenon works neither confirms nor denies the existence of an agency behind it. Ironically, his statement conflates agency and cause in the same way as any run-of-the-mill, naïve pronouncement of a young-Earth creationist: "I don't believe in evolution because God did it." It doesn't matter if you're religious or not; either way, "evolution" is a cause, "God" is an agent, and the naïve creationist who says this is wrong because the two kinds of explanation relate to fundamentally different questions. Atheists who reverse the nouns in this sentence—"I don't believe in God because evolution did it"—are wrong for the same reason.

Any sensible person can legitimately rule out some of the more naïve creation stories that implicate material elements of cause behind life's evolution at various levels. We know, for example, that female humans were not independently created from a male rib, and that terrestrial life was not packed into a wooden ship during a global flood within the last 10 000 years. Ample material evidence of Earth history makes it clear that both stories, while of great metaphorical or aesthetic value to some, are little more than superstition when portrayed as historical fact. However, superstition is not the same as religion, and the evidence of Earth history does not tell us anything beyond the realm of what humans can rationally perceive. In the context of behavioral ecology and evolutionary psychology, we can indeed understand much about the causes of animal social behavior, including some that qualifies as "religious" among humans.[26] But the scope of theology extends beyond the mechanisms behind the evolution of human social behavior. In my opinion, it follows from this that the absence of a scientific proof for God is more indicative of the limits of science than the lack of a deity.[27] No, I cannot prove this to be empirically true, but I believe it is reasonable.

As noted in Chapter 1, evolutionary biology is not about the origin of life or the existence of God. It is about how living things are interconnected through a specific, natural mechanism, one which we can understand through the fossil record, individual development, and molecular biology.

NINE

DNA AND THE TREE OF LIFE

Molecular biology is the study of how very small components in each of your cells interact to generate and regulate living things. On average, for every gram of tissue in your body, there are about one billion cells.[1] That means if you weigh 70 kg (about 154 lbs or 11 stone) you've got about 70 trillion (7×10^{13}) cells comprising your brain, olfactory receptors, eyeball muscles, toes, bladder, immune system, etc. Somehow, all of these are derived by division from just one: an egg fertilized by a lucky sperm in your mother's reproductive tract, nine months before you were born. This process is known as development, or ontogeny. One of the most significant areas of progress in evolutionary biology over the last 50 years concerns the means by which unique aspects of many animals, like ear bones and fish scales, result from subtle alterations in the timing of individual development across generations and geological time. Thanks to many scientists working in molecular biology and development, we know something about how the 70 000 000 000 000 cells came about from division of the single, fertilized egg which once comprised your entire physical being. Molecular biology is a vast discipline, and in this chapter we will examine only two small parts: how molecules are understood to document evolutionary genealogy and generate novel anatomical features of organisms.

As we've previously observed, evolutionary biologists convey their ideas about common ancestry by using diagrams of trees, also called cladograms or phylogenies. The animals represented in these trees that are genealogically close to one another are connected by lines that intersect at a common node, like the kangaroo and bandicoot in Figure 3.1. This means that these two animals are descended from a single common ancestor, and are more closely related to one another than to any of the other animals shown on that tree. From Chapters 3 and 4, you know that the vertebrate Tree of Life (Figure 4.1) places the lancelet (*Branchiostoma*) and jawless fish (lampreys, hagfish) near its base, followed by cartilaginous fish (sharks, rays), ray-finned fish (gars, goldfish), coelacanth, lungfish, amphibians, synapsids (mammals), and diapsids (reptiles and birds). We've reviewed in Chapter 4 how this basic branching pattern was deciphered quite accurately by nineteenth-century biologists like Theodore Gill, Ernst Haeckel, and Thomas Huxley. With only a few differences (like where sturgeons fit, relations within mammals, whether or not hagfish and lamprey are each other's closest relatives, and the possibility that tunicates are closer to vertebrates than the lancelet), recent trees built from very different sources of data are largely consistent with the work of early biologists who explicitly adopted natural selection as the cause behind the emergence of animal and plant diversity over time.

The science of determining how some animals are more closely related than others is called systematics, phylogeny reconstruction, or phylogenetics, and it typically uses data from embryology, anatomy, behavior, and genetics. These days, genetic information at the level of the deoxyribonucleic acid (DNA) molecule is the most popular source of data to build trees for modern groups. For fossils, things are a bit different. Paleontological data usually consist of bones and teeth, i.e., the parts of the animal sturdy enough to withstand fossilization. With the notable exception of certain animals that are relatively close to us in geological time (e.g., neanderthals, woolly mammoths) direct information about the genetic make-up of long-extinct animals is usually lacking, so we use anatomical data to build their evolutionary trees, combining them with other kinds of data (molecules, soft tissue, etc.) when we can.[2]

The methods for building trees from both anatomical and molecular data now draw on rigorous quantitative techniques such as parsimony, Bayesian probability, and maximum likelihood.[3] Furthermore, while some of the anatomical data on development (such as the embryology of mammalian ear bones) have been available since the nineteenth century, new anatomical information is regularly forthcoming. This includes not only the latest fossil discoveries, but also new techniques such as tomographic or magnetic resonance imaging to extract data from animals that we thought were already well known (see Figure 5.6).

Nevertheless, genetic information, or the series of nucleotides A, T, G, and C (for adenine, thymine, guanine, and cytosine) that form the tightly coiled strands of DNA in your cells, has become a tremendous resource for working out the branching pattern on the Tree of Life. For living organisms, DNA sequences have been used extensively in the last few decades to test ideas that had been proposed using anatomy and development (like the basic interrelationships among vertebrates shown in Figures 3.1 and 4.1), and to elucidate new patterns that few had predicted.

The most basic reason why DNA is so useful for phylogenetics is the amazing fact that every living thing on Earth uses the same code to convey information from one generation to the next. All of life shares a common language of inheritance. Since the 1980s, there have been two additional factors contributing to the modern revolution in the utility of molecular data for phylogenetics. First, improvements in laboratory techniques (e.g., polymerase chain reaction, or PCR) have made it much easier to infer DNA sequences at relatively low cost. Second, thanks to the rise of the internet and the availability of genetic databases, most of the DNA sequences generated by scientists are freely available on the web. If you publish a paper in a scientific journal (the bread and butter of any research career) and you've used DNA sequence data, you're usually obliged to upload those data to a web library such as GenBank.[4] In mid-2009, this web resource had about 255 billion nucleotides sampled across 157 million individual organisms, all of which are freely accessible to anyone with a web browser. Ongoing genome projects consisting of the entire sequence of nucleotides in the chromosomes of a given animal are also publicly available. As of this writing, www.ensembl.org is slowly converging on the complete genome of several dozen vertebrates, from alpaca to zebrafish, each of which contains millions of nucleotides distributed across thousands of genes.

It's really hard to overstate the magnitude of the current level of data availability in science, and molecular biology in particular. Take all of the great life scientists who reached retirement age before the advent of the internet and modern sequencing techniques: T.H. Morgan, J.B.S. Haldane, R.A. Fisher, G.G. Simpson, Theodosius Dobzhansky, Sewall Wright, Linus Pauling. Take every unit of data they ever produced, read, stored, or even saw. Add them together. Are you at home with a web browser and a decent broadband connection? Are you in your university library, in an airport, or at Starbucks? If so, then you've got more. At your fingertips is an order of magnitude more data regarding life's genetic make-up than any of these guys ever had.

This is a humbling thought indeed. But all of these data aren't much good unless you know something about how to interpret them, which brings us back to the topic at hand: building an evolutionary genealogy with DNA sequences. You may recall the text quoted in Chapter 4 from T.H. Huxley's 1863 book, *Man's Place in Nature*:

> the embryos of a Snake and of a Lizard remain like one another longer than do those of a Snake and of a Bird; and the embryo of a Dog and of a Cat remain like one another for a far longer period than do those of a Dog and a Bird; or of a Dog and an Opossum; or even than those of a Dog and a Monkey. ... Without a doubt, [humanity] is far nearer the Apes, than the Apes are to the Dog.[5]

In this passage, Huxley is describing the early development of individual animals. By emphasizing humanity's "affinity with the lower world of animal life," and the mechanism of Darwin a few pages later (as quoted in Chapter 4), he is outlining a tree of common descent, one in which humans are "far nearer the Apes, than the Apes are to the Dog." He has made very specific statements about which animals are most closely related based on the data then available to him. Needless to say, he could not have imagined the sheer quantity of information, in particular molecular data, that we can bring to bear on his very testable statements about degree of affinity based on development. Figure 9.1 depicts what Huxley said above in tree format, showing the pairs that he noted: human–chimp, cat–dog, snake–lizard. The more distant relations among birds, opossums, and dogs are derived from the trees published by other nineteenth-century scientists, such as Theodore Gill and Ernst Haeckel, as we've discussed in previous chapters.

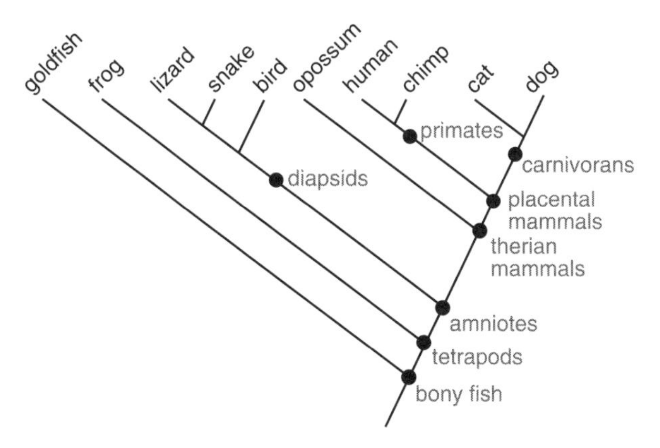

Figure 9.1. Evolutionary tree, or cladogram, consistent with the pattern of genealogical relations among vertebrates as described by T.H. Huxley in 1863 (branch lengths are arbitrary). In addition to Huxley's ideas (as quoted in the text), this cladogram also includes theories of common ancestry from other nineteenth-century investigators such as Theodore Gill and Ernst Haeckel (for example, the placement of birds within diapsids or "reptiles").

Right now, you could use your unparalleled access to molecular data on GenBank to examine the genetic diversity of all of the animals mentioned above and ask a simple question: do molecular data support these assertions about who's closer to whom? Are humans really "far nearer the Apes, than the Apes are to the Dog"? Inferring, as Huxley did, that modern animals derive from common ancestors who have evolved into their current niches via the process of natural selection leads us to expect that two animals with recent common ancestry will share more of their genetic make-up than either will with a third animal with whom that common ancestry is a bit more distant: ((human, chimp), dog), for example.

Another way of thinking about evolutionary tree building is to use it to predict the answer to specific questions. In particular, what would the implications be for anatomical, genetic, paleontological, and geographic similarity if animals evolved from a common ancestor? Having a common ancestor means specifically that different animals alive today—chimp and human, for example—can trace their heritage to a real population of interbreeding animals in the distant past that lived in a specific place at a specific time, and furthermore had a uniform anatomical and genetic make-up. It would be reasonable to conclude that if evolution actually happened, then two animals descended from an immediate common ancestor would show more similarity (anatomical, genetic, paleontological, and geographic) than either would with a more distant relative.

We cannot be too simplistic about measuring similarity because we know that in each case complicating factors exist. Some kinds of similarity are more relevant than others. Features that are shared between chimp and dog that arose from their common ancestor with a frog (e.g., walking on muscular limbs) are primitive; they do not help us identify the closest relatives of chimps and dogs. At first glance this may seem circular: how can I use common ancestry to discount a trait because it's "primitive" when it is common ancestry I'm trying to determine in the first place? This reaction is understandable as a first impression, but it's not accurate. The example of ((human, ape), dog) is one for which the tree has been known for 150 years; we already know what this part of the tree is like, which characters are useful for building it, and that "walking on all fours" is not one of them. For real phylogenetic problems, we don't know in advance which features are primitive (although scientists commonly make educated guesses). In such cases—and there are many—the process of adding more data is iterative, not circular.[6] As you add each trait, without knowing beforehand if it is primitive for most of the nodes in your tree, it helps to test whether or not the previous one defines a pair of closely related animals. Do this hundreds or thousands of times, connect your network of animals to a part of the Tree of Life that's already known, and you'll end up with a scientific estimate of interrelationships for your favorite group of organisms.

Relatedly, differences may accumulate in one animal or plant lineage more slowly than in another, for example leading to a descendant that is similar to its parent and very different from its sibling. Similarity that has nothing to do with common ancestry may result from adaptations to a similar environment (digging, swimming, foraging at night), or the unpredictability of mutation and fossilization. Changes in sea level and mobile continents enable some animals to disperse together to a certain area whether or not they are closely related. If such complicating factors were the rule throughout history, or if evolution had never happened, then anatomical, genetic, paleontological, and geographic similarity should not vary together. More often than not, animals that looked like each other anatomically would be strikingly different genetically, or would live in completely different parts of the world, or would be found at very different points in geological time.

If you recall our discussion in Chapter 4, then you know already that the last point is false. Organisms that are the most similar anatomically tend to be close to one another temporally in the fossil record (Figures 4.1, 6.2). Here, I want to show you that their anatomy also correlates well with their genetic similarity. Broadly speaking, when the pattern generated by one kind of data is consistent with that generated by another, and we have a mechanism (evolution by natural selection) that predicts that this pattern will be so, it is reasonable to regard that mechanism with higher confidence than if it had not been confirmed by multiple bodies of data.

Getting back to Huxley's statement about vertebrate interrelations (Figure 9.1), he didn't know how animals passed on their similarity to their descendants, or that genetic information was stored in the form of DNA located in the chromosomes within your cell nuclei, with a little extra hidden away in cellular organelles called mitochondria. But if his scheme of common ancestry is right, then it should be borne out by the level of similarity present in animal DNA.

BUILDING TREES WITH GENES

Since you've got at your fingertips more information on genetics than did all scientists of the nineteenth and early twentieth centuries put together, you can test whether or not Huxley's scheme was correct, right now, following the directions in Table 9.1. The data to do this are available on GenBank,[7] a website maintained by the National Center for Biotechnology Information, a division of the US National Institutes of Health. This website shares data with other sources such as the European Molecular Biology Laboratory and the DNA Databank of Japan. Thus, taxpayers from many countries worldwide help to maintain this information, and if you're one of them you've got every right to have a look. And it's not just the United States, European Union, and Japan who are footing the bill. This storehouse of information contains data from investigators from almost every university in every country in the world.

One of the first genes that was widely available across many different vertebrate species to build evolutionary trees is called cytochrome *b*. During the early 1990s, scientists interested in phylogeny wrote a lot about this gene, which contains far less information than the molecular datasets now used to build phylogenetic trees. Nevertheless, it makes a good case study to illustrate the science of phylogenetics. Cytochrome *b* is found in the mitochondrial genome, residing not on a chromosome within a cell nucleus, but in the organelle responsible for generating your body's energy from oxygen and sugar: the mitochondrion. This gene contributes to a protein important for energy production, which helps in the process of collecting oxygen. Like any other gene, cytochrome *b* consists of a series of nucleotides, or As, Ts, Gs, and Cs. There are about 1140 of these nucleotides in cytochrome *b* of mammals, fewer in snakes, and more in crocodiles. It is one of about a dozen protein-coding genes in the mitochondrial genome of a vertebrate, which at roughly 17 000 nucleotides in length is much smaller than the genome of your chromosomes contained within the nucleus of each of your cells. In humans, this "nuclear" genome contains well over three billion nucleotides, and can be much bigger in other organisms, such as onions, rice, and grasshoppers[8] which, by the way, are generally not regarded as more complex organisms than humans.

By lining up the cytochrome *b* nucleotides of one animal below those of another, ideally using a sequence editor or at least a simple text application (see the directions in Table 9.1), you can begin to appreciate the similarities and differences across these animals. First, note how the cytochrome *b* gene is not the same length in each of these animals. In the snakes it's rather short; in the crocodile it's a bit longer. We don't have to assume anything about evolution to line up those sequences and infer which pairs show the most similarity, as depicted for a short stretch of this gene in Figure 9.2. Despite some differences, the overall sequence of As, Ts, Gs, and Cs for these animals is quite similar. All we need to do is insert gaps in a few places (mostly in the snake sequences) to maximize the similarity of one series to another, while preserving the "reading frame" of the gene. That is, gaps shouldn't be inserted just anywhere, but ideally in units of three to preserve the way in which a protein-coding gene like cytochrome *b* helps to build a molecule: every three nucleotides contributes one building block (amino acid) of a protein.

Table 9.1. Directions for obtaining sequences of the mitochondrial gene cytochrome *b* from GenBank.

(1) Visit http://www.ncbi.nlm.nih.gov/nucleotide.
(2) For each of the animals listed in the table, the accession number provides the record of its cytochrome *b* gene. Type in this accession number on the top of the page in the "search" field, with "nucleotide" selected in the drop-down menu.
(3) Your result consists of annotations organized into fields labeled in capital letters (e.g,. "LOCUS," "DEFINITION," "ACCESSION," etc.) and the actual DNA sequences toward the bottom, in the ORIGIN field. For each of the species shown above, starting at the LOCUS field and ending at the double-slash ("//"), copy and paste it into a text editor. Some of the accession numbers above refer to an entire mitochondrial genome, others specifically to the cytochrome *b* gene ("*cytB*" as indicated in the "content" column). You can see how much larger the entries are for the mitochondrial genomes, about 17 000 nucleotides instead of 1140–50 for cytochrome *b* alone. When the record consists of an entire mitochondrial genome, look for "cytB" identified in the GENE field and click on the hyperlinked and underlined word "gene" to the left of "cytB." Cut and paste your cytB sequences into a text document, including all of the annotated field data provided by GenBank.
(4) Open-source software such as Mesquite (http://mesquiteproject.org/mesquite/mesquite.html, accessed August 1, 2010) enables reading of GenBank entries like those you've just downloaded. Save your cut-and-pasted cytochrome *b* entries in plain-text format (ensuring that no stray formatting characters or extra text creeps into any of your entries). Open the file in Mesquite to compare different sequences across animals. Alternatively, you could also use a text editor without word-wrap, e.g., by copying all of the characters in the ORIGIN field, up until the double-slash. Delete all of the line breaks and numbers along the left column, and use a uniform font such as courier for all of the nucleotides. The file I used with this information, which generated two of the phylogenetic trees depicted in Figures 9.2 and 9.3, is available here: http://people.pwf.cam.ac.uk/rja58/EB/chap9_cytb.nex

ANIMAL	SCIENTIFIC NAME	ACCESSION	CONTENT
Dog	*Canis familiaris*	U96639	mt genome
Crocodile	*Crocodylus palustris*	FJ173286	cytB only
Rattlesnake	*Crotalus viridis*	AF471066	cytB only
Opossum	*Didelphis virginiana*	NC_001610	mt genome
Cat	*Felis catus*	U20753	mt genome
Fish	*Gobius couchi*	FJ389196	cytB only
Human	*Homo sapiens*	GU990521	mt genome
Green lizard	*Lacerta viridis*	AM176577	mt genome
Orangutan	*Pongo pygmaeus*	NC_001646	mt genome
Python	*Python regius*	AB177878	mt genome
Frog	*Rana nigromaculata*	DQ006267	cytB only
Ostrich	*Struthio camelus*	U76055	cytB only
Thrush	*Turdus philomelos*	AY495411	cytB only

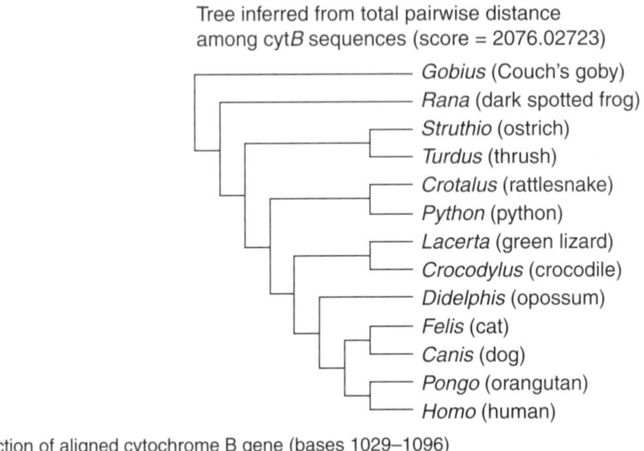

Section of aligned cytochrome B gene (bases 1029–1096)

Taxa		Characters
1	Canis	AGTTGAGCACCCTTTCATCATTATCGGACAAGTCGCTTCAATCTTATATTTCACCATCTTATTGATCC
2	Crocodylus	AGTACAAGACCCATACATTCTAATTGGACAAATAGCCTCCTTCATCTACTTCTTTACCATCCTTATCC
3	Crotalus	AGTAGAACCCCCTTTCACAGAAATTGGTCAACTAGCATCACCCATATACTT---TCTGTTCTTCATGA
4	Didelphis	AGTAGAACAACCCTATATTACCATTGGCCAATGAGCCTCCATTTCCTACTTTACTATTATCATCATCC
5	Felis	TGTAGAACATCCGTTCATCACCATCGGGCAACTAGCCTCCATCCTATATTTCTCAACCCTCCTAATCC
6	Gobius	AGTCGAGGACCCGTACATCGTCATCGGACAACTTGCATCCCTCCTGTACTTCTCTATCTTCCTGGTTC
7	Homo	AGTAAGCTACCCTTTTACCATCATTGGACAAGTAGCATCCGTACTATACTTCACAACAATCCTAATCC
8	Lacerta	AGTTGAACACCCATTCATTATTATTGGCCAACTCGCCTCAACTATCTACTTTATAATCTTCCTTTTTC
9	Pongo	AGTAAGCTACCCCTTTATTACTATTGGCCAAGTAGCATCCGTACTATACTTTACCACTATCCTACTCC
10	Python	AGTAGAACCCCCATACATCATTATCAGCCAAGCAACTGCAACACTATACTT---CACCTTCTTTATCT
11	Rana	AGTCGAGGATCCCTTTATTACCATTGGTCAAATCGCCTCCGGACTTTACTTTCTTATCTTTGTCCTTC
12	Struthio	AGTAGAACACCCCTTCATCATCATCGGCCAAGTAGCTTCCTTCACTTACTTCCTCATCCTATTAGTTC
13	Turdus	AGTCGAACACCCATTCATCATCATTGGCCAACTAGCCTCAATCTCCTACTTCACAATCATTCTAGTCC

Figure 9.2. Aligned segment of the cytochrome *b* mitochondrial gene showing 67 out of about 1200 nucleotides, or As, Ts, Gs, and Cs. All of these DNA sequences are publicly available on the internet (see Table 9.1). Note the presence of "gaps" (dashes) inserted in the sequences of the python and rattlesnake to maximize their overall sequence similarity to the other animals. Alignments such as this form the basis of comparison across species that help build evolutionary trees. The one shown above was built using a raw distance criterion (with arbitrary branch lengths), placing animals together on the basis of simple DNA similarity. Using raw similarity in this way is quick, but more subject to error than other methods (see text and Figure 9.3).

Typically, scientists use a program like Clustal[9] to do this, one version of which is freely available on the web.[10] This procedure is called "alignment"; it is the process by which a molecular biologist interested in making comparisons of DNA across animals lines up sequences of different lengths. It is typically done based on raw similarity and on known properties of genes, such as the groups of three nucleotides that represent amino acids in protein-coding genes, or structural features in non-coding genes. While there are sophisticated analytical approaches to alignment, this process does not change any of the actual nucleotide sequences, and in its most basic form does not require assumptions about evolution itself.

Once you've got a text file with the DNA sequence for the cytochrome *b* gene in these 13 animals (Table 9.1), it's possible to count the number of differences from one animal to another, starting at humans. Of the animals included in this exercise, *Homo sapiens* shows the fewest differences from the cytochrome *b* gene of the orangutan (*Pongo*) and the most with the rattlesnake (*Crotalus*). The rattlesnake, in turn, shows the fewest differences with the python, and the most with the crocodile. There are a variety of means to connect pairs of sequences to one another based on similarity. One option takes the total character distance (counting mismatches as "1" and matches as "0") and connects pairs of species with the fewest mismatches, averaging between them, and connecting more pairs until all are joined together in a "tree," as depicted in Figure 9.2. Although this tree is not quite accurate in all regards, it does support Huxley's statements about closeness: snake and lizard are closer than snake is to bird; dog and cat are closer than either is to bird, opossum, or primate; human and orangutan are closer to one another than either is to dog.

A different, more sophisticated tree-building method is shown in Figure 9.3, based on the principle of parsimony. This principle says that the simplest explanation is the best, and is broadly applicable throughout science. (And, indeed, life. Using the principle of parsimony, it is reasonable to reject the hypothesis that alien abduction is the chief cause of incomplete fourth-grade homework assignments, among other phenomena.) Here, rather than counting all similarities and differences, parsimony examines a large number of competing trees and chooses as "best" that which minimizes the number of required changes, or steps, to get from one branch to another. A step consists of a change from one nucleotide to another at a given position in our aligned cytochrome *b* gene. In the alignment shown in Figure 9.2, many of the nucleotides are invariant in all of the animals. These help us to recognize parts of the sequence that are "the same" across species. Other nucleotides show a difference only in a single animal. Nucleotides that are invariant or unique are useful at some levels, but with parsimony they don't help us to distinguish between competing tree shapes; they don't tell us if it's better to connect the lizard to the snake or to the crocodile, for example. No matter how you rearrange branches, invariant nucleotides will never add a step and unique nucleotides will add only one step. This is one of the differences between distance and parsimony methods for tree-building: distance will group by overall similarity even if observed changes are not uniquely shared by any two animals. Hence, parsimony is not subject to bias due to shared primitiveness. Perhaps for this reason, the cytochrome *b* tree favored by parsimony (Figure 9.3) does a better job of recognizing the affinity of snakes, lizard, and crocodile, and it too supports the tree implied in the above quotation from Huxley.

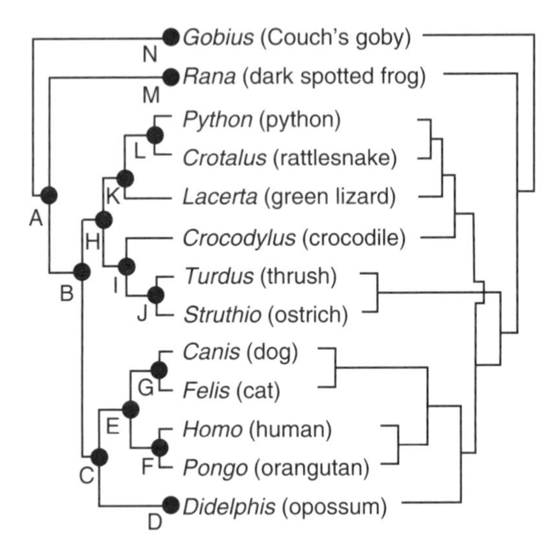

Black circles indicate nodes, as follows:

A, tetrapods	F, primates	K, squamates
B, amniotes	G, carnivorans	L, snakes
C, mammals	H, diapsids	M, amphibians
D, marsupials	I, archosaurs	N, ray-finned fish
E, placentals	J, birds	

Figure 9.3. Phylogenetic trees of vertebrates derived from molecular sequence data (branch lengths are arbitrary). At left is a tree derived from analyses of the nuclear gene RAG-1 (Hugall *et al.* 2007), consistent with others derived from relatively large datasets (e.g., Zardoya *et al.* 2003). At right is a tree derived from a much smaller dataset (the same one used to construct the tree shown in Figure 9.2), consisting of approximately 1150 nucleotides of the mitochondrial gene cytochrome *b*, requiring 2602 steps using the criterion of parsimony (see text). The data used to build these trees are freely available via the National Center for Biotechnology Information website, or GenBank (see Table 9.1). Both trees support the ideas on common ancestry given by T.H. Huxley on pp. 80–1 of his 1863 book, *Man's Place in Nature*. The tree at right differs in the placement of birds and crocodiles because the cytochrome *b* dataset is not sufficiently informative to accurately reflect the pattern supported by much larger DNA datasets such as that shown on the left.

Both distance and parsimony methods applied to our approximately 1140 nucleotides of cytochrome *b* support a genealogy close to those proposed by Huxley (Figure 9.2), Haeckel (Figure 4.2), and Gill (Figure 4.3). The major discrepancy of our brief study of cytochrome *b* concerns the placement of birds at the base of the amniote tree, just inside of the ray-finned fish (*Gobius*) and frog (*Rana*). When we take much larger molecular datasets analyzed by more sophisticated methods, including data from the nuclear genome[11] or larger analyses of mitochondrial DNA (which includes cytochrome *b* plus about a dozen other genes),[12] this discrepancy disappears and the correspondence gets even closer (Figure 9.3). These studies have shown that birds are very close to "reptiles," closer in fact to crocodiles than crocodiles are to lizards or snakes.

Interestingly, a similar result was implied in Haeckel's tree diagram (Figure 4.2) when he placed birds close to turtles and crocodiles, separate from the branch joining lizards and snakes. The closeness of birds to crocodiles reflects the concept of the Archosauria, famous because of the realization that extinct dinosaurs comprise part of this group too. This idea dates to the nineteenth century, but has received its most emphatic support from anatomical data collected by paleontologists since the 1970s.[13] Although our cytochrome *b* dataset is too small to pick up this signal, larger studies of molecular data support the argument that birds, crocodiles, and turtles are the only living archosaurian reptiles (Figure 9.3).[14] Moreover, our brief analysis of cytochrome *b* does recognize most of the other features of the tree hypothesized by anatomy (Figure 9.1).

Hopefully you can appreciate the importance of what we've just accomplished, and will take a moment now to recognize an extremely significant fact. We have just shown that comparisons of a small but important mitochondrial gene support the overall pattern of vertebrate interrelationships as first proposed in the nineteenth century based on anatomy and development. This is exactly what you would expect if Darwinian descent with modification were the mechanism behind evolution.[15] The series of nucleotides present in a gene like cytochrome *b* was not put there by scientists. Instead, using a variety of lab techniques, we simply infer what they are, as independently as any human endeavor gets from our own biases and pet hypotheses.

If the levels of similarity across DNA sequences were driven primarily by some other signal besides Darwinian evolution—for example, by the functional adaptations of the animals in question—you would expect animals with similar habitats (aquatic whales and fish, flying bats and birds, burrowing lizards and moles) to show the greatest genetic affinity. If animals had been created independently by an intelligence using mechanisms unknown to us, we might expect the pattern of affinity shown by different genes to be wholly incompatible, with each analysis of a different gene showing support for very different trees among the billions of possibilities resulting from a phylogeny with a dozen species. Alternatively, maybe the physiology of the animals would direct the shape of the tree, in which case we would expect (for example) warm-blooded birds and mammals to be grouped together.

Indeed, in the early days of molecular phylogenetics, before biologists had a decent body of comparative DNA data, the idea of a bird–mammal group to the exclusion of crocodiles, lizards, snakes, and turtles had been suggested,[16] and was a revived version of an idea discussed sporadically by anatomists, dating to T.H. Huxley himself.[17] Mitochondrial data, even comparisons of the *ca.* 17,000 nucleotides in the entire mitochondrial genome, have proven to be tricky for certain questions, leading some molecular biologists to propose such ideas as the affinity of the lungfish with ray-finned fish (e.g., tuna and carp) to the exclusion of lobe-finned fish (e.g., coelacanths and tetrapods like you and me),[18] the notion that guinea pigs aren't really rodents,[19] or that monotremes are closer to marsupial than to placental mammals.[20] In all of these cases, larger and more comprehensively analyzed molecular datasets have since shown that these results were artifacts: birds really are closer to crocodiles and other "reptiles" than to mammals;[21] lungfish really are closer to land-dwelling animals (tetrapods) than they are to ray-finned fish such as tuna;[22] guinea pigs really are rodents;[23] and marsupials and placentals really are close relatives to the exclusion of monotremes.[24] With increasing information on the DNA structure of various animals, each idea was recognized as a mistake by some of the same investigators who had originally proposed them.[25] These scientists recognized that bigger and better data analyses contradicted their previously published results, a difficult but commendable admission for any professional. The most comprehensive molecular datasets applied to vertebrate phylogeny[26] support the tree shown in Figure 9.3, one in which the basic pattern is very similar to that proposed in the nineteenth century (Figures 4.2, 4.3) based on morphology and development.

We observed in Chapter 4 how the fossil record is also consistent with the tree depicted in Figure 9.3. That is, relatively basal nodes are represented by relatively older fossils (Figure 4.1). So now we've got anatomy, development, the fossil record, and DNA sequences from the mitochondrial and nuclear genome showing us basically the same pattern. We can establish that this pattern is not due to ecology or physiology, and it is clearly not a chance result from the immense number of possible genealogies among sampled organisms. In the parlance of a prosecuting attorney, the observations above are analogous to finding a bloodstain on the victim matching your suspect, in addition to a bogus alibi, fingerprints, eyewitness testimony, and a confession. The ability to predict such correspondence in patterns across very different sources of data doesn't just happen unless you've got a process behind it, and we do: evolution by natural selection.[27]

ANTI-EVOLUTIONISTS AND TREES

Anti-evolutionists claim that phylogenetic study front-loads the assumption of evolution into the analysis before it even starts, and they use this false understanding to portray the entire field of molecular phylogenetics as irrelevant as a proof of evolution.[28] It is true that the mere act of drawing lines between species, as a tree reconstruction computer program is bound to do, is not by itself proof of anything. However, the proof derives not from the lines themselves, but from the fact that the patterns represented by these lines are very similar when using completely different sources of evidence.[29] When criteria such as distance, parsimony, and other methods are applied to the series of As, Ts, Gs, and Cs present in the nuclear and mitochondrial genomes of organisms, they support only one or a few potential branching patterns out of an astronomical number of possible trees. Such trees generated by different genes show many similarities, and few differences, and they are not biased to do so by any of the predominant analytical techniques or computer-based phylogenetic tools commonly used today.

In many of my own scientific publications[30] I've used phylogeny reconstruction programs such as PAUP,[31] NONA,[32] and MrBayes.[33] These programs reconstruct trees based on the given data and a specific method, such as parsimony or distance, as mentioned above. None of them "knows" in advance what an investigator wants to see; all they do is apply the method—such as parsimony—and pick the tree that is judged optimal according to that method. Deceit and data manipulation may indeed occur in phylogenetics, as in any human endeavor. However, the standard for data availability in phylogenetics is high, as is the degree of competitiveness among investigators. This makes data manipulation a potentially career-ending move. Far more likely than active deceit are simple mistakes. Even nice guys make these and, in phylogenetics, they are usually uncovered by your not-so-nice fellow phylogeneticists. These guys do not hesitate to rub your nose in them.[34]

For our example of 13 species (Table 9.1; Figures 9.2, 9.3), there are over 316 billion possible trees that could be used to represent their evolutionary genealogy.[35] Out of this huge number, we identified two using explicit criteria (distance and parsimony), and both trees are very similar to previously made predictions of relatedness using the very different sources of data. The 2007 study of Hugall and colleagues[36] sampled 88 species, for which the number of distinct, bifurcating trees (approximately two followed by 157 zeros) makes the 316 billion in our sample look truly pathetic.[37] Yet using different genes, they honed in on basically the same pattern (excepting birds, as noted above) as we did, and those of the 2003 study by Zardoya,[38] the 1994 study by Hedges,[39] among others.[40] The congruent results obtained by these analyses, and their correspondence with studies of anatomical, developmental, and stratigraphic data, are not an accident. Rather, they comprise powerful support for the idea that organisms share anatomical and genetic information via common ancestry, as predicted by natural selection.

The *Explore Evolution* textbook, to which we referred in the previous chapter, appears to be the nominal replacement for *Of Pandas and People*,[41] a book made infamous by its role in the Dover School District trial of 2005.[42] Two Discovery Institute members (S.C. Meyer and P. Nelson) are identified as contributors in the pages of both books. Pages 52–63 of *Explore Evolution* comprise the chapter on "molecular homology," which contains a discussion specifically on cytochrome *b* and how this gene has been used in phylogenetics. Here's what *Explore Evolution* has to say about using genes to build phylogenetic trees (pp. 53, 56):

> The same kind of sequence similarity we find in proteins is also found in DNA. ... Furthermore, the genes that code for the same proteins in different organisms have remarkably similar sequences. This is just what we would expect if these genes and proteins originated from a common ancestor.

Not bad; I actually agree with this! But as we noted in the last chapter, the layout of the book is explicitly pro and con, and the preceding was the "pro" side. Now comes their "con" (p. 57):

> Molecular similarities may result from common ancestry or they may reflect common functional requirements. ... If Darwin's single Tree of Life is accurate, then we should expect that different types of biological evidence would all point to that same tree. ... Many scientists have argued that ... this is frequently not the case. ... For example, one analysis of the mitochondrial cytochrome *b* gene produced a "family tree" in which cats and whales wound up in the order primates [citation of M.S.Y. Lee 1999]. Yet anatomical analysis says that cats belong to the order Carnivora, while whales belong to Cetacea—and neither of them are Primates. ... Critics point out that the real problem may be that Universal Common Descent is wrong. In other words, maybe the reason the family trees don't agree is that the organisms in question never did share a common ancestor.

This directly contradicts my presentation of the relative consistency of cytochrome *b* with anatomical studies, and I would definitely include the recognition of carnivorans (cats), cetaceans (whales), and primates as among the more robust areas of agreement between anatomical and molecular data. So who is right?

The paper they cite in which "cats and whales wound up in the order Primates" was written in 1999 by the University of Adelaide paleobiologist Michael S.Y. Lee.[43] Incidentally, Lee was also a co-author on the Hugall *et al.* paper referred to above which shows very close agreement to the vertebrate trees proposed by Huxley, Gill, and Haeckel (Figure 9.3). Lee's 1999 paper was a review article about the potential for function to mislead molecular phylogenetics. He did make a rather flippant comment about cytochrome *b* not doing a terribly good job with primates, but his paper was a review, and did not contain original research on cytochrome *b*. The paper concerning this gene to which Lee was referring, and which therefore formed the basis for the above quote in *Explore Evolution*, was published in 1998 by T.D. Andrews and colleagues.[44]

The Andrews paper also concerns the extent to which DNA similarity may be influenced by function. They analyzed cytochrome *b* sequences for seven primates, one whale, one cat, and one rat. Contrary to the claims of the *Explore Evolution* textbook, the discrepancy raised by the Andrews *et al.* paper is not that cytochrome *b* supports cats and whales within primates, but that the tarsier (discussed in Chapter 3) does not form a group with the other six primates they sampled. In fact, the *Explore Evolution* caricature of the cytochrome *b* signal is worse: despite their small sample size (even for 1998 standards), Andrews *et al.* still retrieved an optimal tree that *supports* the close interrelationships of all of their sampled primates except the tarsier, including such externally divergent animals as galago, loris, lemur, squirrel monkey, colobus monkey, and human, just as comparative anatomy has done since the nineteenth century. And even though the tarsier sits off by itself at a basal node near cat, rat, and whale, Andrews *et al.* state that this was not a significant result:

> The tree that represents an acceptable phylogeny [i.e., consistent with comparative anatomy with tarsier close to other primates] ... was among the 1731 trees that are not significantly different from the maximum likelihood tree [i.e., with the divergent position of the tarsier]. ... Only two groupings are clearly resolved by the cytochrome *b* data. These are the simian primates ... and the lorisoid strepsirhine primates. (p. 252).

Stated differently, far from concluding that cats and whales wound up in primates, the only supported groups resulting from the Andrews *et al.* study were completely consistent with those hypothesized by comparative anatomy. At certain levels, the tarsier has indeed proven surprisingly difficult to resolve with molecular data,[45] and mitochondrial DNA in particular has led some investigators to revive an older idea about the association of tarsiers with toothcombed lemurs, or prosimians[46] (see Chapter 3). Nevertheless, these studies still place tarsiers with other primates, not with cats or whales. Moreover, the latest and most densely sampled analyses of molecular data have supported a close affinity of the tarsier with humans, apes, and monkeys,[47] as proposed by most primate biologists over the last century.[48]

There are plenty of controversies in vertebrate phylogenetics, and we will discuss a few specific examples below. However, the basic pattern of vertebrate interrelationships, and the constituents of nearly all modern mammalian orders, are not among them. Concerning the phylogeny of whales, primates, and carnivorans (among other topics, as mentioned in Chapter 8[49]), the poor scholarship of *Explore Evolution* implies a controversy where none exists. There are some exceptions to the generally agreed-upon nature of the vertebrate tree (discussed below). However, the hypotheses of vertebrate interrelationships outlined repeatedly over the past century (Figures 4.2, 4.3, 9.1) receive strong, independent support from modern analyses of DNA.

THE NODES ANATOMY GOT WRONG

In our brief look at cytochrome *b*, I have emphasized the parts of the Tree of Life built by anatomists and paleontologists that have been confirmed by modern molecular biology, in particular the basic branching pattern among vertebrates. There were only a few quibbles; for example, Gill wasn't sure where sturgeons fit, or if fossil heterostracans and osteostracans were closer to bony fish (like guppies) or jawless fish (like lampreys). Once twentieth-century researchers obtained better information on both groups, these ambiguities were not difficult to resolve with anatomical data.

However, certain other parts of the Tree of Life have proven more controversial. For those who specialize in certain groups (e.g., placental mammals) the controversies seem huge, but the vertebrate Tree of Life has remained remarkably stable when viewed as part of the big picture. To make an architectural analogy, the basic edifice of vertebrate phylogeny constructed by nineteenth- and early twentieth-century biologists has remained pretty solid, even though modern biologists have had to do a fair amount of renovation to a number of rooms.

One of the areas of improvement concerns the group on which I've been working for the last two decades: mammals. When I was a graduate student, one of the more interesting symposia I attended on this subject took place at the 1997 meetings of the Willi Hennig Society, a group dedicated to investigating the methods and practice of animal and plant phylogenetics. You might be surprised at how passionately some individuals feel about using a particular methodology for building phylogenetic trees, or the extent to which another methodology can be viewed with derision. For example, at that meeting, I would sooner have performed a few Barbara Streisand numbers from *Yentl*[50] during my talk than presented the distance-based method shown in Figure 9.2. Raw distance was not a respected means for building phylogenetic trees in that crowd, and there are few things worse for a grad student in phylogenetics than to have one of the Hennig Society bigwigs hammering away at some perceived methodological weakness in your presentation. The pain can be particularly acute within this society; one of its by-laws is that the moderator of a session can never cut short the post-talk questioning following a talk, which might therefore more accurately be referred to as a smackdown, shark attack, or human sacrifice. Meetings of the Hennig Society are famous for their animated discussions and passionate feelings about a subject that few outsiders would associate with passion: animal and plant phylogenetics.

One of the presentations at the 1997 meeting was notable even by Hennig Society standards. That year, the University of Houston molecular biologist Dan Graur (then at the University of Tel Aviv) gave a talk about mammal phylogeny, subtitled "morphological schemes crumble under molecular scrutiny," and it was peppered by frequent shouting and jeers from the audience (I remained respectfully silent during the whole thing, of course). The schemes to which he was referring were ideas about where certain mammals belong on the Tree of Life, previously articulated mainly by anatomy and development. His area of focus was DNA sequence data, such as information from cytochrome *b* discussed above. Dan Graur was representative of some molecular biologists who thought that data from anatomy and development were incorrect about many branches of the mammalian tree. Two of the most prominent areas of contention were about groups we've already discussed. He thought living toothed whales (or odontocetes, like dolphins, orcas, and sperm whales) did not form a natural group, and that whales as a whole were very close to even-toed ungulates such as the hippopotamus. In addition, at the time he favored the idea that the guinea pig was not a rodent, and that African mammals like aardvarks, hyraxes, and golden moles were very close to other African groups like elephants and tenrecs. The ideas he mentioned about guinea pigs and toothed whales turned out to be wrong, but the other two (hippo–whale and African mammals) were correct.[51] Here, we'll briefly review some of these new concepts, including what I think is the most radical idea in mammalian phylogenetics published in the last 50 years: the African mammal clade.

AFROTHERIA AND THE NEW TREE OF PLACENTAL MAMMALS

If you had asked me at that 1997 meeting to name my favorite group of animals, I would have said tenrecs. There are about 30 species of them, 27 of which are endemic to the island of Madagascar, off Africa's southeast coast in the Indian Ocean. The other three are the rare, semiaquatic otter shrews (potamogalines) that live in the African rainforest, with a discontinuous range near the equator from Guinea in the west to Kenya in the east. In Madagascar, some tenrecs are fairly common; the biggest (*Tenrec ecaudatus*) resembles a cross between a porcupine, raccoon, and opossum. It weighs a kilo or two, sports some long, defensive quills along its neck and back, and eats just about anything. Other tenrecs, like *Hemicentetes*, *Setifer*, and *Echinops* are smaller but more fully covered with spines. The latter two are very similar in appearance to European hedgehogs, with the capacity to roll up into a defensive ball when threatened. In contrast, *Microgale* tenrecs have fur instead of spines and show an extraordinary similarity to shrews (Soricidae). One species on Madagascar (*"Limnogale" mergulus*) has exploited a semiaquatic niche like its mainland African cousins. Yet another tenrec from Madgascar (*Oryzorictes*) is rather good at burrowing. In short, this is a very diverse group which exhibits a number of similarities to mammals on other continents, like shrews and hedgehogs.

Evolutionary biologists have known since the nineteenth century that tenrecs form a natural group, and that many similarities in their external appearance with non-tenrecs were influenced by ecological convergence, or the evolution of anatomical features due to the influence of similar habitat, rather than common ancestry (Figure 9.4). There are only so many strategies by which a small mammal can get around in a forest and defend itself from predators, and the Malagasy tenrec *Setifer* and the European hedgehog *Erinaceus* have honed in on the same one. We know that the similar external anatomy of these animals is convergent because it does not extend to certain parts of their internal anatomy, such as their teeth. All tenrecs, from the semiaquatic ones to *Microgale*, *Tenrec*, and *Setifer* have the same kind of triangle-shaped upper and lower molars, not seen in any modern shrew or hedgehog.

Back in 1997, after telling you these interesting facts about my favorite animals, I would have also told you that the best hypothesis about their closest relatives included a group of small, burrowing South African animals called golden moles, and that both groups belonged to the "Insectivora," along with European talpid moles, shrews, hedgehogs, and the Caribbean *Solenodon*. We knew that the shrew-like *Microgale* tenrecs were not themselves soricid shrews, but we still thought that shrews were not far off, closer to tenrecs than to other mammals like rodents, bats, or primates. While I was right to say that tenrecs share a close evolutionary relationship with golden moles, I know now that the "Insectivora," a group which dates to the nineteenth century and was generally regarded as true by most zoologists throughout the twentieth, is wrong. Tenrecs and golden moles are related to shrews and hedgehogs only in the general sense that they are placental mammals (as discussed in Chapter 3). Within Placentalia, tenrecs and golden moles have a much closer relation to other African mammals in a group now known as Afrotheria, which also includes aardvarks, sengis (also known as elephant shrews), hyraxes (also known as dassies), sea cows, and elephants.

Yes, elephants. Despite the extraordinary differences in anatomy between a five-ton African elephant and a five-gram Malagasy *Microgale*, and the

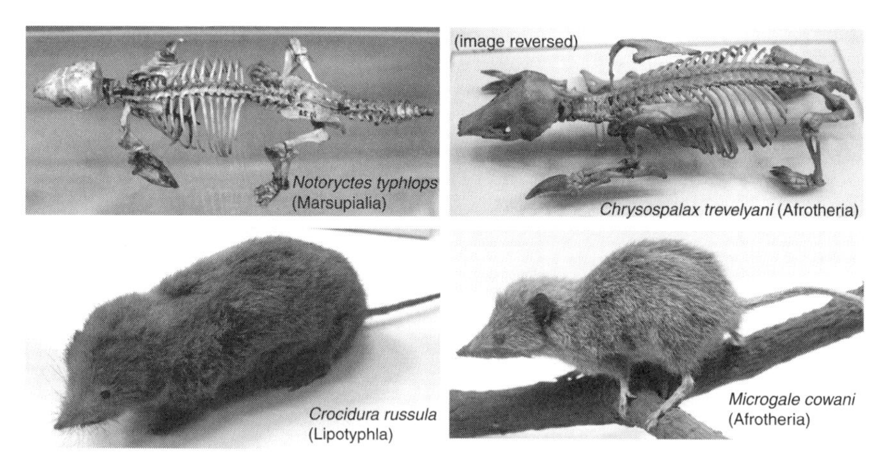

Figure 9.4. Convergence in biological form. Afrotherians *Chrysospalax trevelyani* and *Microgale cowani* (right) occupy similar ecological niches as, respectively, the marsupial *Notoryctes typhlops* and the lipotyphlan *Crocidura russula* (left). *Crocidura* and *Microgale* are both small, insectivorous, nocturnal predators that live in forested or semi-forested habitats. *Notoryctes* and *Chrysospalax* are both blind, burrowing predators of insects and other small invertebrates, and use their hands and heads to dig. Despite their shared habitats, the "moles" (above) and the "shrews" (below) are not closely related to one another. In terms of common ancestry, *Microgale* and *Chrysospalax* are closer to an elephant than either is to *Crocidura* or *Notoryctes*. The great evolutionary distance between lipotyphlan and afrotherian "shrews" has only recently been recognized, thanks largely to our understanding of their DNA. The evolutionary distance between afrotherian and marsupial moles has been recognized for over a century, and confirmed by DNA analysis, thanks in part to the marked anatomical and developmental differences between the two "moles."

many external similarities of the latter with a five-gram soricid shrew (Figure 9.4), the *Microgale* and elephant share a common ancestor with each other to the exclusion of that shrew (Figure 9.5).[52] In fact, the latest and best-supported hypotheses about placental mammal interrelationships place you and I, as members of the order Primates, on a branch that's closer to shrews than those shrews are to *Microgale* or other afrotherians.[53] Afrotheria is now considered to occupy an early branch in the radiation of placental mammals, possibly one shared with another group that is now most widespread on southern continents, the Xenarthra (or sloths, armadillos, and anteaters; see Figure 9.5). Shrews and hedgehogs are part of a group known as Laurasiatheria, along with such diverse creatures as bats, carnivorans, pangolins, and even- and odd-toed ungulates, including whales. As primates, you and I are part of a larger group known as Euarchontoglires,

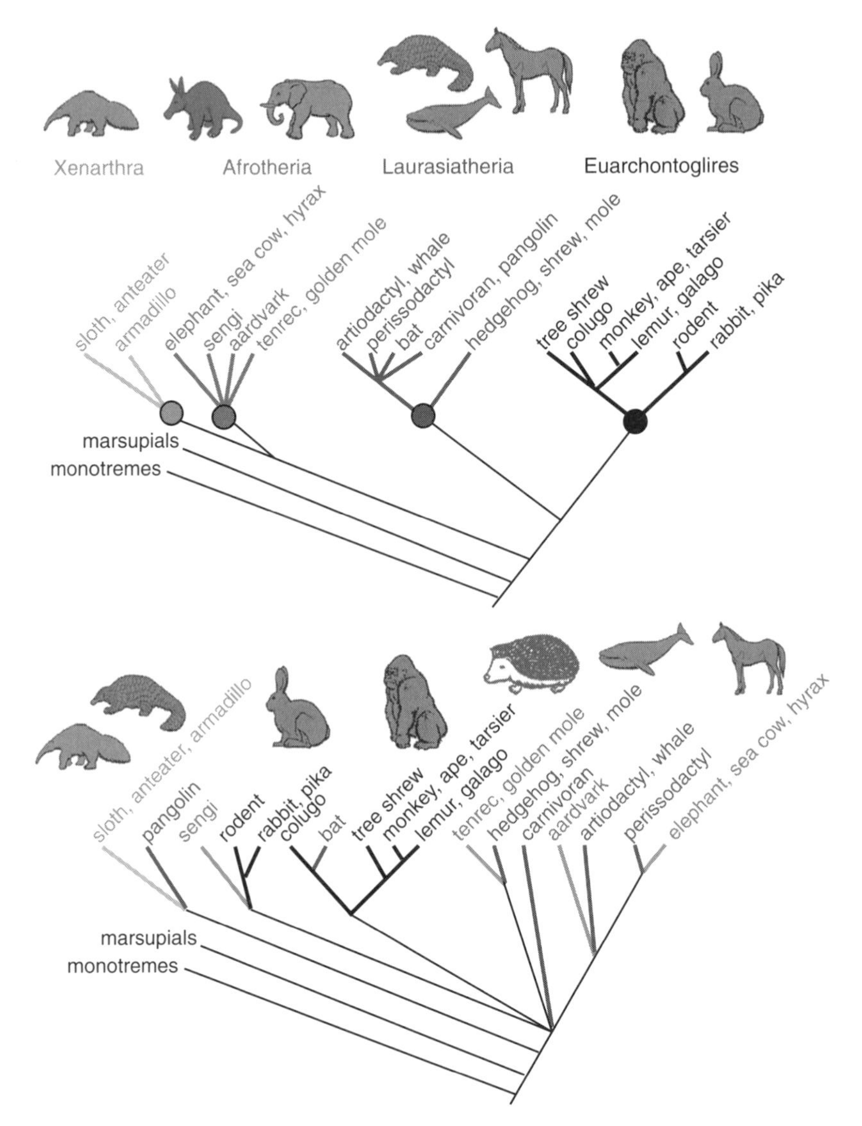

Figure 9.5. Evolutionary genealogies of placental mammals after (above, based on Murphy *et al.* 2001, 2007, reprinted with permission from AAAS) and before (below, based on Novacek 1992, reprinted with permission from Macmillan Publishers Ltd) the revolution in availability of molecular data in systematics (branch lengths are arbitrary). The four groups shown at the top correspond in shades of gray to the animals listed below them. The tree above is based primarily on comparisons of mammal DNA; the one below is derived primarily from anatomical data. While anatomy and DNA agree about many aspects of mammal evolutionary history (e.g., coherence of most orders such as Xenarthra, distinction from marsupials and monotremes), DNA sequences have changed many aspects of the tree, including high confidence in Afrotheria, Laurasiatheria, and Euarchontoglires.

near colugos, tupaiids, rodents, and lagomorphs. Laurasiatheres (including shrews) and Euarchontoglires (including primates) are closer to each other than either is to an afrothere (including tenrecs).

Partial recognition of Afrotheria has a long history among evolutionary biologists, dating at least to 1925, when Sir Wilfried LeGros Clark and Charles Sonntag argued that the aardvark, hyrax, elephant, and sea cow share close common ancestry.[54] Similarly, in 1945 George Simpson gave the name "Paenungulata" to the group consisting of elephants, sea cows, and hyraxes.[55] At various points during the twentieth century, some biologists have disagreed with these authors, arguing, for example, that hyraxes belong on a branch close to horses.[56] Others argued that golden moles were not closely related to northern moles, shrews, or hedgehogs, but then didn't really come up with any better ideas.[57] However, prior to 1997, no one thought that a tenrec or golden mole was anywhere near elephants on the mammalian Tree of Life. This idea began to circulate in 1997–8, when the molecular biology labs of Mark Springer (Riverside, USA) and Michael Stanhope (then in Belfast, Northern Ireland) started publishing analyses of the DNA of golden moles and tenrecs, and found very strong support for Afrotheria, a name coined in the 1998 paper by Stanhope and colleagues.[58]

I devoted parts of my PhD thesis and postdoctoral work to scrutinizing the data published by Stanhope, Springer, and colleagues, along with data I had collected on the anatomy of afrotherian mammals.[59] I suspected that the molecular support for this group might be a methodological artifact, or perhaps the result of not being able to sample any of the numerous, potentially relevant mammalian fossils. As we pointed out earlier in this chapter, DNA sequences are available primarily for living species only. Molecules, like muscles, nerves, or other soft tissues, are generally not amenable to fossil preservation. For anything older than about 0.5 million years, we do not yet know how to directly reconstruct their genetic diversity. But we do know a lot about the anatomy of fossil mammals. I was initially very skeptical of Afrotheria, and by including this information, alongside that known for living animals, I thought I might come up with a different resolution of the branching evolutionary history of afrotherian mammals.

I didn't, at least not concerning their general position in the placental mammal tree.[60] Despite using a substantially different tree-reconstruction methodology, and including information on morphology that my molecular biologist colleagues had previously ignored, the strength of the DNA support for Afrotheria was very strong. Some aspects of the trees I obtained in my postdoctoral work did indeed differ from those of Springer, Stanhope, and colleagues. However, their essential conclusion that tenrecs and golden moles were part of the African clade was untouchable. They were right.

Within a few years of those first papers sampling tenrecs and golden moles in the late 1990s, the DNA database grew a lot. The hard work of these molecular biology labs was made public on GenBank (see Table 9.1), a testament to the high standards of scientific practice in evolutionary biology. In 2001, Springer, Stanhope, and several other colleagues published additional papers[61] based on much larger datasets (compared to their work from 1997–8) that, with only a few subsequent tweaks, have set the standard for what we now know about the placental mammal Tree of Life (Figure 9.5). This standard includes several ideas that no one had previously expected based on anatomy alone, such as Afrotheria including tenrecs and golden moles.

They determined that the branching pattern within Placentalia consisted of four main groups (depicted in Figure 9.5), as mentioned above: Xenarthra, Afrotheria, Laurasiatheria, and Euarchontoglires. One of the several other surprising results that has turned out to be extremely well supported includes the removal of bats from a group that anatomists had thought included them along with colugos, primates, and tree shrews. "Colugos" are gliding mammals from southeast Asia, also known as dermopterans or flying lemurs (despite the fact that they're not lemurs and they don't fly). Previously, there had been a consensus that bats and colugos were each other's closest living relatives, but the new DNA evidence indicated strongly that this was not so.

Prior to the vast increase in DNA data availability during the 1990s, confidence about relationships among placental mammal orders, i.e., primates, bats, whales, perissodactyls, pangolins, sengis, aardvarks, etc., was never really at a comparably high level. There were several ideas with a fairly strong basis, including the associations of elephant–sea cow–hyrax, rodent–rabbit, primate–tree shrew[62] (Figure 9.5), and primate–colugo,[63] but even some of these were not accepted by prominent mammalian biologists.[64] For much of the twentieth century, the operative word for many ideas of placental mammal common ancestry was "uncertainty," and for some reason, anatomical data on their own seemed unable to conclusively nail down questions such as the immediate sister group of aardvarks, insectivorans, and even primates.

This was frustrating, since anatomy had done such a good job at both higher and lower levels in the tree. Based on anatomy, we knew that mammals consisted of marsupials, placentals, and monotremes; we knew that mammals were amniotes along with birds, crocodiles, turtles, lizards, and snakes; we knew that amniotes shared a common ancestry with tetrapods to the exclusion of other vertebrates. The signal of common ancestry was also clear at lower levels. We knew, for example, that placental mammals consisted of about 20 different orders (like primates, rodents, sea cows, and bats) and that each descended from a single common ancestor to the exclusion of any other animal. With only a few exceptions (e.g., insectivorans and hippo–whale), the new data from DNA that started to accumulate in ever-greater quantities during the 1990s confirmed these results based on anatomy. For some reason, however, disentangling the interrelationships between most mammalian orders seemed more difficult with anatomical data.

The reasons why anatomy by itself didn't yield high confidence at this particular level in the tree (between orders), but did so at others, are still unclear. Probably they derive from the fact that modern mammalian orders diverged from one another relatively quickly (on a geological scale) toward the end of the Cretaceous in places of the world that have not yet yielded an abundant fossil record (e.g., the African and southern Asian tropics). Nevertheless, while anatomy alone has had a hard time making sense of the evolutionary branching pattern among mammalian orders, in combination with molecular data, it has been shown to increase our confidence at specific nodes in the newly discovered tree (Figure 9.5). That is, adding data from bones and teeth into a molecular matrix (similar to the one shown in Figure 9.2, but much bigger) increases the strength with which nodes are supported, including those underpinning the four major placental groups (Afrotheria, Xenarthra, Laurasiatheria, Euarchontoglires) shown in Figure 9.5.[65]

Some of the hyperbole from those 1997 meetings of the Willi Hennig Society, represented by Dan Graur's subtitle "morphological schemes crumble under molecular scrutiny," has thankfully subsided. From my perspective as a grad student during the 1990s, fascinated by fossils and the anatomical data they contributed to understanding mammalian evolution, molecular biologists like Dan Graur seemed extremely bold about their radical new ideas, and maddeningly dismissive of the fossils they ignored while building their trees. This agitated me and many other biologists to question claims about what the mammalian Tree of Life really looked like. Thanks to the data that molecular biologists like Mark Springer, Michael Stanhope, and many others have produced over the last 20 years, along with the integrative analyses of these data by biologists interested in tree-reconstruction methods, paleontology, and other aspects of evolutionary science, the result is a resounding success. The tree shown in the top of Figure 9.5 represents a hypothesis about the evolutionary branching pattern within placental mammals that gets stronger as more data are added to it.[66] Not every node is 100% certain, and there may be a few more tweaks in the coming years. However, compared to where we were just 15 years ago, this tree represents a substantial improvement in our understanding of the natural world.

At the same time it is essential to keep things in perspective. Did any "morphological schemes" actually crumble during the 1990s? Within mammals, yes, a couple. Nevertheless, the novel ideas of the last decade do not substantially conflict with the trees of vertebrates as published by nineteenth- and early twentieth-century evolutionary biologists (Figures 4.1, 4.2, 9.1). Most of the orders are still the same, and mammals are still understood to occupy a twig of the Tree of Life among amniotes and other vertebrates. As we've already discussed, the basic edifice of vertebrate evolutionary history hypothesized many years ago has been supported, not scrambled, by the new data from DNA.

Building the Tree of Life, or the discipline of phylogenetics, is one relatively narrow application of molecular data in the life sciences. It's not usually the one most people think of when they hear the phrase "molecular biology." Perhaps more common is the notion that molecules control the development of phenotype, or your patterns of size, shape, anatomy, and behavior. The information in your DNA governs what you look like as a species. Hence, we now turn to the subject of genetic regulation of phenotype across species, and a few specific ways in which this "information" has been shaped by evolution.

TEN

DNA AND INFORMATION "CREATION"

Between my home and office, I've got a very good bike path. For the most part it takes me through farm fields, completely removed from any automotive competition. So listening to audio while biking does not, fortunately, greatly increase my risk of running into any traffic surprises. This has enabled me to listen to quite a few lectures and debates on the intersection of religion and science, the "controversy" surrounding intelligent design (ID), how some atheists view Charles Darwin as antagonistic to religion, and how certain theists view evolution as caustic to their beliefs.

In one such debate, part of an ongoing podcast called *Unbelievable* produced in the United Kingdom by Premier Christian Media, an advocate of ID articulated his view that Darwinian evolution cannot "create information":

> Randomness has never produced a single clump of information in the history of the universe. ... Have we ever seen a new structure, or a single gene, come into existence by natural selection? I know of no empirical evidence, and I've been asking biologists for 10 years now.[1]

I've admittedly spliced together two parts to this quote (on either side of the ellipsis), but in doing so I don't believe I've misrepresented the speaker's views. This quote conveys his belief about two things concerning evolution: (1) it is essentially random, and (2) it cannot produce novel biological structures at the anatomical or genetic level. For him, the analogy of the hurricane blowing through a junkyard and assembling a functional jet represents what evolution is all about. As we've already discussed in previous chapters, and will discuss further in the next, evolution is not random. The above quote is based on the faulty premise that it is. Besides this misunderstanding, the speaker makes the empirical claim that after "ten years" of asking biologists, he's never heard of a case in which natural selection is causally responsible for the appearance of biological novelty, or in his words, "a new structure, or a single gene." His conclusion in that bit of the *Unbelievable* podcast is that not just genes but novel anatomy in general is beyond the capacity of the mechanisms articulated in contemporary evolutionary biology.

This is a major rallying cry of the ID movement. Evolutionary biology, they say, cannot account for novel information. I want to convey to you in the coming pages some specific examples that show how this view is wrong. Very wrong. Many examples of evolution "creating" what the above speaker claims to be impossible exist: opsin genes and color perception,[2] *hox* genes and novel vertebrate body plans,[3] pancreatic ribonuclease and leaf-eating in mammals,[4] proteins such as Ectodysplasin and novel mammalian tooth patterns.[5] Quite of few of these and other examples are well documented in the literature, including popular accounts by John Avise,[6] Sean Carroll,[7] Jerry Coyne,[8] Ken Miller,[9] Neil Shubin,[10] Marc Kirschner and John Gerhart,[11] Ken Weiss and Anne Buchanan,[12] among others.[13]

In case you'd like a summary of a few of the latest examples, Table 10.1 lists a few dozen peer-reviewed, scientific publications from 2009–10 that document how features of morphology or behavior in organisms have been influenced by natural, genetic mechanisms. These publications are ones that I've happened to notice in the tables of contents of journals I often read, and there are other cases out there that I've missed. In addition, the examples described in the coming pages on viral DNA, color vision, and skull shape were first documented before 2009; these and many others are not included in Table 10.1. I don't know who the ID advocate quoted above has been asking in his search for the natural mechanisms behind the evolution of biological novelty, but he and his friends have got a lot of reading to do (Table 10.1).

Before we discuss specific examples of how natural mechanisms of mutation, duplication, and selection have led to novel phenotypes (meaning aspects of your anatomy and behavior), let's consider what it means for a natural mechanism to explain novelty. To do this, recall our discussion about agency and cause from Chapter 1. We can ask how something works, or what the agency is behind it, but the answer to one of these questions generally does not rule out the possibility of the other. Thus, to say that structure X "was designed" is fine as far as it goes, but it doesn't address the mechanism by which structure X actually functions, designer or not. This mechanism is what evolutionary biologists are trying to uncover.

Table 10.1. Partial list of cases in which genetic and phenotypic "novel information" is shown to be linked to natural processes (for example gene duplication, mutation, and selection) as demonstrated in papers published during 2009–10. This is not an exhaustive list. Viral resistance, color vision, and skull shape are a few such examples published before 2009, not listed here but discussed in the text.

SUBJECT	SOURCE
Evolution of hermaphrodism in nematodes by lowering expression of *tra-2* and *swm-1*	Baldi *et al.* 2009
Accumulation of beneficial mutations to *E. coli* bacteria at regular pace	Barrick *et al.* 2009 See also http://rationalwiki.org/wiki/Lenski_affair
Directed evolution experiments show high frequency of neutral and postitive mutations in many protein molecules	Bloom and Arnold 2009
Expression of genes *Tbx1* and *Brn4* in embryonic tissues of inner ear help regulate length and extent of inner ear coiling	Braunstein *et al.* 2009
Variation in FGF3 expression mirrors ancestral tooth shape in rodents and primates	Charles *et al.* 2009
Genes orthologous to vertebrate odorant receptors in the lancelet at base of chordate tree	Churcher and Taylor 2009
Evolution of eukaryotic protein transport from bacterial membrane proteins	Clements *et al.* 2009 See also www.sciencemag.org/cgi/content/summary/327/5966/649
Oral teeth in jawed vertebrates co-opted genetic regulatory network of pharyngeal teeth in jawless vertebrates	Fraser *et al.* 2009
Recently duplicated genes show high levels of adaptive selection	Han *et al.* 2009
Evolution of pituitary hormone (*PRL, GH, SL*) and mammalian placental lactogen (*PL*) genes via gene duplication	Huang *et al.* 2009
Point mutations in *FOXP2* confer potential for different language capacity in chimps and humans	Konopka *et al.* 2009
Experimentally transplanted *Tbx4/5* gene in lancelet (basal vertebrate) can initiate formation of limbs in mice. Evolution of genes regulating expression of duplicated *Tbx4* and *Tbx5* genes in non-lancelet vertebrates comprises example of evolutionary novelty (paired limbs) via gene duplication	Minguillon *et al.* 2009

Table 10.1. (*cont.*)

SUBJECT	SOURCE
Fgf4 gene duplication via retroposed mRNA followed by selection, responsible for short stature in chondrodysplastic dogs	Parker *et al.* 2009
Duplication of *hox* genes in vertebrates and cartilaginous fish and identification of conserved noncoding regions potentially responsible for tissue-specific regulation of Hox expression	Ravi *et al.* 2009
Scale phenotype in zebrafish and carp due to gene duplication and subsequent mutation of *fgfr1*	Rohner *et al.* 2009
Summary of evidence for genetic and developmental basis of evolution of animal eyes and vertebrate limbs	Shubin *et al.* 2009
Violet pigment sensitivity caused by point mutation to *SWS1* in ray-finned fish	Tada *et al.* 2009
Variation in dental morphology results from variation in conserved gene networks. "Tinkering" with expression/inhibition of elements in pathway leads to novel tooth morphology both in the lab and during evolution	Tummers and Thesleff 2009
Shift in embryological digit positions from 1–3 to 2–4 in birds by experimental manipulation of *shh* signaling	Vargas and Wagner 2009
Shift in embryological digit positions from 1–3 to 2–4 in skinks by experimental manipulation of *shh* signaling	Young *et al.* 2009
Gene duplication and subsequent mutation responsible for sexual organ identity in flowering plants	Airoldi *et al.* 2010
HAS2 genetic control of skin folding in domestic dogs	Akey 2010
Genetic basis of timing differences in bone formation in Antarctic fish	Albertson 2010
Multiple origins of electric organs in skeletal muscle of teleost fish by gene duplication and natural selection	Arnegard *et al.* 2010
Mutations in *EPAS1* gene lead to lower hemoglobin concentration among high-altitude dwelling humans and results in reduced susceptibility to hypoxia-related diseases	Beall *et al.* 2010
Duplication as source for novel genes relevant to early zygotic development in mosquito	Biedler and Tu 2010

SUBJECT	SOURCE
Morphological change in stickleback fish caused by regulatory mutations in *Pel* that modify expression of *Pitx1*	Chan *et al.* 2010 See also www.nature.com/news/2010/100217/full/463864a.html
Small tandem repeat variation during slipped-strand mispairing in DNA replication influences mutability in bacteria	Chen *et al.* 2010
Hox code in snakes shows some loss of redundancy, explaining in part their reorganized skeleton. Rib-suppressing activity retained only by *Hoxd10*; termination of body elongation retained by *Hox c12* and *c13*	Di-Poï *et al.* 2010
Proteins in a given area of genotype space contain a spectrum of functions. Proteins with a similar genotype can perform disparate functions; increasing sequence divergence corresponds with functional divergence	Ferrada and Wagner 2010
CAPN[11] protease in placental mammals shares common CAPN ancestor with other vertebrates, and shows increased level of selection relating to reproductive physiology	Macqueen *et al.* 2010
More frequent gene duplications within Pancrustacea, a group with greater variety of eye morphology than other eukaryotes	Rivera *et al.* 2010
Chimeric genes with substantial phenotypic effects produced by partial chromosome fusion in *Drosophila*	Rogers *et al.* 2010
Variation of tooth developmental parameters and gene networks enables approximation of mammalian tooth shapes in nature	Salazar-Ciudad and Jernvall 2010
Correspondence of scale patterns between fossil and modern fish based on variations in genes such as *FGF* and *Eda*	Schmid and Sánchez-Villagra 2010
LIM homeobox genes with influence on neural patterning diversified in metazoans via gene duplication	Srivastava 2010
Vertebrate heart tissue has common origin with tunicate pharyngeal tissue, influenced by expression of COE transcription factor	Stolfi *et al.* 2010
Evolution of yeast ribosomal protein regulators via gene duplication	Wapinski *et al.* 2010

Second, you might be very interested in the origins of the most basic processes of life, or of the first living molecule. However, the origin of life is not what Darwin's mechanism for evolutionary biology is about, as he himself wrote in the *Origin of Species*. Complaining that Darwinian evolution can't explain life's origin is like complaining that your Mercedes can't fly. It wasn't supposed to do that in the first place. So why not focus on what it is supposed to do? In the case of Darwin's theory of evolutionary biology, this is providing a causal mechanism by which organisms like newts, monkeys, tuna, spiders, and ostriches attained their current diversity. That's a substantial bit of explanatory value, undiminished by the fact that we've got a long ways to go before explaining the creation of life from non-life. This is not to say that scientific investigation into the origin of life, or abiogenesis, has to invoke supernatural causes, or that there won't be a definitive scientific case for it sometime in the coming years. However, it is very important to realize that studies of abiogenesis comprise a distinct field of science, one that does not draw on the same mechanisms relevant to Darwinian evolutionary biology.

Third, if you hear someone say that "evolution can't explain novel information," you should ask them to define exactly what they mean by "information." What is the novelty that they feel evolutionary biology has got to explain? If it's the origin of life, see the preceding paragraph. I interpret novel information to mean, biologically, the basis for some feature or behavior that distinguishes one organism from another. For example, I have two legs that are good at traversing long distances; a chimp has legs that are less efficient bipedally. I've got three small bones that convey sound in my ear; an iguana has just one. A pitbull has a long snout without much of a forehead; a poodle has a pretty big forehead. A mouse has very differently shaped teeth and a coiled inner-ear cochlea; in most lizards the teeth look fairly homogeneous and its cochlea is just a bulge. Some mosquitoes in southern France are resistant to insecticides; others are not. A theory that provided a mechanism by which each of these differences came to be, how the "information" inherent in each novel feature arose, would be very impressive, whether or not it simultaneously tells us how life originated.

WHAT DOES DNA DO?

In Chapter 9 we noted that the nucleus and mitochondria of each of your cells contain DNA, from your eyeballs to your big toe. The nuclear genome consists of tightly wound coils of DNA represented by chromosomes, and is far larger in size than the mitochondrial genome—nearly 200 000 times larger in a human. We noted that DNA itself consists of four bases, or nucleotides: A, T, G, and C. So far we've described how series of these letters are comparable across organisms, and how these comparisons can be used to build evolutionary trees which usually match estimates based on development and anatomy. What we've not yet done is describe what this DNA actually does, or how it contributes to the phenotype (i.e., the morphological and behavioral attributes) of a given organism.

The DNA in your cells is a repository of information. It doesn't do anything unless acted upon by other features of the cell. Its most basic function is to provide the information with which proteins, or small molecules consisting of strings of amino acids, are constructed. If you recall from Chapter 9, the identity of a given amino acid is determined by one of a small number of sets of three bases (A, T, G, or C), also known as a triplet or codon. Hence, a protein-coding "gene" typically refers to a stretch of these codons, the particular sequence of which determines the sequence of amino acids, which in turn determines the identity of a specific protein.

Protein synthesis, or the process of turning nucleotide codons into a protein, involves many additional steps. In order to bring the information in codons (among other things) out of the cell nucleus, we require the activity of ribonucleic acid (or RNA), a molecule similar to DNA but with one strand of nucleotides instead of two. "Messenger" RNA (mRNA) copies the genetic code of DNA within the nucleus, a process known as transcription. mRNA then leaves the nucleus with this information and brings it to a cellular structure called the ribosome. Another form of RNA, called transfer RNA (tRNA), then brings to the ribosome specific amino acids in the correct order for a given protein. As we've already noted, this order is determined by the nucleotide codons carried by mRNA from the DNA molecule.

Each codon represented by mRNA identifies one of 20 amino acids. Interestingly, these codons are redundant, i.e., more than one combination of three nucleotides may code for the same amino acid. This is readily apparent when you remember that there are only four nucleotides (A, T, G, and C). Combining these in all possible sets of three results in 64 different combinations ($4^3 = 64$); yet just about all organisms use the same codons for only 20 different amino acids. Usually, it is the third nucleotide in the codon that shows the most variability. For example, the amino acid "valine" is represented in the DNA within your chromosomes by any codon that begins with "GT" (GTA, GTT, GTG, GTC).

Once put into place in the ribosome thanks to the tRNA, the amino acids are bound together to form a protein in a sequence determined by the mRNA in a process called translation. The unique sequence and capacity of these amino acid chains (or polypeptides) to fold into different shapes gives each one a specific identity as a protein or component thereof. Proteins comprise the substance of every cell and are its workhorses; DNA carries the information that makes and regulates them.

Certain genes help regulate the activity, amount, and timing of transcription and expression of proteins. This is one way all of our diverse tissue types can retain their distinctive identities, making sure muscle cells remain muscle and liver cells remain liver. Remember that cells in your body have a shorter lifespan than you do; the whole really is greater than the sum of its parts! Cells are constantly going through cycles of division and death, and each time one of them divides, it has to duplicate all of the DNA in its chromosomes. Regulatory genes provide the essential task of ensuring that the right kind of protein is synthesized for a given tissue type. One of the more impressive facts about the information stored in your cells is that in every one of them it's all there. Cells in your teeth, brain, and small intestine contain the same genetic instructions as every other cell in your body. Thanks to regulatory genes they don't act on them all, but the information is there.

Some of the segments in a strand of DNA are transcribed into mRNA, but not assembled into chains of amino acids. Some of these contribute to the structure of ribosomes themselves. There are many DNA sequences in your chromosomes that are not transcribed at all, and in many cases their functions are not yet known.

PSEUDOGENES

Certain genes retain some of the codons relevant for building a given protein, but then contain nonsense series of nucleotides that stop and/or interrupt the codons that tRNA needs to form the right series of amino acid for that protein. We discussed examples of such pseudogenes in Chapter 7: the degraded *AMBN* and *ENAM* genes that code for enamel proteins in toothless, baleen whales.

Importantly, it is not only baleen whales that exhibit dental pseudogenes. Other vertebrates without teeth (such as birds and anteaters) also exhibit nonsense mutations characteristic of reduced function in their enamel-related genes.[14] In fact, many other pseudogenes exist relating to a plethora of other features across species.[15] Jerry Coyne's book, *Why Evolution is True*,[16] has a nice discussion of pseudogenes, describing, for example, the vestigial genetic pathway of vitamin C metabolism in apes (including humans), monkeys, and a few other mammals.

Humans have a very small nose, a weaker sense of smell than most other mammals, and a correspondingly large number of pseudogenes that in our primate ancestors once coded for odorant reception.[17] Olfactory genes also show high rates of reduced functionality among aquatic mammals,[18] which resemble humans in their emphasis on modes of sensory input besides smell. Animals which have an absent or reduced sense of vision (e.g., subterranean mole rats) have relatively more optic pseudogenes.[19] Giant pandas are unusual as vegetarian members of the Order Carnivora; accordingly, they exhibit nonsense mutations in a gene that regulates taste sensitivity to an amino acid commonly found in meats and other animal-derived food.[20]

While pseudogenes are a powerful demonstration of the hereditary connections among different species (e.g., a blind, burrowing rodent retains its degraded eye lens crystallin gene from an ancestor that could see), the fact that pseudogenes may provide a template for natural selection, and evolve novel functions, has been recognized as a source of evolutionary novelty for decades.[21] For example, some pseudogenes still yield RNA transcripts. This means that the pseudogene DNA within the cell nucleus is still copied and delivered into the cell cytoplasm by messenger RNA (mRNA). Although the sequence of amino acid codons represented in that mRNA has accumulated mutations that make it a pseudogene, and although it therefore cannot deliver the same polypeptide as its functional cousins, the presence of such mRNA outside of the nucleus can still influence the mRNA derived from functional copies of that gene. This phenomenon is known as "RNA interference" and it means that pseudogene mRNA, or a sequence derived from it, has the potential to interfere with the synthesis of functional mRNA in the process of assembling amino acids into a protein.[22] RNA interference is another means by which the activity of a given gene may be regulated, potentially giving an important function to certain genes (or derivatives thereof) that have partially or wholly lost their ability to code for specific proteins.

VIRAL DNA AND FUNCTION

Pseudogenes are not the only example of how function in DNA may be altered over the course of evolution. Parts of your genome contain stretches of DNA that have been inserted by a virus. At least initially, viruses are nothing more than parasitic bits of DNA or RNA seeking to use your cellular machinery to replicate themselves. Every time you've had the flu or some other viral infection, you get sick because a virus interferes with your ability to synthesize a standard complement of protein from your own DNA. Among the many kinds of viruses, one category known as retroviruses (which includes HIV) reverses part of the transcription cycle. These viruses manage to transmit their own sequence of nucleotides from mRNA to the DNA inside the cell nucleus, and in so doing change the host's genetic code to varying degrees. If the infected cells include gametes (i.e., sperm or eggs), then the host could pass on the viral genetic information to its offspring.

This gives you an idea of the dynamic nature of DNA and replication. From one minute to the next, many of the foreign viruses present in our environment could conceivably change your genetic make-up, and ultimately that of every organism descended from you.[23] It doesn't matter how many prizes you've won, papers you've written, or friends you've made; your most lasting contribution to future life on Earth might be derived from that retroviral infection, residing (probably latently) within your DNA as you read this.

Importantly, not all viral genetic material makes you sick. It is possible to transmit elements of viral DNA via your sperm and eggs without fatal consequences for the health of your child. As a matter of fact, primates, mammals, and organisms generally have been doing just this throughout geological time.[24] Traces of viral DNA may be exceedingly old in animal and plant genomes. We've already discussed at length how the similarity between species' DNA corresponds with ideas about their common evolutionary history based on other lines of evidence, such as anatomy, development, and the fossil record. This similarity extends to the genetic material inserted by viruses. Humans share more retroviral signatures in their genomes with other primates, particularly apes, than they do with non-primates, as you would expect if the common ancestor of these animals was transmitting these sequences from a once-infected germ-line.[25]

Being infected with a virus has understandably negative implications. Yet there is substantial evidence that the retroviral sequences left behind in animal genomes actually have been selected during the course of evolution to provide important functions.[26] For example, modern mice possess a gene aptly called "Friend-virus-susceptibility-1," or *Fv1*. This gene makes an mRNA transcript that interferes with the ability of a retrovirus to influence the DNA of a host cell, in a similar fashion as the pseudogene-derived RNA interference described above. It does so in a manner that's based on its close similarity to foreign retroviruses, binding to viral RNA while it is still in the cell cytoplasm and thereby impeding its ability to infect the DNA inside of the cell nucleus. Furthermore, the composition of *Fv1* indicates that it has itself originated from genetic material of a retrovirus, inserted via an aboriginal infection into the germ cells of an ancestral mouse.[27] In other words, *Fv1* evolved from one retrovirus and now helps to reduce the ability of other retroviruses to replicate in hosts with *Fv1*.[28] Surveys of the DNA composition of other genomes have uncovered many genes, such as the mouse *Fv1*, with retroviral origins.[29]

MUTATION AND COLOR VISION

In English, there is an understandable association of the word "mutation" with serious deviation from the norm, or at least substantial loss of function.

"Mutants" are popularly associated with creepy science-fiction characters, Chernobyl cows with one eye, or glowing rabbits who live too close to the local nuclear-waste depot. In reality, genetic mutations in animals and plants are very common and not necessarily dramatic. For example, blue eye color,[30] freckles, and red hair[31] are all results of genetic mutations, ones that most frequently occur among (and are inherited by) humans of European origin.

Another example is color blindness, which (for reasons that we'll describe below) is fairly common among human males. Sensitivity to various parts of the visual spectrum throughout vertebrates comprises a well-studied phenomenon, both in terms of its anatomy and genetics.[32] Humans, other apes, and Afro-Asian monkeys (collectively referred to as catarrhines) typically have sensitivity across the visual spectrum, which can be divided into "short," "middle," and "long" wavelengths (roughly corresponding to blue, green, and red). The genetic potential for this sensitivity resides, in part, along genes that code for three proteins known as "opsins": those for middle and long wavelength sensitivity are located on the X-chromosome, one of the two sex-specific chromosomes present in mammals. The opsin for short wavelength sensitivity is on one of our non-sex (or autosomal) chromosomes. In its most common genetic form, color blindness results from a mutation to one or both of the opsin genes that reside on our X-chromosome.

Humans generally have 46 chromosomes in all cells of the body except for the gametes (sperm and egg). You know already that genetic material comes in roughly equal halves from your mother and father. In order to ensure that you don't get too much DNA, gametes have half as many chromosomes, or 23, enabling offspring to recombine maternal and paternal DNA into a new, standard chromosomal complement of 46. The sex-determining Y-chromosome in human males is very small; it is the only one that does not duplicate all of the genetic material present in the complementary chromosome from the mother. While size usually doesn't matter, on occasion it does, and any function dependent on a gene residing on the X-chromosome, such as the middle- and long-wavelength color sensitivity influenced by our two X-chromosome opsin genes, has a higher chance of abnormal expression in males. This is why color blindness is more common in men. Unlike females, all we've got is a tiny Y-chromosome which, while very masculine, does not have an opsin gene copy. It therefore cannot do the job of providing a back-up for genes on our single X-chromosome, should their function change via mutation.

All it takes to change the color-spectrum sensitivity enabled by opsins are very specific mutations of one amino acid to another within their polypeptide chain. Of the approximately 360 amino acids that make up each of the three catarrhine opsin genes, changes at sites 180, 277, and 285 affect sensitivity to longer wavelengths.[33] It has been demonstrated that a gene duplication event of the long-wavelength, X-chromosome opsin genes in the common ancestor of catarrhines, combined with simple point mutations at these amino acid positions, was the source of our middle-wavelength, X-linked opsin gene that enables trichromatic vision in humans, great apes, and Afro-Asian monkeys.[34] In fact, gene duplication leading to trichromatic vision in primates has happened more than once, since at least one species of South American primate (the howler monkey, or *Alouatta*) also has two X-linked opsin genes, including a middle-wavelength opsin with subtle differences indicating that it emerged independently of that present in catarrhines.[35]

Most other mammals, including South American monkeys, lemurs, galagos, and lorises, have just two opsin proteins, one on a "regular" (or autosomal) chromosome and the other located on their X-chromosome. Usually, this facilitates sensitivity to short and long regions of the visible spectrum, but not the middle region. Thus, your pet retriever is unlikely to be able to distinguish reds from greens as well as you can. Importantly, non-catarrhine mammals (i.e., most lemurs, cows, dogs, and others besides apes and Old World monkeys) also show considerable diversity in their color sensitivity, and a different configuration of opsin genes enables some of them to also perceive middle-wavelength light. In some South American monkeys and lemurs, for example, females possess slightly variant copies of opsin genes on each one of their X-chromosomes (one maternal, one paternal). With the subtle differences in visual-spectrum sensitivity enabled by single amino acid mutations, which do occur in some individuals, a female would be able to better discriminate across the visual spectrum than other members of her own species. Such individuals have better color perception than any male, since males lack the second X-chromosome on which they'd need a slightly mutated copy of the long-wavelength opsin gene to perceive middle-wavelength regions of the visual spectrum.

How do we know that gene duplication is responsible for two versus three opsins in various species of primates? First, we know because the mechanisms of DNA transcription and replication are good, but they're not perfect.[36] Categories of genetic duplication include retroposition, tandem duplication, and polyploidy, and each differs in terms of the amount of genetic material copied.[37] Retroposition copies small amounts of genetic material from RNA back into the DNA of the cell nucleus. Single genes or even parts thereof can be copied multiple times into the DNA of a cell by retroposition. Tandem duplication can result from unequal crossing over, when during the course of cell division, a small part of a chromosome does not divide evenly in the process of providing daughter cells with their own complement of genetic information. Polyploidy involves duplication of large tracts of genetic material, potentially involving the entire genome itself. All of these forms of gene duplication are common in the natural world.[38]

The nucleotides in the two X-linked opsin genes of humans and other catarrhines provide important clues that reflect history.[39] Comparisons of the single X-linked opsin gene in New World monkeys, such as a marmoset, with the long- and middle-wavelength opsin genes in humans and other catarrhines, show a close similarity in all of these genes for at least part of their sequences. In particular, the marmoset X-linked opsin shows close similarity to the catarrhine (including human) long-wavelength opsin for its entire sequence. Similarity to the catarrhine middle-wavelength opsin is also present, but is much more pronounced for the first third of the sequence. After that point, the similarity decreases, and this decrease has been interpreted to reflect the gene duplication event from an ancient X-linked opsin shared between Old World and New World monkeys.[40] As noted above, catarrhine primates are not the only ones with two opsin genes on their X-chromosome which provide sensitivity to middle- and long-wavelength light: South American howler monkeys (genus *Alouatta*) have them too. However, the sequence similarity of their middle-wavelength opsin to the X-linked opsin of a marmoset is higher, consistent with the interpretation that the howler monkey opsin duplication is an independent and more recent event than that of catarrhines.[41]

While humans and other apes have richer color vision than male lemurs or dogs (for example), our color perception is rather impoverished compared to that of other vertebrates. Many birds, lizards, fish, and even a few mammals have sensitivity to ultraviolet (UV) wavelengths, which are very short and—for the most part—beyond the capacity of the human short-wavelength opsin. Such animals also possess opsin genes, but may have more than three copies which (among other adaptations, such as an optic medium that differs from our UV-absorbing cornea) enable perception of parts of the color spectrum including UV.

Color perception in many animals, such as birds and guppies,[42] plays a critical role in mate recognition. Females of these species may be selective of the males with whom they mate, based in large part on unique color patterns that they judge to be particularly appealing. Note, therefore, how important changes to the amino acid sequence in an opsin gene, demonstrably linked to changes in perception of the color spectrum, can be for the capacity of an individual to recognize potential mates. Given an interbreeding community and thousands of generations, such mutations are neither deleterious nor rare. Opsin mutations enable a slightly different sensitivity to the color spectrum across multiple individuals, and such mutations have the potential to contribute to the reproductive isolation of parts of that population.[43] With isolation comes reduced interbreeding, enabling other kinds of genetic and anatomical differences to accumulate and, eventually, paving the way for the emergence of a phenotypically distinct species.[44]

"TUNING KNOBS" AND SKULL SHAPE

Perhaps you've seen the alarm clock used by the main character in episodes of the animated series *Wallace and Gromit*. Upon ringing, it sets off a series of gears, trap doors, mice on treadmills, springs, all interconnected to get Wallace out of bed, start the shower, squeeze the toothpaste, pour the coffee, etc. His alarm clock represents a Rube Goldberg contraption,[45] named for the early/mid-twentieth century cartoonist who liked to draw such things, and it serves as an apt metaphor for genetic regulation in terms of how regulation and expression of multiple genes influence the development of a given adult structure.

Although we're only a few years after the decoding of the first large animal genome (*Homo sapiens*) and our appreciation of genomic diversity across organisms is still in its infancy, many aspects of the genetic control of phenotype have been decoded, like the individual point mutations that change the spectral sensitivity of opsin genes discussed above.[46] Another kind of variation frequently seen in the genes of animals that influences phenotype, and which is demonstrably the subject of natural selection over time, concerns short stretches of repetitive DNA, sometimes called short sequence repeats or microsatellites.[47]

Microsatellites occasionally appear as a stretch of DNA that's transcribed by RNA into a protein. They consist of a repeated series of nucleotides that may show substantial variation in length, and influence how a given protein is expressed at certain points during the growth of an organism. For example, the gene *Runt-Related Transcription Factor 2* (*Runx2*) plays a large role in the development of bone cells. In mammals, this gene contains a microsatellite which codes for a repeated series of the amino acid glutamine (represented by "Q") followed by a single glutamic acid (or "E"), followed by a series of alanines (or "A"). Different species of mammals show variable Q/A ratios in this specific region of the gene. This ratio is important because it is along this segment of *Runx2* that other molecules have the capacity to bind—effectively, partially, or not at all—depending on the configuration of repeats.[48] This region is therefore a "binding site" of *Runx2*.

In 2004, University of Texas molecular biologists John Fondon and Harold Garner discovered that the Q/A ratio of *Runx2* showed an interesting correlation with skull shape in domestic dogs.[49] They noted that individuals with a longer snout and flatter forehead tended to have more glutamines relative to alanines (a higher Q/A ratio) coded within their microsatellite than those with a shorter snout and more conspicuous forehead. Even over the course of a few dozen generations within a single breed, they observed differences: a purebred bull terrier from the 1970s shows a less conspicuous forehead and longer snout than one from the 1930s, and has a higher Q/A ratio. The difference is subtle, but it's there, and it has occurred within 40 years of breeding.

In addition, Fondon and Garner showed that despite the inbreeding typical in purebred dogs, which normally reduces the genetic variation in a given species and which is why inbreeding is taboo in human cultures, this particular sort of genetic diversity was still quite abundant. A kind of mutation that may be responsible for this high degree of variation is known as slippage. As noted above, every time a cell divides (which is happening constantly in millions of your cells as you're reading this), it has to replicate its complement of DNA. "Slippage" refers to a minor error in the duplication process that leads to the repetition of a brief series of nucleotides. Thus, microsatellites throughout the genome (not just in *Runx2*) can easily become shorter or longer in one individual relative to another, even within a single, inbred population of human pets. This kind of variability makes DNA evidence so valuable to forensic and police work. Each one of us has a unique signature in our DNA due in part to our own peculiar patterns of (for example) microsatellite mutations.

The causal mechanism joining Q/A ratio with skull shape relates to the capacity of *Runx2* to affect the process of bone development by controlling the expression of other genes. Different Q/A ratios at the *Runx2* binding site influence the downstream activity of other molecules, making more or less of a given RNA, and ultimately protein, that influences bone growth. The hypothesis that *Runx2* influences downstream molecules as part of this cascade was confirmed in a recent study by University of Illinois molecular biologist Karen Sears.[50] In their paper, Sears and colleagues investigated the activity of a target molecule influenced by *Runx2* proteins with varying Q/A ratios, and found that, as predicted, the higher the Q/A ratio, the more activity they observed in their downstream target. Many other genes like *Runx2* have similar binding sites consisting of microsatellites, and these too have been linked with downstream activity of other molecules.[51]

One of the potential problems with variation in the *Runx2* binding site as a mechanism of natural diversity in morphology, and that of many other potential sources of genetic variation, concerns effects of a given mutation beyond the benign differences seen in skull shape of domestic dogs. "Pleiotropy" is the formal term that signifies the influence of one genetic change on many aspects of phenotype, with potentially destructive consequences. In fact, there is a human disorder called cleidocranial dysplasia that results in part from mutations to the *Runx2* gene. This condition is associated with many detrimental symptoms, including not only some of the aspects of skull shape observed by Fondon and Garner in domestic dogs, but also failure of skull bones to completely ossify, lack of clavicles, lack of testicular descent in males, and delayed eruption of adult teeth.[52]

Those readers who came of age during the 1980s will recall the phenomenon of the Rubik's cube. Pleiotropy is similar: adjust one bit of your genome, and a whole suite of changes follows whether you like it or not.

Purebred dogs are helped through their development by attentive owners who are so pleased by some feature of their "creation" that they overlook such defects as hip problems, inability to naturally give birth, and early onset of cancer. Hence, an initial concern with a study like that of Fondon and Garner[53] was that the variation they observed would be a special case of artificial breeding by humans, not representative of the natural world. Sears and colleagues addressed this possibility by asking whether or not *Runx2* Q/A ratios in wild-occurring members of the order in which dogs are classified, Carnivora, co-varied with skull shape. They did, particularly in members of the group more closely related to dogs than to cats, such as foxes, raccoons, badgers, and bears (caniforms). Sears and colleagues found that the longer the snout on a naturally occurring carnivoran species, the higher their Q/A ratio. They observed that the microsatellite variation observed in domestic dogs, i.e., the repeated series of amino acids within the *Runx2* molecule that influences the capacity of the molecule to regulate others via its binding site, is also present in naturally occurring species of carnivorans. These are healthy animals, capable of reaching reproductive age, and do not suffer from debilitating diseases, despite exhibiting substantial variation in their *Runx2* microsatellites. Stated differently, it is now clear that *Runx2* microsatellite variation does not necessarily come with an undue level of pleiotropy; the Rubik's cube metaphor is not necessarily applicable in this case.

While the *Runx2* example in carnivorans is compelling, evolutionary biology has a long way to go before the molecular control of skull shape is fully understood. As a matter of fact, I and several close colleagues are now actively involved in exploring this correlation in vertebrates besides carnivorans, and the story shows additional complications, such as a weaker correlation between *Runx2* Q/A ratio and skull shape in non-carnivorans.[54] Nevertheless, the kind of variation seen among microsatellites at the *Runx2* binding site occurs widely throughout the genome, and this variation is demonstrably linked to other aspects of phenotypic diversity besides skull shape.[55]

The continuous nature of variation in microsatellites has led to the metaphor of the "tuning knob."[56] In the same way that a circular knob on your radio enables a continuous gradient of control over its volume, subtle changes in microsatellite composition are causally linked to correspondingly subtle changes to the regulation of downstream molecules. Because some microsatellite variation in important genes such as *Runx2* does not necessarily impede healthy functioning of a wild-occurring animal, and because this variation is frequently occurring in nature and linked with many different morphological and behavioral phenotypes,[57] microsatellites comprise a compelling and very specific case of how natural selection operates on genetic variation to "create" novel phenotype. In other words, microsatellite mutations may enable subtle changes in one or few traits while not greatly influencing others, matching the kind of gradual, continuous, and heritable variation envisioned by Darwin in the *Origin*.

ORIGIN OF INFORMATION AND "MACROEVOLUTION"

The above examples show how novel "information" may be generated by natural processes, and contradict the claim of creationism and ID that biological novelty is beyond the reach of Darwinian evolutionary biology. As we've discussed, there are many such examples (e.g., Table 10.1 and Chapter 11). Mechanisms of genetic mutation, variation, and selection, easily observable within the scale of a human lifetime, are demonstrably connected to viral immunity, color perception, and skull shape, among other phenotypes. In none of these cases does their inherent complexity require anyone to sacrifice a real understanding of evolutionary cause. It is difficult for many of us to extrapolate short-term effects of mutation, drift, and selection over the grand scale of biodiversity through geological time. However, this difficulty is a result of our own short lifespans and mode of perception, not the facts of genetics, development, and paleontology.

Most creationists and ID advocates will acknowledge the role of Darwinian selection in such obvious phenomena as microbial resistance to antibiotics, insect resistance to pesticides, or short-term breeding in animals. Probably such individuals would classify the scientific research described above into this category of "microevolution." However, as discussed in Chapter 4, the line between "micro" and "macro"-evolution is for the most part arbitrary. Recent publications regarding, for example, the genetic control of segmentation,[58] of color patterns in birds,[59] and jaw and ear development[60] show why both are equally within the explanatory reach of natural phenomena we can observe today. The relationship behind the "microevolutionary" dynamics of *Hox* genes and the "macroevolutionary" differences among animal body plans (among other issues) has already been described in a very accessible manner by the University of Wisconsin developmental biologist Sean Carroll in his book *Endless Forms Most Beautiful*. I would like to close this chapter with another example illustrating how the mechanisms of biology we can observe today can help explain large-scale patterns of evolution.

GENETICS, DEVELOPMENT, AND THE MAMMALIAN EAR

In Chapter 5 we discussed in some detail how the hearing system of a mammal differs from that of bird, lizard, or crocodile. Recall that mammals have three bones connecting the eardrum to the organ of hearing, or cochlea, inside the inner ear. Most reptiles, birds, and amphibians have just one. In addition, the cochlea itself has a very peculiar shape among mammals: it is coiled (Figure 10.1). Coiling enables the organ of hearing to become elongate without taking up an undue amount of space within the skull, thereby comprising a spatially efficient means of increasing acoustic sensitivity. The longer your cochlea, the wider the range of auditory signals you can interpret. Importantly, coiling is not the only means of enlargement observed in nature; birds have a large, uncoiled cochlea which provides at least as effective acoustic sensitivity as the coiled cochlea of mammals. For whatever reasons, all living marsupial and placental mammals have increased the size of their cochlea by tightly coiling it (Figure 10.1).

However, mammals do not start out embryonic life with a coiled cochlea. Between the initiation of development, when your head showed recognizable visceral arches (Figure 5.5), and birth, your inner-ear cochlea changed from a small bump below your incipient organ of balance to the fully coiled "snail" you've got today. This is true for all therian mammals investigated so far, such as the mouse shown in Figure 10.1. The only exceptions to this pattern are the two egg-laying mammals we discussed in Chapter 3: the platypus and echidna. Both have a reasonably effective hearing mechanism, and a somewhat elongate cochlea, but the extent of coiling is much less than in other mammals (Figure 10.1). This is because the mechanisms of differential growth during their early development does not result in quite as much cochlear turning as in, say, a mouse or human.

The genetic control mechanisms behind ear development are slowly becoming better understood. For example, several lines of evidence point to differential expression of a number of important genes that help determine ear morphology, such as *Tbx1*, *Brn4*, and *Pax2*, particularly in regards to cochlear growth and coiling. Mice bred with defects to these genes (known as "transgenic" mice) show abnormal extension and incomplete coiling of the cochlea, suggesting that they are active in the regulatory cascade involving the cell proliferation and death that, ultimately, would result in a coiled cochlea in a normal individual.[61]

We owe some of our understanding of inner-ear development to Evan Braunstein and colleagues at Yeshiva University in New York City. These

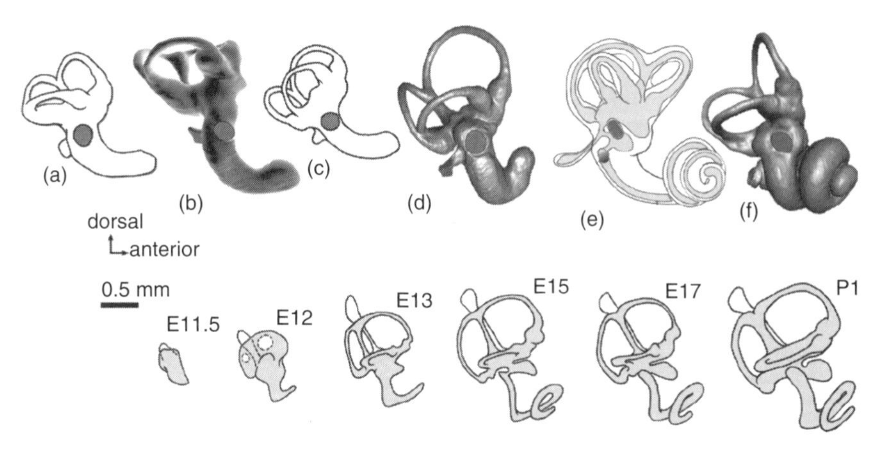

Figure 10.1. Inner ears of fossil and modern mammals (top row) compared with development of the middle ear in embryonic and newborn stages of the mouse (bottom row). In both rows, the organ of balance, or semicircular canals, comprise the top half of each image, and the organ of hearing, or cochlea, comprises the bottom half. The solid oval dot in the middle of each image on the top row represents the oval window, or the site at which the hearing bones of the middle ear connect to the fluid-filled chamber of the inner ear. The top row consists of *Morganucodon* (a: late Triassic/early Jurassic), platypus (b: modern), multituberculate (c: fossil mammal common in the Cretaceous and Tertiary), *Dryolestes* (d: Jurassic), human (e), and opossum (f: modern). Note how the degree of coiling is much more pronounced in the human (e) and opossum (f) than in the other adult mammals. Note furthermore how the degree of coiling increases during the growth of the mouse, represented by individuals of known age in days postconception ("E") or postnatal ("P"). As in many aspects of vertebrate form, variation exhibited during growth provides a range of phenotypes that may be selected for during evolution, for example by extending or truncating growth of specific features, such as cochlear coiling. Furthermore, while the genetic basis of cochlear coiling is not fully understood, specific genes (e.g., *Pax2*, *Otx1*) are known to influence the degree of inner-ear development in mice and humans. Evolution of these same genes help to explain the patterns of morphology seen in both fossils and living mammals. (Images redrawn from Luo *et al.* 2011; figure 3, printed by permission of The Royal Society.)

scientists have focused on two genes, *Tbx1* and *Brn4*, and have found that the genes interact with one another to influence the extent of cochlear turns in transgenic mice. Individuals with altered expression of either *Brn4* or *Tbx1* alone exhibited relatively few changes to the number of cochlear turns. But strains of mice bred for defects in both genes exhibited significantly reduced cochlear coiling, not exceeding a single turn. They furthermore determined that expression of these two genes early in development in a specific tissue of the inner ear, known as the periotic mesenchyme, influences the activity of a substance present in developing embryos that is known to inhibit cell growth: retinoic acid. Different levels of expression in these genes help to control the activity of retinoic acid, which in turn influences the location and extent of coiling in the organ of hearing, the cochlea.[62] This discovery of Braunstein and colleagues represents part of the control cascade behind the uniquely mammalian, coiled form of the inner ear.

As you may recall from Chapter 5, the evolution of the mammalian ear has a lot to do with the form of the mammalian jaw. Mammals have just one bone in their jaw (the dentary), which connects to the skull at the squamosal bone, whereas other vertebrates have several jaw bones, with the joint formed by the quadrate and articular. We've already described in some detail how during the course of synapsid evolution the tooth-bearing jawbone (dentary) becomes larger at the expense of the posterior jaw bones (Figure 5.3), including the articular, to the point where some fossil synapsids actually have two different jaw joints. One of these is squamosal–dentary (like modern mammals) and the other is articular–quadrate (like modern reptiles).

As is the case with coiling of the inner-ear cochlea, several groups of scientists are now deciphering the precise genetic mechanism(s) behind jaw and ear bone development. For example, developmental biologist Abbie Tucker from King's College London has examined gene expression patterns during growth of the jaw and ear bones of the mouse.[63] She found that at key stages during early growth, a gene called *Bapx1* is strongly expressed in the embryonic precursors of the malleus, incus, and the tympanic ring. These bones are derived from the first visceral arch of the embryo (see Figure 5.5), and all of them are part of the hearing apparatus of modern mammals, as discovered by Karl Reichert in the 1830s and as discussed in Chapter 5. If the derivation of mammalian hearing bones from the jaw bones of a common ancestor shared with reptiles is true, you would expect these jaw bones in non-mammals to show similar genetic expression patterns. And that's exactly what Tucker and colleagues found. Even though the articular and quadrate form the jaw joint of a bird and fish, and are not located in the ear or intimately related to their capacity to hear, these animals exhibit expression of *Bapx1* during the development of these posterior jaw elements, revealing that they are genetically and developmentally very similar to mammalian ear bones.[64]

Tucker and colleagues also found a couple of surprises in the genetic regulation of ear bone development, such as the more pronounced effect of *Bapx1* on the width, but not length, of one of these bones (the malleus), and the independence of *Bapx1* on their presence. *Bapx1* mutant mice still have small ear bones, but they show important differences in size and shape relative to normal mice. Tucker and colleagues found that *Bapx1* contributes to development of the first arch ear bones in mammals and the jaw in non-mammals as a regulator within a potentially large genetic control cascade, not as a simple on/off switch that influences first arch elements directly.

And herein lies the challenge for evolutionary biologists: to figure out every step of this control cascade, or Rube Goldberg contraption, that directs formation of first arch elements, whether as ear ossicles in mammals or as the jaw joint in non-mammals. The same challenge applies to development of cochlear coiling. Several of these steps are now known, such as the contributions of *Brn4* and *Tbx1* in regulating the expression of retinoic acid in specific tissues of the developing inner ear, thereby helping to determine the extent of cochlear coiling in a mammal.[65]

In cases where the entire genetic control cascade of a given complex has not yet been 100% deciphered, including both ear and jaw development in mammals, evolutionary biologists make the assumption that there are mechanisms in this cascade by which cochlear coiling or size and shape of ear and jaw bones can be altered without undue, detrimental effects on other aspects of their growth, i.e., without too many of the Rubik's-cube-like effects of pleiotropy. This assumption is a reasonable one because we already know that continuous mechanisms of genetic control exist, such as the microsatellites present in *Runx2* of both domestic and wild carnivoran mammals that correlate with aspects of their skull shape.[66] We know that point mutations exist among opsin genes that can have major effects on the capacity of an animal to discriminate regions of the color spectrum,[67] and that stretches of DNA on our X-chromosome that encode color sensitivity can be mutated and duplicated to enhance color vision without detrimental effects to other aspects of animal phenotype.[68]

We also know that a genetic mechanism to form an "intermediate" mammal-like ear exists because we have the end-products of such a mechanism: a living platypus or echidna, and a 150-million-year-old *Dryolestes* mammal with an essentially mammalian ear but with an uncoiled cochlea (Figure 10.1)[69] or other fossil mammals that show middle-ear bones like those of modern ones, but which are still attached to the jaw in adults via an ossified first arch cartilage.[70]

By applying the principles of uniformitarianism discussed in Chapter 2, i.e., that the processes observed today are applicable to explaining the phenomena of the past, it is not difficult to recognize how anatomical features of major groups (reptiles versus mammals, for example) can be explained by a causal process such as natural selection. The genetic processes of slippage, retroposition, or tandem duplication (for example) and their effects on the natural variation of growth and development, combined with the non-random force of selection acting on many generations across geological time, give us an understanding of a macroevolutionary event: the evolution of mammals with their three ear bones, single jaw bone, and coiled cochlea from a common ancestor they share with birds, lizards, turtles, and crocodiles. In terms of their basis in natural processes evident to us today, the genetic and developmental processes outlined above are not fundamentally different from those that explain so-called microevolutionary phenomena such as antibiotic and pesticide resistance.

Of the vertebrates now available on genome browsers,[71] only a handful are relatively complete and fully annotated. Without knowing what their DNA sequences actually contain, it is obviously much more difficult to disentangle the genetic control cascades behind the shapes and behaviors of organisms. Even so, we're making progress (Table 10.1). The negative claim that certain kinds of complexity are impervious to natural explanation is a cop-out, and an unjustified one at that. This claim is useless to the biologists who are doing the hard work to figure out the mechanisms by which biological complexity has arisen.

ELEVEN

BIOLOGY AND PROBABILITY

Much in the popular debate about life's evolution revolves around our own perception about what is likely, or what could and could not happen in nature without the intervention of a human-like, purposeful agent. It's therefore worth having a look at probability, and to consider the kinds of things that you and I could agree upon as probable, or not, under certain circumstances.

Take five six-sided dice and roll them. The odds that you'll come up with a 1, 2, 3, 4, and 5 in a single roll are 6^5, or 1 in 7776. If it took four seconds each time you gathered up the dice and rolled them, on average you'd get only one such result every 8.6 hours of continuous rolling. In the scheme of a single game involving dice, this is an improbable event, but given lots of humans spending many hours rolling dice, a result such as this would actually occur quite frequently. Hence, something that seems at first glance to be very improbable may be boringly common when a little time and extra protagonists are added.

Let's take a much more improbable event: rolling 1, 2, 3, 4 and 5 three times consecutively. This amounts to the previous probability, 1 in 7776, multiplied by itself three times: 7776^3, or 1 in 470,184,984,576. On average, you would achieve one such combination about every 522,427,761 hours (61,000 years) of continuous rolling, again at four seconds per roll. The participation of many humans rolling dice simultaneously would substantially increase the probability of obtaining this result within a set period of time; after all, there are nearly seven billion of us and with everyone at work someone would probably make this unlikely set of rolls without too much delay. But all you have to do is start increasing the number of successive occurrences of 1, 2, 3, 4, 5 and even a world full of eager humans rolling dice would be ever more unlikely to hit upon the magic sequence.

I'm going to roll my five dice three times, right now: (2, 2, 4, 5, 5), (3, 5, 5, 6, 6), and (1, 4, 5, 5, 6). Had I told you in the last paragraph that I was about to roll these specific numbers in succession, you would have rightfully expressed skepticism and noted that the odds of such an event are 1 in 470 billion—no different than the odds for rolling 1, 2, 3, 4, 5 three times consecutively. You might have said that I'd achieve such a thing only once every 61 000 years and wagered your life's savings that I would fail to obtain these numbers in the order I said I would—but there they are, and you would have lost the wager.

Similar kinds of improbable events happen all the time. For example, in my second-year class on vertebrate biology, I lecture to about 80 students. Each one of them developed from a unique combination of one of about 400 maternal eggs and a couple hundred billion sperm generated by each parent over their lifetimes. If you want to be impressed with improbability, go ahead and figure out (1) the odds of each of those students developing from one particular pair of egg and sperm, and (2) the odds of all of those 80 improbable individuals meeting together in one lecture theatre, out of the many thousands of others who might otherwise have been there. I won't bother, except to note that the potential genetic combinations for each student, combined with the potential combinations of 80 such individuals attending a lecture in one room in Cambridge, exceed the number of atoms in the visible universe.

CONTINGENCY

These examples illustrate the difference between contingency and chance in history. Contingency is the retrospective appreciation of how what happens is one possibility out of many, one which is dependent upon what else has happened. Some combination of numbers has to result with every roll of the dice, although each combination, by itself, is improbable. The occurrence of some series of numbers after a roll of the dice is not only probable, but certain, and is not worthy of awe on our part. The same goes for any single group of 80 students attending a lecture. I work in a university, and providing lectures for students is our bread-and-butter. Some combination of students in the audience has to happen every time. Each individual aggregation of numbers or students may be rare, but they are contingent upon

Figure 11.1. Instances of botanical contingency in Newnham College, Cambridge (left) and the Oranjezicht neighborhood, Cape Town (right). The number of possible shapes for these plants is vast; the number that would fit into holes in a wall or through a fencepost is in comparison infinitesimally small. Yet no sensible person would argue that the improbability of the plant having just the shape that matched its substrate means that the wall or fencepost was specifically constructed to support these individual plants. Rather, the shape of each plant was contingent upon the already-existing substrate; in each case the plant grew and took shape around a medium by which it was constrained. Analogously, the physical and mathematical properties of the universe, while unique and incredibly improbable, long predated the existence of life, which grew upon its cosmic substrate as a creeping vine grows into and around a fence or wall. Therefore, interpreting the anthropic principle to mean that the cosmos was designed specifically to support humanity is unconvincing.

other events that are not necessarily rare (rolling dice, students attending university), and which are guaranteed to produce some kind of result.

Figure 11.1 illustrates contingency with a botanical analogy. Is it not incredibly improbable that the plants shown just happen to have forms that match the stone and metal substrates upon which they are growing? Strictly speaking, the shapes these plants have are very improbable, but of course you recognize immediately that the substrates on which the plants are situated were there first. The brick wall and the metal fence determined the shape of the plants, which were contingent upon the shape of their substrate; they were not the reason why the wall and fence were built. This is similar to the example of architectural spandrels discussed in Chapter 4.

Contingency is why you should not be impressed when someone tells you how incredibly improbable the conditions for life are in this universe, thereby implying that someone or something out there planned our existence all along. The "anthropic principle" is the observation that, for example, had the relative strengths of gravity and electromagnetism been a tiny bit different following the Big Bang, or had the mass ratio of different subatomic particles been ever so slightly larger or smaller, our planet, galaxy, universe, etc. wouldn't exist. Therefore, it is argued, these constants have been purposefully arranged to pave the way for our particular kind of life.

If you look at the constants of the universe in the same way that we looked at the improbability of predicting my three dice rolls, a 1 in 470 billion chance without any contingency, then sure, it seems "finely tuned" for our existence, as if some human-like consciousness had made everything for us from the beginning. Many theistic evolutionists with whom I agree about biological evolution use the anthropic principle to argue for the existence of a Creator at a cosmic level. I don't find this argument compelling, and regard such fine-tuning as a consequence of contingency, rather than the purpose-driven agency implied by the anthropic principle. It is still reasonable to believe in a creator (see below and Chapter 1), but in my view the anthropic principle is not a compelling argument to do so, and no such belief has much to do with the kind of probability relevant to evolutionary biology.

The anthropic principle is unimpressive as an argument for a human-like "god" because had the initial conditions of the universe been different immediately following the Big Bang, something else would have existed, perhaps with conditions amenable to another form of consciousness. It would probably not have been carbon-based, but we cannot rule out alternatives that might have arisen from other potential cosmic starting points.[1] It is the beginning of either universe, who or what got them started, and the consistency of natural law that are worth considering as evidence for a "god." However, the fact that any single universe is unique and improbable is, by itself, no more profound than the observation that plants growing on a wall have a unique and improbable shape. Had the wall or fence shown in Figure 11.1 been shaped differently, the plant too would have exhibited a different appearance, one that would be equally improbable. Unlike the plants, however, the alternative universe resulting from different starting conditions is unimaginable from our perspective, dependent as it is on the particular starting points of our "finely tuned" universe.

Our existence is certainly improbable, but such improbability is by itself not a compelling argument directing us toward the action of a human-like force, priming the cosmic way for our existence. The improbability of our world is contingent on a series of events that could have unfolded in another way, and in so doing would have resulted in some other, equally improbable course of cosmic history.

CONVERGENCE

I don't know if or how the universe was constrained to exhibit the particular features of physics, chemistry, and mathematics that it now has. Because the origin of the cosmos is a far more remote event than those relevant to evolution on Earth, it is often difficult to collect and evaluate the data relevant to this problem. Moreover, as an evolutionary biologist, concerned with life on Earth after its origin, I freely confess that the issue of cosmic origins is far beyond my expertise. Much more certain is the fact that life on Earth has been constrained, as the well-documented phenomenon of convergence makes clear. One need not be impressed by the anthropic principle to recognize the limited directions in which biological evolution on Earth has occurred. My paleontologist colleague Simon Conway-Morris has persuasively argued that, in biology, not every conceivable kind of form is possible.[2] If life were to start over again, evolution would be constrained by many factors that, probably, would lead to animals and plants taking on forms not too different than what we've got today.

If you want to take advantage of solar radiation for energy, you've got to collect it with something that efficiently utilizes surface area, like a leaf; if you want to extract oxygen from your environment, you've got to maximize exposure to the relevant medium with structures like gills in water or lungs in air; if you want to swim, fly, or dig, there are only a limited number of shapes that allow you to do this in an efficient manner. This is why, as Conway-Morris and many other biologists have shown, animals and plants from completely different parts of the Tree of Life can converge on the same solutions for common ecological problems. For example, soricid shrews and afrotherian microgales are small mammals with an extremely similar reliance on smell and tactile sensation in foraging for insects and small vertebrates; marsupial moles and afrotherian golden moles are very similar in the way they burrow (Figure 9.4). These animals show strikingly similar aspects of anatomy in several regards, even though they share only a distant common ancestor (Figure 9.5). Few of the possible designs we can imagine to handle the tasks observed in nature have actually been realized, and this is due not only to the raw material with which evolution has to work, but also to the limited ways in which a given ecological problem (such as hunting insects in the forest or burrowing under the earth) can be solved.

EVOLUTION IS NOT RANDOM

The phenomenon of convergence reflects an important fact: evolution by natural selection is not random. If it were, and the constraints of development and environment had less impact on how animals evolved, we would expect to see far more differences in form and function of anatomical and genetic features across animals and plants. Once you imagine shapes and substances that could conceivably help some creature make a living, you quickly realize how limited in material and form living things really are. As we noted in Chapter 4, not a single vertebrate animal takes advantage of stone or steel to form its skeleton; not one of them has developed wheels to help them move on land or in water (Figure 11.2). In contrast, they all use the same language of heredity (DNA), and have the same tissue types (such as bones, teeth, and muscle) and cellular components. A truly random process of generating animal diversity would show more variety than that.

In Chapter 4 we also mentioned an oft-used, misleading analogy for evolution: a hurricane sweeping through a junkyard and assembling a functional jet. Like rolling 1, 2, 3, 4, 5 with six-sided dice a million times in succession, such an event is exceedingly improbable. However, neither has much to do with evolution. Evolution by natural selection is not a random process, but a biased one. Although I've disagreed with Richard Dawkins elsewhere in this book, his discussion of probability in chapter three of *The Blind Watchmaker*[3] is right on the mark. As Dawkins and other biologists[4] and philosophers[5] have pointed out, you've got to realize that natural selection is not a random process in order to appreciate its creative power.

Let's have another look at our dice. Instead of trying to get a 1, 2, 3, 4, 5 from scratch every roll, you could aim for such an outcome by keeping from each roll all of the relevant results and re-rolling any duplicates or sixes. I've just done this right now: my first roll was 1, 2, 4, 4, 5. I took the duplicate four, rolled it again and got a six. Rolling that same dice again got me the missing three. So in three rounds I got my result: 1, 2, 3, 4, 5. Rather than a 1 in 7776 chance of obtaining this result, likely to occur once about every 8.6 hours of constant rolling, it took me three rolls in about 12 seconds. You would be unlikely *not* to get 1, 2, 3, 4, 5 with more than just a few rolls using this process, which although it entails a random element (rolling dice), is more accurately understood as a biased, non-random one (selection).

Similarly, in chapter three of *The Blind Watchmaker* (pp. 46–9), Richard Dawkins describes how the biased—not random—process of selection succeeds in a few seconds on a vintage, 1980s home computer in generating the phrase "Methinks it is like a weasel" from letters that are randomly generated. As metaphors, Dawkins' exercise and my biased dice rolling are of course distinct from natural selection, but they are far better representations of selection than analogies consisting of hurricanes and junkyards.[6]

An objection made to Dawkins' and my artificial analogies for natural selection, i.e., keeping partially formed words from random letter generation or re-rolling dice by picking out the "right" ones, is that both require

Figure 11.2. Mammalian skeletons on display at the Galerie d'Anatomie Comparée, Muséum Nacional d'Histoire Naturelle, Paris. Few museum exhibits display with such simplicity and clarity the highly conserved layout of the vertebrate skeleton, one which is nevertheless the structural basis for thousands of distinct species. The animals represented in this exhibit are diverse, but consider that not a single one incorporates steel, stone, or wheels into its structure. This is because evolution by natural selection is not a random process, but one constrained by the material and development shared by all of these animals with their biological ancestors, linked by common descent to one or few primordial forms of life on Earth.

the foresight of a "designer" to work. I know ahead of time that I'm searching for a series of 1, 2, 3, 4, 5 on my dice, and Dawkins knew the phrase he wanted to produce from the first generation. In contrast, selection in nature would have no such capacity for forward planning—or so it is claimed.[7]

It is certainly true that you and I would be hard pressed to explain why a natural process would prefer syllables like "wea" followed by "sel" and others that comprise the phrase "Methinks it is like a weasel." That particular phrase, as opposed to "No it's a skunk," clearly involves the premeditated foresight of Richard Dawkins' intelligence behind the software described in his book. However, objecting to this metaphor on this basis is taking it too literally. The identity of the numbers or syllables you pick for the "target" is irrelevant to the real point: biased processes can generate complexity far more easily, and frequently, than random ones. Strictly speaking, the differential survival inherent in natural selection responds to stimuli only in the present. However, when those stimuli are consistent over time, as may very well occur with selective pressure resulting from predation (among other possibilities), they comprise an analogous kind of "foresight" as that evident in metaphors of language or dice rolling. Stated differently, although natural selection obviously cannot plan ahead or anticipate in the same way a human can, the results of selection can greatly resemble the results of human-like foresight. Hence, metaphors of dice rolling and word generation can indeed accurately reflect some ways by which natural selection causes the evolution of complexity via a biased process.

SELECTION AND STICKLEBACK FISH

Natural selection favors individuals with features that help them to survive and reproduce, and thereby contribute genetic material to the next generation. Sometimes selection occurs in an intuitive way that is easily understood. If a little more of something good increases your ability to contribute genes to future generations, then selection has something concrete on which to work, a process which is not so different than the artificial selection applied to series of dice-rolls. In terms of probability, this is not a problem.

A dramatic example of the influence of natural selection on animal form concerns stickleback fish, composed of several closely related species a few centimeters in length, native to nearshore seas, rivers, and lakes in North America and Eurasia (Figure 11.3). Marine stickleback species tend to have superficial armor plates along their flanks, a robust and bony pelvis, and spines extending from their backs and bellies, whereas in many freshwater species the spines, pelvis, and armor are much reduced. Over the past few decades, the work of many scientists[8] has established that the extent and size of armor, spines, and other pelvic bones correlates well with the presence or absence of certain predators, such as certain fish, birds, and insect larvae.[9]

Long spines on a stickleback make it more difficult for predatory fish and birds to ingest them, since they're limited in how wide they can open their mouths. Indeed, when you examine gut contents of cutthroat trout, for

Figure 11.3. Differences in the dominant form of threespine stickleback over time in Lake Washington, Seattle. The dark body plates represent dermal armor, with which this fish protects itself from predators such as trout. The change in the population from mostly unarmored individuals to mostly armored ones took place over the course of about ten years, from the late 1960s to the late 1970s, representing about ten generations of stickleback fish. The unarmored variant did not disappear from Lake Washington, but by 2005 was greatly outnumbered by the armored variant. (Reprinted from Kitano *et al.* 2008, with permission from Elsevier. Scale bars = 10 mm.)

example, ingested sticklebacks tend to have no or relatively small spines.[10] Similarly, when the sticklebacks possess an extensive series of bony plates along each flank (Figure 11.3), predatory fish find it more difficult to capture and eat them. Other environments in which sticklebacks occur have fewer predatory fish but more insect larvae, against which spines and body armor are not so effective and may even be a liability. For several decades, correlations between predation and stickleback anatomy have been the subject of a great deal of scientific investigation,[11] and it is demonstrably the case that body armor and long spines are inherited across generations and have been acted upon by natural selection. Indeed, spines and body armor have repeatedly been lost in sticklebacks that have re-colonized freshwater habitats from a marine ancestry.

One recent study by Jun Kitano[12] of Tohoku University in Japan, formerly of Seattle's Fred Hutchinson Cancer Research Center, has demonstrated the opposite trend: a gain of armor in descendants of freshwater sticklebacks. Kitano and colleagues observed an increase in body armor since the late 1960s in Lake Washington, an urban freshwater lake in Seattle, located about two miles east of the Discovery Institute, home to the anti-Darwin lobby we mentioned in Chapter 8. Kitano proposed that an increase in water transparency during the late 1960s and 1970s enabled visual predators such as the cutthroat trout to see farther, leading to high levels of predation on sticklebacks with less armor. This greatly biased the kinds of stickleback that survived to produce offspring in the initial generations as underwater visibility increased.

Natural selection applied to sticklebacks during the late 1960s and 1970s resulted in a very different morphology in a very short period of time. Kitano *et al.* reported that in the course of about ten years, the most common plate number among Lake Washington sticklebacks increased from about 7 to over 30 (Figure 11.3) in roughly ten generations. In the course of about 50 years, the percentage of fish with little armor decreased from 91% in 1957 to 16% in 2005. Samples collected from Lake Washington during the 1950s and early 1960s showed no or few stickleback individuals with more than a dozen armor plates along the fish's side, whereas the most common number after the mid-to-late 1970s was over 30. Furthermore, Kitano's genetic data showed that the heavily armored sticklebacks were not simply marine immigrants from nearby Puget Sound, but were direct, lineal descendants, within just a few generations, of their relatively less-armored ancestors from Lake Washington.

In addition, the level of armor plating in these fish is correlated with the activity of a specific gene called *Ectodysplasin*, or *Eda* for short. Confirming the results of a previous study,[13] Kitano and colleagues found that their fish with the dominant form of this gene showed relatively more armor plating than those with a different form of *Eda*. This is representative of the genetic variation upon which natural selection works in nature. It has major anatomical consequences for this little fish, a resident of a lake just down the road from an anti-Darwin lobby whose members claim or imply that the complexity of these animals exists because of human-like, supernatural intervention.

To be sure, the study of Kitano *et al.* did not focus on the actual mechanism by which *Eda* synthesizes a protein that, when expressed at the right place and time during the growth of a stickleback fish, contributes to the development of an individual with lots of armor plating along its side. In addition, water transparency is unlikely to be the only factor influencing the re-evolution of body armor in Lake Washington sticklebacks, as other lakes with an appreciable level of water clarity have not witnessed a similar phenomenon. Mineral concentration is not uniform across lakes, and the relative physiological expense of developing bone in such a small animal undoubtedly plays another important role in the evolution of stickleback armor. Nevertheless, the research of Kitano[14] and other evolutionary biologists[15] has reasonably demonstrated not only that a natural mechanism for the appearance of stickleback armor exists, but also that it is reversible. Individuals with little armor plating can, over time, give rise to descendants with lots of armor plating, as in Lake Washington,[16] or the reverse, as typically found in freshwater descendants of marine populations.[17]

Recently, another genetic mechanism has been articulated to explain not the presence of armor plating of sticklebacks, but the pelvic bones and spines that project downward on either side of the fish's midsection (Figure 11.3). Typically, these two morphologies are associated: freshwater sticklebacks tend to show small or absent spines and few armor plates; marine sticklebacks tend to show longer spines and more armor plates. In a series of papers over the past few years,[18] biologists like Michael Shapiro from Stanford University and the University of Utah have shown how changes to the pelvic bones of stickleback fish result from changes to the regulation of a gene called *Pituitary homeobox 1* (or *Pitx1*).

Certain segments of *Pitx1* show a very high level of conservatism across different species, including mice and humans. When the scientists compared the *Pitx1* sequences of stickleback populations with pelvic spines to those without, and to other vertebrates like mice and humans, they made a startling discovery: the sequences were nearly identical. Not only was the *Pitx1* gene virtually the same in the two sticklebacks with different pelvic morphologies, but the fish genes were very similar to those of humans and mice.[19]

So if this gene is essentially the same in animals with and without a pelvis, how is it supposed to be relevant to pelvic reduction in these little fish? Biologists have known since the 1970s[20] that genes may influence anatomy and other features of phenotype not by changes to the actual proteins they make, but by changes to other parts of the genome that control the timing and expression of those proteins. Scientists have recognized that *Pitx1* has an important influence on the development of pelvic anatomy in vertebrates, yet it showed virtually no differences in its sequence of nucleotides among such diverse creatures as fish, mice, and humans. This led them[21] to take the next logical step: to search for regulatory sequences that control the expression of *Pitx1* in the pelvic skeleton, and find out how they differ in animals with and without pelvic spines.

During the course of this search, the scientists found that their mechanism for the reduction of the pelvic skeleton seemed to be the same across not only different kinds of stickleback fish (e.g., *Gasterosteus aculeatus*, *Pungitius pungitius*), but appeared consistent even with the anatomical data they collected for a mammalian species without a pelvis: the manatee.[22] The fish are, however, easier to study, since there are closely related species that vary in their possession of pelvic spines, and unlike manatees they fit under a microscope without too much trouble. In early 2010 a group of scientists led by Frank Chan published their discovery regarding the genetic regulation of *Pitx1*.[23]

Using a series of genetic mapping experiments, Chan and colleagues found that a stretch of DNA located about 30 000 nucleotides before the site containing the actual *Pitx1* gene could influence the activity of *Pitx1* specifically in the pelvis. They named this segment, logically, *Pel*. While the DNA sequences of *Pitx1* were the same in fish with and without a pelvis, those of the controlling region *Pel* were not. In populations of reduced-pelvis sticklebacks, Chan and colleagues found various changes in a specific region of *Pel* that were associated with loss of expression of *Pitx1* in the developing pelvic region. Such changes were not present in sticklebacks that matured into adults with a full pelvis.

In addition, by manipulating different combinations of the *Pel* regulatory sequence with *Pitx1*, they conducted a clever series of experiments to try and recover the full-pelvis morphology in sticklebacks that normally lack a pelvis. To accomplish this, they combined a functional *Pel* sequence plus the DNA encoding the *Pitx1* protein, and injected this artificial construct into the fertilized eggs from a pelvis-reduced population. As they expected, the pelvis developed in individuals into which this artificial DNA had been inserted, but not in control specimens that lacked the functional *Pel–Pitx1* combination. Once present in the genome of a fish that would normally lack a pelvis, *Pel* turned on the expression of *Pitx1* at a critical moment in that fish's development, leading toward the formation of pelvic bones and spines.[24] Perhaps most impressive was their discovery that the activity of *Pel* was tissue-specific. Whether or not *Pitx1* was active in the pelvis, it would still express itself in the fish's head (for example).

The discovery of *Pel* is consistent with the idea that mutations in regions of genes that code for proteins (such as *Pitx1*) are not necessarily required for major evolutionary change to take place. Indeed, due to the typical function in protein-coding genes across many different body structures and processes (the phenomenon of pleiotropy we discussed in Chapter 10), it has been argued that changes to gene regulation via non-coding sequences such as *Pel* are more important sources for evolutionary novelty because they reduce the effects of pleiotropy. As Chan and colleagues demonstrated, changes to the sequence of their newly discovered *Pel* regulator affect the expression of a protein like *Pitx1* in the pelvis, but not in the head.

Opinions on the relative importance of mutations in regulatory versus coding DNA sequences for large-scale evolutionary patterns vary,[25] and the two kinds of changes are not mutually exclusive. They may co-occur, and we know already that at least some aspects of animal form (such as primate color vision, discussed in Chapter 10) are due to mutation and selection in coding regions of the genome.

Contemporary biologists have a pretty good understanding of the genetic mechanisms that influence the stickleback's body armor and pelvis. Variable expression of *Eda* and *Pitx1*, the latter through the regulatory activity of *Pel*, are now known to be responsible for important aspects of the skeleton in not only stickleback fish, but other vertebrates as well.[26] Hence, these examples would have been right at home in Chapter 10, where we more explicitly focused on the genetic mechanisms upon which natural selection works to produce novel phenotypes.

I included the discussion of these small fish in this chapter not only to briefly summarize how these mechanisms work, but also to show that the evolutionary pressures on stickleback fish are analogous to the "foresight" apparent in dice rolls or vowel–consonant combinations discussed above. Just like dice rolls, mutations leading to changed expression of *Eda* or *Pitx1* do entail random elements. But the survival of one variant or another across generations can be precisely skewed in a non-random direction over long periods of time. Although selection cannot anticipate the future, the bias of the present may coincide with that of the future. When this happens, bias resembles foresight. In the case of the stickleback fish in Lake Washington during the 1960s and 1970s, the bias was toward fish with more body armor—the ones that larger, predatory fish found most difficult to ingest. The same kind of predator pressure is known to influence the morphology of pelvic spines. In each example, there is a consistency to the bias that persisted for more than one or a few generations. For the fish, this is the likelihood that, relative to armored sticklebacks, unarmored individuals with no pelvic spines will be preyed upon more frequently. The transformation of Lake Washington sticklebacks during the late twentieth century, or of any number of freshwater descendants of marine species over many thousands of years, resulted from a biased process, not a random one.

PROBABILITY AND ANTI-EVOLUTIONISM

The evolution of stickleback body armor is a well-studied phenomenon that is demonstrably linked to natural selection.[27] But there are other cases in which the connection between morphology and natural selection isn't so obvious. One of Darwin's early critics, George Jackson Mivart, claimed in 1871 that "incipient stages of useful structures,"[28] including eyes, baleen, and mammary glands, were not useful to an animal and that, therefore, natural selection couldn't possibly act on them in evolutionary time.

Darwin was acutely aware of this issue himself. Because of it, he penned the all-time favorite sentence of creationists: "If it could be demonstrated that any complex organ existed, which could not possibly have been formed by numerous, successive, slight modifications, my theory would absolutely break down."[29] Mivart seized on this apparent concession. His objection to the role of natural selection in evolution might be phrased as follows: what use is half a wing for powered flight? By most engineering standards, little indeed, and thus it seems natural selection reaches a stumbling block. Here is the argument in Mivart's own words:

> Complex and simultaneous co-ordinations [of eye and ear development] could never have been produced by infinitesimal beginnings, since, until so far developed as to effect the requisite junctions, they are useless. But the eye and ear when fully developed present conditions which are hopelessly difficult to reconcile with the mere action of "Natural Selection."[30]

Mivart's objection amounts to the concept of "irreducible complexity," as defined by intelligent design advocate Michael Behe 125 years later:

> By irreducibly complex I mean a single system composed of several well-matched, interacting parts that contribute to the basic function, wherein the removal of any one of the parts causes the system to effectively cease functioning. An irreducibly complex system cannot be produced directly (that is, by continuously improving the initial function, which continues to work by the same mechanism) by slight, successive modifications of a precursor system, because any precursor to an irreducibly complex system that is missing a part is by definition nonfunctional.[31]

Both Mivart and Behe present such complexity as an insurmountable challenge to evolution by natural selection. Furthermore, Mivart and the contemporary ID crowd have used probability metaphors to try and convince their readers how ineffective natural selection really is. Here is one metaphor from 1871, cited at length by Mivart in his *Genesis of Species* book:

> There are improbabilities so great that the common sense of mankind treats them as impossibilities. It is not, for instance, in the strictest sense of the word, impossible that a poem and a mathematical proposition should be obtained by the process of shaking letters out of a box; but it is improbable to a degree that cannot be distinguished from impossibility; and the improbability of obtaining an improvement in an organ by means of several spontaneous variations, all occurring together, is an improbability of the same kind.[32]

Although Mivart never saw an airplane, here is one of the first misapplications of the hurricane-sweeps-through-junkyard-makes-jet analogy, as if it were relevant to natural selection. It's not. The fact that things of great complexity do not appear by pure chance is simply irrelevant. As already discussed above and elsewhere,[33] complexity can come about via a biased process such as natural selection.

ID advocates continue to cite such improbability as a problem for evolution: "If you were able to observe the evolution of life, you'd probably see mutations and changes coming along that really defied probability, that were so uncommon you would have no statistical right to expect them. ... It's like throwing a string of dice, sixes, a million times in a row."[34] Here's another example:

> Many scientists and mathematicians have questioned the ability of mutation and selection to generate information in the form of novel genes and proteins. Such skepticism often derives from consideration of the extreme improbability (and specificity) of functional genes and proteins. A typical gene contains over one thousand precisely arranged bases. For any specific arrangement of four nucleotide bases of length n, there is a corresponding number of possible arrangements of bases, 4^n. For any protein, there are 20^n possible arrangements of protein-forming amino acids. A gene 999 bases in length represents one of 4^{999} possible nucleotide sequences; a protein of 333 amino acids is one of 20^{333} possibilities. ... [P]roteins represent highly isolated and improbable arrangements of amino acids ... far more improbable, in fact, than would be likely to arise by chance alone in the time available.[35]

The author of this quote is Stephen Meyer, whose perspective on uniformitarianism we discussed in Chapter 2. In the very next paragraph following this quote, Meyer acknowledges that "Of course neo-Darwinians do not envision a completely random search through the set of all possible nucleotide sequences." This belated acknowledgment does not deter him in the rest of his paper from treating the mechanisms of mainstream evolutionary biology as if they were tantamount to fundamentally random processes. For example, toward the end of his paper he states "Natural selection can favor new proteins, and genes, but only after they perform some function. The job of generating new functional genes, proteins and systems of proteins therefore falls entirely to random mutations."[36] Meyer clearly implies that a large number of such random events must happen simultaneously for novel protein function to emerge, and that novel function is an all-or-nothing event, rather than quantitatively varying.

Such implications are patently false.[37] Occurrence of genetic mutations such as duplication, slippage, and point mutations are indeed subject to randomness. However, as evolutionary biologists have repeatedly stressed, selection can favor intermediate forms between one protein and another;[38] such intermediate pathways between one protein function and another may involve some or many neutral steps;[39] and the multiple layers of selection, redundancy, and constraint that mediate the evolution of novelty are not random.[40]

When pressed by those knowledgeable about evolution, ID proponents such as Meyer generally qualify their comparisons of evolutionary biology with pure chance, as quoted above. However, they tend not to inform their non-scientific, popular audiences about this potential misunderstanding, one which seems to be positively encouraged in their public lectures and movies. Take this example from ID advocate Paul Nelson, the invited speaker for a September 2006 lecture at Calvary Baptist Church in Albuquerque, New Mexico:

> If [a robot] were here in the lobby tonight ... the question that would naturally occur to you is who designed that? Who constructed that thing? The one answer that you would not accept ... that would be irrational to even entertain is that "well we put the engine, and the parts, and the frame in a box and we randomly agitated it for 100 years and this is what popped out." To offer that kind of answer to the question "who built this?" would invite psychiatric attention. Now why is it, when we look at [a dog], why is it that modern biology gives us a different answer, one that does not point to intelligent design?[41]

Paul Nelson is a very polished and entertaining speaker, and I'm glad the staff at Calvary Baptist decided to make his talk available on the web. But his is a fundamentally misleading portrayal of evolutionary biology. Nelson's audience at Calvary Baptist Church presumably consists of hard-working people who do not regularly consult the biological literature. They're busy with their lives and trust professional speakers, like Paul Nelson, to give them a fair accounting of biological thought. Unfortunately, Nelson did them a disservice by implying that the main explanatory force of evolutionary biology is akin to "randomly agitated" parts in a box, repeating rather closely Mivart's objection from 1871, as quoted above. Rhetorical flourishes and hyperbole aside, evolutionary biologists do not maintain that pure chance is how a dog or any other organism is "built."

EXAPTATION

So, what good is half a wing for flight? Are Mivart and Behe correct in saying that a gradual, adaptive climb toward the function of many complicated biological structures does not exist? In the sixth and last edition of the *Origin* (published 1872), Darwin responded to this objection by noting how "incipient stages of useful structures" (in Mivart's words) or "irreducibly complex" structures (in Behe's words) are only problematic for evolution if one presumes conservation of function. Darwin himself stated "This subject is intimately connected with that of the gradation of characters, often accompanied by a change of function."[42] He also wrote "structures thus indirectly gained, although at first of no advantage to a species, may subsequently have been taken advantage of by its modified descendants, under new conditions of life and newly acquired habits."[43] In other words, half a wing won't help you fly, but it might help you run uphill,[44] and its feathers will keep you warm. This potential for change in function is called exaptation,[45] and it enables natural selection to have a profound influence on animal form. The term derives from "adaptation"; it is a feature of anatomy or behavior "extended" or "exapted" for purposes beyond those for which it was originally selected.

Cell biologist Ken Miller of Brown University has shown how exaptation has toppled one of the more popular examples of "irreducible complexity."[46] The bacterial flagellum is the propulsive device of microorganisms such as those in your gut, *Escherichia coli*. This is the hair-like projection in *E. coli* which spins rapidly and propels the organism on its way. The flagellum has been presented by ID as one of those particularly complex structures which are supposed to lack useful subcomponents, and which therefore could not have been the subject of natural selection. Recall Behe's definition of this concept from his 1996 book: "A single system composed of several well-matched, interacting parts that contribute to the basic function, wherein the removal of any one of the parts causes the system to effectively cease functioning."[47]

Bacterial flagella come in many different forms, and it's not really accurate to talk about "the flagellum" as if it were an invariant structure.[48] Nevertheless, all of these different forms appear to share a few dozen proteins arranged in a particular way to effect function.[49] If the ID concept of "irreducible complexity" is to be relevant to natural selection in evolution, then it is not enough that the flagellum be inoperable without its constituent parts. Those parts should also lack structures of potential use to its owner. In other words, if it is indeed "irreducibly complex" in a way that is relevant to evolution, the parts of the bacterial flagellum have to be ineffective as potential subjects of natural selection.

They're not. Miller[50] and other scientists[51] have pointed out that several flagellar subcomponents actually do have functions that occur in nature. One of these is called the "type III secretory system," or T3SS, present in bacteria such as *Yersinia pestis*, a nasty little bug responsible for bubonic plague. It uses its T3SS as a syringe to inject toxins into host cells. Miller shows that the proteins that make up the T3SS comprise a subset of those in the flagellum, exhibiting a similar sequence and structural arrangement. So as a matter of fact, you can take certain parts away from the flagellum without obliterating function. Now this function doesn't necessarily have to do with moving the bacterium around, but it is good for something else (e.g., toxicity or adhesion), and as such is potentially the subject of natural selection. This is an example of exaptation. As for the alleged irreducible complexity of the flagellum, "the contention that the flagellum must be fully-assembled before any of its component parts can be useful is obviously incorrect. What this means is that the argument for intelligent design of the flagellum has failed."[52]

Interestingly, Behe has responded to this critique by emphasizing his insistence on conservation of function in the definition of "irreducibly complex":

> Miller shifts the focus from the separate function of the intact system itself to the question of if we can find a different use (or "function") of some of the parts. ... What's more, the functions that Miller glibly assigns to the parts ... have little or nothing to do with the function of the system ... so they give us no clue as to how the system's function could arise gradually.[53]

On the contrary, biologists such as Ken Miller have proposed a means by which natural selection could have acted gradually upon the parts of the flagellum: exaptation.[54] While we do not yet have a definitive, precise, step-by-step understanding of how the particular flagellum of *E. coli* has evolved using intermediate protein structures of some use to precursor generations (although some possibilities have been suggested[55]), the meaningful claim of "irreducible complexity" was more general: subcomponents of certain complex systems cannot have functions amenable to natural selection. By refusing to consider the possibility of a change in function during the course of natural selection, e.g., constructing a flagellum with some components originally favored due to their role in other processes, Behe has made his concept irrelevant to biology. Sure, he can restrict his view of "irreducibly complex" to one in which function cannot change, but natural selection is not so restricted, and thus the debate degrades into one about semantics rather than biology.

Another objection to critiques of the "irreducible complexity" of the flagellum has to do with the possibility that the T3SS represents a degraded flagellum itself. Maybe the syringe in certain microbes (such as *Yersinia*) is simply a remnant of an ancestral flagellum present at some point in its evolutionary history: "Phylogenetic analyses of the gene sequences ... suggest that flagellar motor proteins arose first and those of the [T3SS, or pump] came later. In other words, if anything, the pump evolved from the motor, not the motor from the pump."[56] Not all investigators agree with this scenario,[57] but there is indeed genuine scientific debate about the microbial Tree of Life,[58] and figuring out how various groups of bacteria are interrelated is very challenging.

We do not yet know for sure if a flagellar-like structure was present in the ancestor or descendant of a T3SS-equipped bacterium. However, at issue is not any particular scenario of what-evolved-from-what, but the assertion that certain structures are "un-evolvable."[59] Such an assertion depends on the claim that subcomponents of a structure like the flagellum are not useful in their own right, or part of other systems that could conceivably have played intermediate functional roles, potentially subject to natural selection. Concerning the flagellum, we know that actually many of its parts qualify on both counts.[60] This is true whatever competing hypothesis for the bacterial evolutionary tree currently discussed in the literature proves to be true.[61] Consider, too, the irony of using an evolutionary argument about the flagellum, "the pump [T3SS] evolved from the motor [flagellum],"[62] to prop up the claim that it cannot itself evolve. The components of the bacterial flagellum are no more impervious to exaptation and natural selection than those of any other complex cellular or anatomical structure, however the shape of the bacterial Tree of Life is reconstructed.

The debate about the bacterial flagellum has helped to focus the energy of evolutionary biologists on deciphering specific mechanisms of how complexity at the cellular level exhibits the properties of evolution by natural selection.[63] These efforts have successfully decapitated the concept of "irreducible complexity," for example by using similar observations of changed function, or exaptation, just as Darwin did in his response to Mivart in the early 1870s.

TWELVE

EVOLUTION, EDUCATION, AND CONCLUSIONS

I grew up in a conservative town in western New York state, about 350 miles northwest of Manhattan. During the early nineteenth century, the rural counties between Buffalo and Syracuse NY were known as the "burned over district" due to their revivalist religious zeal. Among the more famous evangelists of nineteenth-century New York state was Joseph Smith, founder of the Mormon Church. In terms of their more recent political history, western New Yorkers have sent such conservative Republicans as Tom Reynolds and Bill Paxon to represent them in the US Congress. In the US presidential elections of November 2008, several counties of western New York state gave the Republican ticket of John McCain and Sarah Palin a substantial majority, reaching close to 20 percentage points in some cases, compared to the national result in which Barack Obama won the popular vote by a "landslide" of 7%. Evangelical Christianity has a strong following in this part of the United States and played a very important role in my own youth.

At present, I'm a full-time academic at an elite British university and I've paid taxes to two different European governments over the past decade. By all accounts, I should have little in common socio-politically with the people of my hometown, but the fact is I do. For example, like many of my western New York neighbors, I can't stand the idea of voting for political candidates sympathetic to trial lawyers who have litigated the US health care system into the world's most expensive and least efficient. A lot of the men and women in my hometown are proud of the fact that they served in the US military, and have tried to foster democracy in Iraq and Afghanistan. I'm proud of them too. Politics are not one-dimensional, and the stereotypical dichotomy of rural Republican versus ex-pat liberal does not trump the down-to-earth pragmatism that characterizes the part of the United States where I grew up.

There are many in my hometown who disagree with me on a variety of political issues. However, if some of the conservative teachers, dentists, nurses, police, carpenters, and other members of small-town America in which I was raised regard evolutionary biology as suspect, they do so only because they've been busy with their own concerns and have not yet had time to seriously consider the evidence behind it. I don't recall the specifics of my own exposure to biological evolution as a western New York high-school student during the mid-1980s. We covered at least some material on fossils and common descent, but it did not lead anyone in my hometown to complain that such information would endanger the minds or morality of us kids. Far more worrying to the parents of my school district was our interest in *Dungeons and Dragons* and hardcore punk.

The extent to which high-school biology should delve into the details of vertebrate paleontology, plant and animal development, or molecular phylogenetics is limited by all of the other subjects kids should learn by the time they're 18. Personally, I think it would be great for the average student to know something about the fossil elephants and whales we discussed in previous chapters, and to have "heard" the fascinating story behind the evolution of their own ear bones. But there are a lot of other subjects to cover, and if students are going to understand something about history, chemistry, literature, and geometry, then at least some of the details on evolutionary biology may have to wait until college. That's okay, and local and state school districts, parents, and teachers have the right to make and/or influence the decisions about the educational priorities that determine local curricula. However, the freedom to influence your child's education does not equate with freedom to ignore or distort the facts behind that education. We all want to enjoy "academic freedom," but remember that science is not a democratic endeavor. Some ideas are right, others are wrong, and such accuracy is not a matter of politics or voting.

ACADEMIC FREEDOM

In his 2006 book, *The Making of the Fittest*, Sean Carroll describes how academic persecution in Soviet biology led to the downfall and murder of the esteemed Russian botanist Nikolay Vavilov, a member of the Russian Academy of Sciences and former president of the International Congress of Genetics. His death in a Soviet prison in 1943 was a direct result of his scientific criticism of the infamous Trofim Denisovich Lysenko, who in 1940 replaced Vavilov as the head of the Soviet Institute of Genetics. Unfortunately for Vavilov, Lysenko acquired an extraordinary amount of power within the Soviet Union due to his political skills and association with Joseph Stalin, and was able to stifle opposition to his unproven (and economically disastrous) ideas about agriculture. Vavilov viewed biology in the framework of Mendel and Darwin; Lysenko did not, and for many years Lysenko successfully kept the evolutionary biology practiced in the rest of the world out of the Soviet Union. Soviet biology began to recover in the 1960s, after Lysenko's immunity from criticism disappeared along with the regimes of Stalin and Khrushchev. Once it did, the extent of scientific fraud committed by Lysenko became apparent in the Soviet scientific community and, eventually, in the Soviet media. Lysenko died in 1976, regarded by his peers and successors as a cruel and fraudulent failure.

We all agree that such persecution is horrible, so when new cases of academic intolerance happen, we should be very concerned. One case of alleged persecution took place in the state of Minnesota during the 1997–8 school year, when high-school teacher Rodney LeVake was in charge of a tenth-grade biology class at Faribault High School. As the year progressed, Mr. LeVake expressed reservations to his colleagues about teaching evolution to his students. Court documents[1] state that he marginalized the chapter relating to evolution because he was "not allowed to cover the criticisms and weaknesses in the theory." After it became clear to his fellow teachers and administrators that Mr. LeVake was not covering parts of the state-mandated curriculum, he was reassigned to a ninth-grade general science course the following year.

In terms of real human suffering, this case does not come close to what happened to Nikolay Vavilov in the early 1940s. No one was murdered, sent to the gulag, or even fired. Furthermore, this was about a high-school teacher, not an internationally renowned scientist and member of a national academy. Nevertheless, freedom of expression is applicable at every level in society, and Mr. LeVake evidently felt very mistreated by his school district. He was careful not to mention creationism or God in his classes; he even said he would teach evolution to his students, as long as he could also include his perspective on "some of the holes in Darwin's theory."[2] So in 1999, he sued the district for infringement on his academic freedom and right of expression and sought to regain his position teaching biology. The case was heard by a Minnesota state court, which found LeVake's suit baseless on all counts. This decision was appealed at the state level in

2001 and to the US Supreme Court in 2002. Both higher courts refused to hear the case.

And there the story would have ended, except for the fact that among their January 2010 episodes of their "ID-The-Future" podcasts, the Discovery Institute used LeVake's experience as the basis for a series entitled "Rodney LeVake: Expelled Science Teacher."[3] Here is a representative sample of how the Discovery Institute interviewer presented his case:

> [Some] school districts want to teach evolution very dogmatically and they want to censor from students any scientific challenges to Darwinian evolution, even to the extent where they would falsely accuse teachers like you who are teaching the curriculum properly, of refusing to teach evolution, when really all you've committed was a thought crime. You didn't even get a chance to teach these problems that you wanted to talk about; the worst that you did Mr LeVake was have the wrong thoughts inside your head. … This highlights one of the reasons why academic freedom legislation is very important, because it protects the rights of teachers, like you, to teach about scientific strengths and weaknesses of evolution, without having to fear about being reassigned to a different class, or without having to fear getting fired or other negative repercussions to your job.[4]

"Thought crime" and "censorship" are serious accusations, and this interviewer clearly makes the district seem very heavy handed. But what was it, exactly, that Mr. LeVake wanted to critique about evolution, and why did his fellow teachers at Faribault High School question his ability to teach it? Why did the courts refuse to support the claims he made? Mr. LeVake's interpretation of paleontology gives us a clue:

> The fossil record for years, when I was growing up and in school, and even in college … was almost put on a pedestal as "the proof" for … evolution. And when you get right down to it, studying the fossil record, actually it … not only doesn't support evolution it really kinda flies in the face of evolution. … You would expect, if macroevolution, changing from one cell to larger animals, were true, you would expect in the lower rock levels, the Cambrian levels in this case, you would have simple organisms, and that … as you would increase layer upon layer … of rock, you would expect to see more complex organisms … until you came to more present day fossils. The truth of the matter is just the reverse, almost upside down. In the Cambrian layer, in fact they have a name called the Cambrian explosion … there's many many fully formed complex creatures already on the … very low levels of the rock layers. And there's very little change throughout the rock layers. … There's kind of an observed stasis of animals, as they progress up the layers of rock. And so the whole idea of change over time,

> starting with simple organisms going to complex, is not borne out at all by the fossil record.[5]

These are his own words, transcribed verbatim from his Discovery Institute interview. Mr. LeVake is apparently a nice guy and generally popular teacher, and coverage of the 1999 court case was at pains to make this clear.[6] However, in this interview, he demonstrates a profound ignorance of biological history. He may as well insist that the moon is made of cheese as maintain that since the Cambrian, "there is very little change throughout the rock layers." In Chapters 4 through 10 we discussed at length how the ever-expanding body of data from stratigraphy, development, and molecular biology point toward a similar pattern of common descent. In the case of vertebrates, we discussed how the main branching points of the Tree of Life, as recognized on the basis of anatomy, development, and molecular biology, correspond very closely to the ages of relevant groups in the fossil record (Figure 4.1). There are no rabbits in the Devonian, and there is a genuine, stratigraphic signal about the relative ages of ancestral and descendent groups that has been recognized by scientists and laypeople alike for over 150 years. The evidence for this signal has gotten much better, not worse, since the nineteenth century.

Based on his above statement, Mr. LeVake is manifestly unaware of this pattern.[7] As such, he is simply unqualified to do the job asked of him by his school district—teach evolution. So was his school district "censoring" his views, or exercising quality control? I hope you'll agree with me that their decision was a legitimate case of the latter. Most high-school teachers, and apparently Mr. LeVake in his other classes,[8] know what they're talking about. But just as it would be unreasonable for me to sue my educational institution for refusing to let me teach ancient Greek (because I know nothing about it), it is unreasonable for Mr. LeVake to sue his school district for not letting him teach evolution. He doesn't know what he's talking about.

It's important and good that a legal recourse exists for parents, teachers, and students to scrutinize possible instances of censorship and/or intolerance in public education. In the case of LeVake *v.* Independent School District #656, the system worked. By reassigning Rodney LeVake away from teaching evolution, his school district upheld its obligation to provide quality instruction based on a state-mandated curriculum. Its students receive a better education in biology because of this decision. Rodney LeVake was not unfairly persecuted; his case bears little resemblance to that of Nikolay Vavilov, an internationally acclaimed geneticist who really was persecuted, and eventually murdered, because of his adherence to scientific principle over political dogma in Stalinist Russia. The democracies in which I've lived have not, thank goodness, descended to the depths of academic intolerance seen in Lysenko's Russia.

Although I find no merit in Rodney LeVake's lawsuit about persecution by his school district, I am sympathetic to the motivation that is likely at least in part behind his hostility to evolution. He probably shares his anti-Darwin sentiment with William Jennings Bryan, the prosecuting attorney in the 1925 Scopes trial, as discussed in the Prologue. Both fear that a recognition of the animal and material nature of humanity, as implied by evolution by natural selection, is caustic to morality. If we are animals, the reasoning goes, then no one has any reason to behave otherwise. "People have a soul; you can't put them on the same level as animals. To believe in evolution would mean that death is the last word."[9] This is a conviction expressed by those hostile to Darwin. No wonder so many have tried for so long to keep such an apparently threatening force out of their children's biology classes.

The anti-evolutionary zeal of William Jennings Bryan and many of his successors is misdirected, but well-meaning. His was an ultimately noble sentiment, one that affirms a broad responsibility of humans to future generations of our own and other species, based on the conviction that life has reason and purpose derived from a loving God. To the extent that you believe that a certain way of thinking would lead the mass of humanity to deny our responsibility to be moral, to care for one another, and be stewards of the Earth for our grandchildren, then of course you'd want to reject that way of thinking.[10] One of the challenges for biology educators today is to make clear that awareness of biological evolution, and our connection with the natural world, does not lead to a denial of this responsibility, whether in a religious or moral form. In the same way that awareness of gravity does not justify dropping bombs, awareness of evolution does not justify hedonism or selfishness, or lead to "death" in a real or spiritual sense. William Jennings Bryan and many in the anti-Darwin community have misunderstood this point and confused history for politics; they have confused "is" for "ought." One reason for this confusion is that Bryan and some of his well-meaning followers have been concerned with the great questions of morality and purpose in life, and they've mistakenly assumed Darwin was too. But the conclusions of *On the Origin of Species* were not at all a prescription about morality. What Darwin sought to explain was much more narrowly defined.

WHAT DID DARWIN EXPLAIN?

In his 2009 book, *Signature in the Cell*, ID advocate Stephen Meyer makes the case that human-like intelligence is the "best explanation of the origin of information necessary to build the first living cell."[11] He doesn't discuss the evolution of humans, frogs, flies, or any other animal, but he does refer frequently to Darwin, and acknowledges that "in the *Origin*, Darwin did not try to explain the origin of the first life." Meyer nevertheless goes on to impart a sense of failure on Darwin's part for not delving into the issue of life's origins: "Darwin had little more than vague speculations to offer as to how the first life on earth had begun" and Darwin's theory "assumed rather than explained the origin of the first living thing."[12] Consider also the blurb on the back cover of Meyer's book:

> Stephen Meyer has written the first comprehensive DNA-based argument for intelligent design. As he tells the story of successive attempts to unravel a mystery that Charles Darwin did not address—how did life begin?—Meyer develops the case for this often-misunderstood theory using the same scientific method that Darwin himself pioneered.

As a naturalist who never wrote more than a few sentences on the subject in the entirety of his published work, Charles Darwin seems an odd focus for a book on the origin of life. In order to find Darwin's guesses on how life may have started, one is obliged to comb through his personal correspondence,[13] letters which he never intended to be publicly scrutinized.

Should Darwin be cast as a failure for his lack of detail on life's origins? Of course not. Evolution by natural selection is not about the origin of life, but what happened after it first appeared. As we discussed in Chapter 10, recognizing that the scope of evolutionary biology does not include the study of life's origins is not a concession that the latter is impervious to a natural, causal explanation. I have little doubt that further progress will be made toward scientific theories of abiogenesis during the coming years. But if you want to bemoan the current uncertainty about how life began, you should criticize someone else besides Charles Darwin. He did not discuss black holes or atmospheric carbon in his writings, either. Yet no sensible person should view a theory of biological evolution for the worse because its author does not simultaneously come up with a theory for stellar origins or global warming. Darwin regarded life's origins as important and interesting, as any intellectual would both then and now; but this issue was independent of his main interest—explaining how life attained its current diversity after it began.

Awareness of the scope of a given problem is an integral feature of contemporary science. The most successful scientists today are those that focus on fairly specific topics, and are willing to say "I don't know" (at least temporarily) to many other issues. Darwin was fully aware of this point himself. Starting in the third edition of the *Origin*, Darwin explicitly stated that "it is no valid objection [to evolution by natural selection] that science as yet throws no light on the far higher problem of the essence or origin of life."[14] Similarly, regarding critiques of his theory by the German naturalist Heinrich Georg Bronn (the first to translate the *Origin* into German), Darwin remarked

> [Bronn] ... seems to think that till it can be shown how life arises, it is no good showing how the forms of life arise. This seems to me about as logical (comparing very great things with little) as to say it was no use in Newton showing laws of attraction of gravity & consequent movements of the Planets, because he could not show what the attraction of Gravity is.[15]

Casting Darwin as a failure for not explaining life's origins betrays a misunderstanding of evolutionary biology and a naïve view of science generally. The implication that, because evolution by natural selection is not about the origin of life, Darwin was a failure, is wrong and has no bearing on what his theory does explain: the appearance of biodiversity after life started.

CONCLUSIONS

I would like to bring this book to a close by drawing your attention to the distinction between agency and cause, concepts introduced in Chapter 1. These phenomena, under various names, have been discussed by philosophers for centuries. We previously noted that, in the thirteenth century, Thomas Aquinas wrote how ridiculous it would be to debate if fire itself produces heat, or if "god" does so but only in the presence of fire.[16] Does a carpenter hammer in a nail, or is it his tool doing the work? Aquinas made the point that to believe these forces are competing with and antagonistic to one another is absurd. I believe Charles Darwin understood this, but many in the contemporary polemic about evolution have not.

The notion that agency and cause in biology must compete with one another, with God as a tinkerer external to life, which is in turn foreign and even antagonistic to God, comes from a profoundly anthropomorphic way of thinking. This has often been recognized by past and current philosophers and theologians. Consider this excerpt from a recent lecture by the former Bishop of Durham (England), N.T. Wright:

> All this business about God intervening is not a language that I use, because that implies God is outside the process, and occasionally reaches in and stirs the pot and then goes away again. Whereas in the Psalms ... God is the one who "feeds the ravens when they call upon him." What does he throw them tidbits from heaven? No, it's a way of describing the fact that God is active within the world.[17]

Wright is saying that in this passage from Psalms,[18] it is not a toga-clad, Charlton Heston-like figure reaching down into His creation from beyond and scattering magical bread crumbs. Rather, the natural processes of biology, behind which God himself is the author, answers the ravens' call.

Archbishop of Vienna Cardinal Christophe Schönborn, once justifiably criticized for an opinion piece he wrote in 2005 for the *New York Times*,[19] made a far more insightful observation along these lines in a February 2008 lecture at the Dominican School of Philosophy and Theology in San Francisco:

> I think the main mistake, which is perhaps behind the whole debate, also around intelligent design, is the understanding of an extrinsic intervention of the creator, as if he were a foreigner in his world, in his creation, that he had to intervene like a [mechanic] who repairs a car, who is exterior to the car. But God is not exterior to his creation; this creation is not exterior to him.[20]

Without endorsing or committing ourselves to any specific cause, it is rational to believe that an entity beyond our comprehension was the agency by which something was derived from nothing at the beginning of time. But it is not rational to assume a human mode of operation for that entity, or to assume that "He" performs miracles among us like a magician conjures tricks. The point made above—that God is not exterior to our cosmos, that he is not to our world as a mechanic is to a car—means that the actions of natural law themselves comprise "His" activity as we perceive it.[21] Relatedly, it means that the term "miracle" is simply a placeholder for our own ignorance.

Some call this perspective "deism," implying that it's not really God doing the work but only the forces of nature that he created long ago, and abandoned, to carry on under their own inertia. But such a perspective also unjustifiably assumes an antagonism between God and nature, and entails the baseless assumption that if God is "active," this must somehow resemble human activity. Yet who are we to claim that any natural process is bereft of divinity, or that it cannot be intimately connected to a creator, one who is every bit as active as the gods of theism? People tend to assume that their god would act like a human, one who apparently has no use for processes we would call "natural." Again, this is unjustified. As quoted in Chapter 1, Darwin wrote in the *Origin of Species* "Have we any right to assume that the Creator works by intellectual powers like those of man?"[22] In our vanity it is very difficult for us to escape the assumption that a deity must act with human-like regard for time and effort, since we don't really have any other model to go by. Nevertheless, given the clear magnificence of the cosmos and all of its stunning natural depth—which is obviously beyond the capacity of anything remotely human—it is entirely rational to try and move beyond this anthropocentric view of God's creativity.

As now understood using data from embryology and development, the succession of life in the fossil record, and the molecular basis of heredity, the mechanisms that drive biology through time do not indicate or require the existence of a human-like deity. Yet our awareness of these mechanisms cannot be used to rule out the possibility of a deity because our understanding of creative expression is entirely limited to one kind of intelligence—our own. We do not know if or how some other "intelligence" might be behind the material processes that we find in nature. The scriptures of Abrahamic religions can be anthropomorphic, but even they do not completely shackle God with such a limited, human-like capacity to create. On a practical note, the good news is that human understanding of natural cause is possible whatever you think about the nature of God, whether or not you believe any agency at all is behind the cosmos. As long as a given religion does not insist on a scripturally literal, anthropocentric, or otherwise superstitious form of natural action, whether it's stars perched upon a metal firmament, a geologically young Earth, or a god who wills biology into existence without a mechanism, and as long as science recognizes its inherent limitations within the domain of human perception and rationality, this view of God as the agency underpinning natural cause means that science and religion are compatible.

Scientific rationality applied to the past, for example in evolutionary biology, does indeed contradict superstition and literal belief in the scriptures of many religions. Unfortunately, some conclude from this incompatibility that atheism inexorably follows from science. As we've just noted, it doesn't because most science, including evolution, is about cause, and the core of most religion concerns the potential of agency behind such causes. The false association of evolutionary biology with atheism has led to a growing interest across the literate world in fringe misinterpretations of this particular historical science. It is my hope that the information presented in the preceding chapters will help you to better recognize these misinterpretations, along with the factual basis of evolution, and the extent to which evolutionary biology is about explaining natural cause, not divine agency.

Biological evolution, as outlined by Darwin and Wallace in the mid-nineteenth century and elaborated upon and tested by thousands of scientists worldwide ever since, is not about the origin of life or the existence of God. It has little to do with either topic, and those who present Darwinism as an alternative to religion are wrong. With evolutionary biology emancipated from its misappropriated religious implications (Chapter 1), it is easier to see how the evidence in favor of evolution by natural selection as a biological cause is remarkably abundant (Chapter 2). We have looked at numerous, specific examples of how certain living organisms exhibit intermediate morphologies between other modern groups, like the tarsier, bandicoot, platypus, and coqui frog (Chapter 3). We examined in some detail paleontological evidence (Chapter 4) documenting the evolution of mammalian hearing bones and the mosaic appearance of features now found in all mammals (Chapter 5). You've seen a basic outline of the fossil record of elephants (Chapter 6), terrestrial and baleen whales (Chapter 7), and you know that these groups are only a small part of the very diverse fossil record of vertebrates that document how extinct life has been constrained by the same natural mechanisms of development and morphology as extant life (Chapter 8 and Table 8.1).

You know that the basic outline of the vertebrate evolutionary tree, first made on the basis of comparative anatomy and development in the nineteenth century, has been emphatically confirmed not only by the sequential appearance of major vertebrate groups in the fossil record (Chapter 4), but also of the similarities present in DNA sequences of living organisms (Chapter 9). And you have read about examples of natural molecular change, such as point mutations, gene duplication, and microsatellite variation, that can play a causal role in the generation of such animal attributes as color vision, skeletal shape, and hearing (Chapters 10 and 11; Table 10.1). You've read that evolution by natural selection is not a random process, but a biased one that can lead to biological complexity by building on natural variation over time (Chapter 11). Finally, you know that biological evolution is not a process constrained by what humans imagine as "probable" or not. Function of an anatomical or molecular structure does not have to be conserved over the course of evolution, and indeed the co-option of pre-existing structures for novel functions, or exaptation, plays a major role in the natural evolution of complexity (Chapter 11).

I have made the case in this book that there is ample evidence for the mechanism of natural selection as a major, driving force behind the biodiversity of life that exists on this planet today. If you approach this topic with awareness of what evolutionary biology seeks to explain, an understanding of how explanations of agency and cause are independent and not mutually exclusive, an appreciation of some aspects of paleontology, development, comparative anatomy, and molecular biology, and an open mind, I hope you'll agree with me. Darwinian evolution really does explain quite a bit about life on our planet. Those who remain skeptical about this extraordinarily well-supported theory need to spend more time grappling with one or more of the above issues. Perhaps most of all, they need to realize that understanding humanity as a natural part of life on Earth does not dictate how we reach our value judgments about ourselves, our planet, or our future. Knowing what *has* happened does not dictate what *should* happen.

An understanding of biological cause neither supports nor denies the existence of a deity behind life. Speaking for myself, I accept this deity based on my own intuition, and I make no pretension that this intuition has any scientific basis. However, although I acknowledge my belief to be non-scientific, it is entirely rational. Science is a subset of rationality; the former has a narrower scope than the latter. To ignore rationality when it does fall beyond the scientific enterprise would be an injustice to both reason and humanity. Long before any human ever recognized evolution or anything else, something out there expressed itself in the matter that now surrounds us. We now understand that in our universe, 2 + 2 = 4, planets have a predictable gravitational pull, and DNA is the language of life on Earth. Humanity has the extraordinary potential, and hope, of understanding some of this universe. This fact brings me to my knees every time.

NOTES

NOTES TO PROLOGUE (Pages xiv to xxii)

1. Padian, K. 2009. Truth or consequences? Engaging the "truth" of evolution. *PLoS Biology*, 7(3): e1000077.
2. Creationism in its current incarnation as a pseudo-scientific enterprise started in 1961 with *The Genesis Flood* by John Whitcombe and Henry Morris. In the century prior to 1961, the only author to try to make a scientific case for a geologically young Earth, compatible with a literal reading of Genesis, was the Canadian Seventh-Day-Adventist George McCready Price (see Numbers, R. 2006. *The Creationists*, Cambridge, MA: Harvard University Press, chapter 5).
3. Hunter, G.W. 1914. *A Civic Biology*. New York: American Book Co., p. 263.
4. W.J. Bryan led the prosecution of John Scopes for violating the Tennessee Butler Act, the 1925 law forbidding the teaching of evolution in that state. Bryan's team won the trial, and anti-evolution statutes such as the Butler Act remained on the books in some US states until the 1960s.
5. Numbers, 2006, *The Creationists*, pp. 59–60.
6. Gould, S.J. 1991. William Jennings Bryan's last campaign, *Bully for Brontosaurus: Reflections in Natural History*, New York: Norton, ch. 28.
7. Weiss, K.M. 2007. The Scopes trial. *Evolutionary Anthropology*, 16: 126–31.
8. Using the methods and principles of evolutionary biology, science has established, for example, that caucasian, non-African populations are more closely related to some African groups than many African groups are to one another. See, for example, Betti, L. *et al.* 2009. Distance from Africa, not climate, explains within-population phenotypic diversity in humans. *Proceedings of the Royal Society B:*

Biological Sciences, 276(1658), 809–14; Manica, A. *et al.* 2007. The effect of ancient population bottlenecks on human phenotypic variation. *Nature*, 448: 346–8.

9. See, for example, Book of Mormon, Ether 9: 19, 1 Nephi 4: 9, 3 Nephi 6: 1, Alma 18: 12.
10. Weinstock, J. *et al.* 2005. Evolution, systematics, and phylogeography of Pleistocene horses in the new world: a molecular perspective. *PLoS Biology*, 3(8): e241.
11. Luís, C. *et al.* 2006. Iberian origins of New World horse breeds. *Quaternary Science Reviews*, 97(2): 107–13.
12. Pool, C.A. 2007. *Olmec Archaeology and Early Mesoamerica*. Cambridge and New York: Cambridge University Press. Some Mesoamerican cultures did use wheels for decoration or toys, but not in transport.
13. http://www.fairlds.org/Book_of_Mormon/AshHorse (accessed May 23, 2009).
14. Coyne, J. 2009. *Why Evolution is True*. Oxford and New York: Oxford University Press.
15. Padian, 2009, Truth or consequences?
16. Coyne, J. 2009. "Seeing and believing: the never-ending attempt to reconcile science and religion, and why it is doomed to fail." *The New Republic* February 4, 2009.
17. http://www.buffalocurse.com (accessed April 28, 2010).
18. This part of the chapter was written before April 2010, when the Sabres made it to the playoffs, but lost to Boston in the first round. We lost again in the first round versus Philadelphia in 2011. Sigh.
19. See http://www.nytimes.com/2008/04/14/sports/baseball/14jersey.html (accessed November 26, 2010). The jersey was then put on display in Boston to benefit a Red Sox charity in support of cancer research. An apparently irrational act thus had a happy ending, and may provide a surprisingly profound metaphor for how "religious" motivation can ultimately benefit society.
20. Tanner lecture, November 2003, Harvard University. Available at: http://richarddawkins.net/article,2783,The-Science-of-Religion-and-the-Religion-of-Science,Richard-Dawkins-Steven-Pinker-Harvard-University (accessed November 7, 2009).
21. Krauss, L. and Dawkins, R. 2007. Should science speak to faith? *Scientific American*, 297(1): 88–91.

NOTES TO CHAPTER 1 (Pages 2 to 26)

1. See Miller, K.R. 1999. *Finding Darwin's God: A Scientist's Search for Common Ground between God and Evolution*, New York: Cliff Street Books, p.189; Schneider, R.J. 2005. "Essay V: evolution for Christians." http://community.berea.edu/scienceandfaith/essay05.asp (accessed August 2, 2009).
2. Darwin, C. 1859. *On the Origin of Species*, London: John Murray.
3. Gray, A. 1876. *Darwiniana: Essays and Reviews Pertaining to Darwinism*, New York: D. Appleton, p.36.

4. For example, Darwin, C. 1860. *On the Origin of Species by Means of Natural Selection* (second edition), London: J. Murray, p.481.
5. See Sober, E. 2008. *Evidence and Evolution*, Cambridge and New York: Cambridge University Press, pp.187–8.
6. As further support of this interpretation, Darwin noted in an 1860 letter to Jeffries Wyman how much he appreciated the interpretations of the *Origin* by theistic evolutionist Asa Gray: "No other person understands me so thoroughly as Asa Gray. If I ever doubt what I mean myself, I think I shall ask him!" (quoted in Weber, B.H. 2011. Design and its discontents. *Synthese*, 178: 279).
7. Aquinas, T. *ca.* 1260. *Summa Contra Gentiles*, book 3, chapters 69–70, translation by V.J. Bourke (Aquinas, T. and V.J. Bourke. 1957. *Summa Contra Gentiles*, New York: Hanover House. Available at http://www.op-stjoseph.org/Students/study/thomas/ContraGentiles3a.htm (accessed December 27, 2009)).
8. See also Phipps, W.E. 1983. Darwin, the scientific creationist. *Christian Century*, 1983, 809–11.
9. Miller, K.R. 1999. *Finding Darwin's God*, pp.255–6.
10. Owen, R. 2009. Vatican says evolution does not prove the non-existence of God. *Times Online*, March 6, 2009. Available at: http://www.timesonline.co.uk/tol/comment/faith/article5859797.ece (accessed November 7, 2009).
11. Phipps, 1983. Darwin, the scientific creationist; Carroll, W.E. 2000. Creation, evolution, and Thomas Aquinas. *Revue des Questions Scientifiques*, 171(4), 319–47; Burrell, D. 1993. *Freedom and Causation in Three Traditions*, Notre Dame, IN: University of Notre Dame Press.
12. Mitchell, C.E. 2009. It's not about the evidence: the role of metaphysics in the debate, in Schneidermann, J.S. and Allmon, W.D. (eds.) *For the Rock Record*. Berkeley, CA: University of California Press, pp.93–116; Miller, K.B. 2009. The misguided attack on methodological naturalism, in Schneidermann, J.S. and Allmon, W.D. (eds.), *For the Rock Record*. Berkeley, CA: University of California Press, pp.117–40.
13. Moreland, J.P. and Reynolds, J.M. 1999. *Three Views on Creation and Evolution*, Grand Rapids, MI: Zondervan, pp.31–2.
14. Gould, S.J. 1992. Impeaching a self-appointed judge. *Scientific American*, 267(1): 118–21.
15. Johnson, P. 1991. *Darwin on Trial*, Westmont, IL: Intervarsity Press.
16. Gould, 1992, Impeaching a self-appointed judge.
17. http://www.arn.org/docs/orpages/or151/151johngould.htm (accessed July 25, 2005).
18. Early inspirations for the ID movement include Dean Kenyon and Charles Thaxton; see Dembski, W. 1998. The Intelligent Design Movement, originally published in *Cosmic Pursuit*. Available at: http://www.leaderu.com/offices/dembski/docs/bd-idesign.html (accessed November 7, 2009).
19. Forrest, B. and Gross, P.R. 2007. *Creationism's Trojan Horse: The Wedge of Intelligent Design*, Oxford and New York: Oxford University Press.
20. Behe, M.J. 1996. *Darwin's* Black Box: *The Biochemical Challenge to Evolution*, New York: Free Press; Behe, M.J. 2007. *The Edge of Evolution: The Search for the Limits of Darwinism,* New York: Free Press.

21. See, for example, chapter 3 or Appendix 2 of Dawkins, R. 1986. *The Blind Watchmaker*, New York: W.W. Norton; or Dawkins, R. 1996. *Climbing Mount Improbable*, New York: W.W. Norton; also see Sober, E. 2008. *Evidence and Evolution: The Logic Behind the Science*, Cambridge and New York: Cambridge University Press, p. 125.
22. Gould, S.J. 1977. *Ever Since Darwin: Reflections in Natural History*, New York: W.W. Norton, p. 126.
23. Gould, S.J. 1982. In praise of Charles Darwin. *Discover*, 3(2): 20–5.
24. Wieland, C. 1992. Darwin's real message: have you missed it? *Creation*, 14(4): 16–19.
25. Gould, 1992, Impeaching a self-appointed judge.
26. Simpson, G.G. 1967. *The Meaning of Evolution*, New Haven, CT: Yale University Press, pp. 344–5.
27. Simpson, G.G. 1961. 100 years without Darwin are enough. *Teachers College Record*, 60(1961): 617–26.
28. The "blind watchmaker thesis" from Richard Dawkins' 1986 book of the same name implies an atheistic portrayal of Darwinian evolution, and is an example of an unjustified negation of agency due to the understanding of cause. Inference of natural selection as a mechanism behind biological evolution neither supports nor denies a potential agency behind it.
29. http://www.arn.org/docs/orpages/or151/151johngould.htm (accessed July 25, 2005).
30. Dawkins, 1986, *The Blind Watchmaker*.
31. Another example is an interview given by Discovery Institute fellow and ID advocate David Berlinski on the Christian talk-show program "Stand to Reason" on October 18, 2009; for example, consider the segment about 20 minutes from the end of the podcast, available at http://www.strcast2.org/podcast/weekly/101809.mp3 (accessed November 14, 2009).
32. See Miller, 1999, *Finding Darwin's God*, p. 189: "The giddy irony of this situation is that intellectual opposites like [creationists and materialist atheists] actually find themselves in a symbiotic relationship – each insisting vigorously that evolution implies an absolute materialism that is *not* compatible with religion. ... Each validates the most extreme viewpoints of the other."
33. Sober, 2008, *Evidence and Evolution*, p. 148.
34. Sober, 2008, *Evidence and Evolution*, p. 143. See also the historical discussion by Weber (2011, Design and its discontents: 279), including acknowledgment of a similar point made by Louis Ezra Hicks in 1883, who "warned of the ambiguity of conflating design, as instantiated in humanly created contrivances, with intent or purpose resulting from the action of natural law. He saw defenders of design unjustifiably gliding from the one usage to the other in many of their arguments. ... Further, Hicks claimed that many of the arguments presupposed their conclusion in their major premise (design supposes a designer). He agreed with Whewell's argument in *The Philosophy of the Inductive Sciences* that a final cause is not deduced from the phenomena associated with 'organised bodies' but rather is assumed."

35. Sokal, R.R. and Rohlf, F.J. 1995. *Biometry: The Principles and Practice of Statistics in Biological Research*, New York: W.H. Freeman.
36. Darwin, 1859, *On the Origin of Species*, p.188
37. Incidentally, Francis Collins in *The Language of God* (2007, p.166) quoted the same text and left this part out too.
38. While the extent of such persecution is claimed to be substantial (e.g., in the 2008 film *Expelled* with Ben Stein), actual cases seem to have been substantially exaggerated (see www.expelledexposed.com, accessed December 19, 2010). One such claim is discussed further in Chapter 12, in the section "Academic freedom."
39. Smith, L. and Henderson, M. 2008. "Royal Society's Michael Reiss resigns over creationism row," *The Times*, September 18.
40. Miller, 2009, The misguided attack on methodological naturalism.
41. Miller, 2009, The misguided attack on methodological naturalism; Scott, E.C. 2009. *Evolution vs. Creationism: An Introduction,* Berkeley, CA: University of California Press.
42. DeVries, P. 1986. Naturalism in the natural sciences: a Christian perspective. *Christian Scholars Review*, 15: 388–9. Cited in Miller, 2009, The misguided attack on methodological naturalism.
43. Aquinas, *ca.* 1260, *Summa Contra Gentiles*; Numbers, R.L. 2003. Science without God: natural laws and Christian beliefs, in Lindberg, D.C. and Numbers, R.L. (eds.), *When Science and Christianity Meet*, Chicago, IL: University of Chicago Press, pp.265–85.
44. Pennock, R. 2007. God of the gaps: the argument from ignorance and the limits of methodological naturalism, in Petto, A.J. and Godfrey, L.R. (eds.), *Scientists Confront Intelligent Design and Creationism*, New York: W.W. Norton, pp.309–38.
45. Interestingly, Darwin seemed to regret this theistic invocation in an 1863 letter to J.D. Hooker (as quoted in Sober, 2008, *Evidence and Evolution*, p.111). This is in spite of the fact that in the later editions of the *Origin*, three of which postdated this 1863 letter, he kept the theistic implication of his final sentence by retaining the term "Creator."
46. Miller, 2009, The misguided attack on methodological naturalism.
47. "On the one hand, having a natural explanation for something does not logically require denying that God exists or plays a role in bringing forth and sustaining the cosmos. ... Methodological naturalism requires that science be practiced as if it is without a priori metaphysical commitments, just methodological ones, so that the same science can be done and debated by theists, agnostics, and atheists. Philosophers and theologians, on the other hand, can and should consider the metaphysical implications of the science and engage in dialogue with scientists." (Weber, 2011, Design and its discontents: 285).
48. Plantinga, A. 2006. On Christian scholarship. Audio from http://www.redeemingreason.org/archives/audio/aplantinga_2006.MP3 (accessed May 13, 2010) (quote with *ca.* 56 minutes remaining).
49. Dawkins, R. 2006. *The God Delusion*, Boston, MA: Bantam Press.

50. Johnson, 1991, *Darwin on Trial.*
51. Schneiderman, J.S. and Allmon, W.D. (eds.) 2009. *For the Rock Record: Geologists on Intelligent Design*, Berkeley, CA: University of California Press, pp. 187–9.
52. Phipps, 1983, Darwin, the scientific creationist.
53. Owen, 2009, Vatican says evolution does not prove the non-existence of God.
54. Reuters, 1988. Pope believes contraception unacceptable, even With AIDS. November 12. Available at: http://www.aegis.com/news/ads/1988/AD880025.html (accessed January 7, 2010).
55. http://news.bbc.co.uk/1/hi/world/europe/7950671.stm (accessed April 9, 2009). In November 2010, 30 years after the beginning of the human AIDS epidemic and millions of deaths later, a Roman Pope seems to be partially coming around to the idea of condoning (but not advocating) condom use to prevent the spread of infection (see http://www.bbc.co.uk/news/world-europe-11821422, accessed November 23, 2010).
56. Hearst, N. and Chen, S. 2004. Condom promotion for AIDS prevention in the developing world: is it working? *Studies in Family Planning*, 35(1), 39–47.
57. Rosenbaum, J.E. 2009. Patient teenagers? A comparison of the sexual behavior of virginity pledgers and matched nonpledgers. *Pediatrics*, 123: e110–20.
58. Hitchens, C. 2007. *God is Not Great: How Religion Poisons Everything*, New York: Twelve.
59. As summarized regarding the penal system in the Islamic Republic of Iran; see http://www.iran-bulletin.org/political_islam/punishmnt.html (accessed August 19, 2010).
60. CNN and Mohammed Jamjoom, 2009. Top Saudi cleric: OK to wed young girls. January 17. Available at: http://edition.cnn.com/2009/WORLD/meast/01/17/saudi.child.marriage.
61. "We believe that the land of Israel was given to us Jews by the Lord. Eighty percent of Israelis are Jewish..." Ariel Atias, Israeli Minister of Housing. BBC, July 24, 2009. Available at http://news.bbc.co.uk/2/hi/middle_east/8164755.stm.
62. Dobson, J. 2008. A look at modern day Israel. May 15. Available at: http://listen.family.org/daily/A000001195.cfm.
63. CNN, 2001. Falwell apologizes to gays, feminists, lesbians. September 14. Available at http://articles.cnn.com/2001-09-14/us/Falwell.apology_1_thomas-road-baptist-church-jerry-falwell-feminists, (accessed 26 December 2011).
64. http://www.cornwallalliance.org/articles/read/an-evangelical-declaration-on-global-warming (accessed September 5, 2010).
65. http://www.arn.org/docs/orpages/or151/151johngould.htm (accessed July 25, 2005).
66. Kant, I. 1788. *Kritik der praktischen Vernunft*. JF Hartknoch: reprinted 1984 Erlangen.
67. Warraq, I, 2002. Virgins? What virgins?. *The Guardian*, January 12. Available at http://www.guardian.co.uk/books/2002/jan/12/books.guardianreview5 (accessed December 31, 2009).

68. Alexander, D. 2008. *Creation or Evolution: Do We Have to Choose?*, Oxford: Monarch Books, pp. 319–20. Also see Morris, T. and Petcher, D. 2006. *Science and Grace*, Wheaton, IL: Crossway Books, p. 146.
69. Shanks, N. and Green, K. 2011. Intelligent design in theological perspective. *Synthese*, 178: 315.
70. Alexander, 2008, *Creation or Evolution*, p. 164.
71. Genesis 2: 19, New International Version English Bible (italics added).
72. I first heard reference to this verse in a speech by Lawrence Krauss during a demonstration against the *Answers in Genesis* "museum" of creation: http://www.youtube.com/watch?v=LkrJsvxZyYo (accessed August 17, 2009).
73. Genesis 1: 7–14.
74. Exodus 14: 15–31.
75. Joshua 10: 12–13.
76. Lamoreux, D.O., undated. Evolutionary creationism. Available at: http://www.ualberta.ca/~dlamoure/3EvoCr.htm (accessed November 9, 2009).
77. For example, listen to *Guardian* columnist Madeline Bunting's answer when asked this question by Richard Dawkins, on the *Guardian* website. Available at: http://www.guardian.co.uk/commentisfree/audio/2008/feb/14/richard.dawkins.
78. Faraday Institute seminar, Cambridge: "Can a scientist believe the resurrection?" May 15, 2007. Available at: http://www.st-edmunds.cam.ac.uk/faraday/Multimedia.php.
79. Faraday Institute seminar, Cambridge: "The language of God: a believer looks at the human genome" September 15, 2007. Available at: http://www.st-edmunds.cam.ac.uk/faraday/Multimedia.php.
80. Van Biema, D., 2006. God vs. Science. *Time*. November 5. Available at: http://www.time.com/time/magazine/article/0,9171,1555132,00.html.
81. Faraday Institute seminar: "The language of God."
82. Faraday Institute seminar: "The language of God."
83. Roberts, M.D. 2007. *Can We Trust the Gospels? Investigating the Reliability of Matthew, Mark, Luke, and John*, Wheaton, IL: Crossway Books.
84. Radin, M. 1931. *The Trial of Jesus of Nazareth*, Chicago, IL: University of Chicago Press.
85. Roberts, 2007, *Can We Trust the Gospels?*
86. C.S. Lewis, chapter 2 in Hooper, W. 1994. *God in the Dock: Essays on Theology and Ethics by CS Lewis*, Grand Rapids: Wm B Eerdmans.
87. As quoted and discussed further in Chapter 12, Cardinal Christophe Schönborn, Archbishop of Vienna, expressed a similar sentiment in a February 2008 lecture at the Dominican School of Philosophy and Theology, San Francisco. Available at: http://www.bringyou.to/CardinalSchonbornChancePurpose.mp3 (accessed June 29, 2010).
88. Darwin, C., Letter to J. Fordyce, May 7, 1879. Available at: http://www.darwinproject.ac.uk/darwinletters/calendar/entry-12041.html (accessed November 7, 2009).

NOTES TO CHAPTER 2 (Pages 27 to 41)

1. Darwin, C. and Wallace, A.R. 1858. On the tendency of species to form varieties; and on the perpetuation of varieties and species by natural means of selection. *Proceedings of the Linnean Society, Zoology*, 3: 45–62.
2. Padian, K. 2008. Trickle-down evolution: an approach to getting major evolutionary adaptive changes into textbooks and curricula. *Integrative and Comparative Biology*, 48(2): 175–88.
3. Zalmout, I., *et al.* 2010. New Oligocene primate from Saudi Arabia and the divergence of apes and Old World monkeys. *Nature*, 466: 360–4.
4. Portrayed by Johnny Depp in the 2004 film, *The Libertine*.
5. Gould, S.J. 1987. *Time's Arrow, Time's Cycle: Myth and Metaphor in the Discovery of Geological Time*, Cambridge, MA: Harvard University Press.
6. See the critique of "flood geology" as applied to the Grand Canyon in Prothero, D. 2007. *Evolution: What the Fossils Say and Why it Matters*, New York: Columbia University Press, pp. 65–73.
7. Lambeck, K. 1980. *The Earth's Variable Rotation: Geophysical Causes and Consequences*, Cambridge and New York: Cambridge University Press.
8. Falkowski, P.G., *et al.* 2005. The rise of oxygen over the past 205 million years and the evolution of large placental mammals. *Science*, 309: 2202–4.
9. McElhinny, M.W. 1973. *Palaeomagnetism and Plate Tectonics*, Cambridge: Cambridge University Press.
10. Clack, J.A. 2002. *Gaining Ground: The Origin and Evolution of Tetrapods*, Bloomington, IN: Indiana University Press.
11. Denlinger, R.P. and O'Connell, D.R.H. 2010. Simulations of cataclysmic outburst floods from Pleistocene glacial lake Missoula. *Geological Society of America Bulletin*, 122(5–6): 678–89.
12. Pennock, R. 1999. *Tower of Babel: The Evidence Against the New Creationism*, Cambridge, MA: MIT Press.
13. For example, his interviews with "Focus on the Family" leader James Dobson on August 1, 2005 (available at http://www.discovery.org/a/2635, accessed February 25, 2008); on the Christian Radio talk show "Stand to Reason" on August 9, 2009 (available at http://www.str.org/site/PageServer?pagename=Radio_Archives, accessed September 21, 2009) or his talk at the Heritage Foundation on June 23, 2009 (available at http://apologetics315.blogspot.com/2009/06/signature-in-cell-mp3-audio-by-stephen.html, accessed January 24, 2010).
14. Paley, W. 1802. *Natural Theology*. 2009 reprint. Ann Arbor: University of Michigan Press.
15. See also chapter 6 in Alexander, D. and White, R.S., *Beyond Belief: Science, Faith and Ethical Challenges*, Oxford: Lion Books.
16. Plantinga, A. 2006. On Christian scholarship. Available at http://www.redeemingreason.org/archives/audio/aplantinga_2006.MP3 (accessed May 12, 2010).
17. "Stand to Reason" on August 9, 2009 or his talk at the Heritage Foundation on June 23, 2009.

18. "I just want to say one word to you, just one word. Are you listening? Plastics." from *The Graduate*, 1967. Directed by Mike Nichols. Embassy Pictures. For the relevant clip, see http://www.youtube.com/watch?v=DHGCvJjat1E (accessed May 16, 2010).
19. Felsenstein, J. 2007. Has natural selection been refuted? The arguments of William Dembski. *Reports of the National Center for Science Education*, 27(3–4): 20–6.
20. Negoro, S., *et al.* 2005. X-ray crystallographic analysis of 6-aminohexanoate-dimer hydrolase. *Journal of Biological Chemistry*, 280(47): 39644–52; Yomo, T., *et al.* 1992. No stop codons in the antisense strands of the genes for nylon oligomer degradation. *Proceedings of the National Academy of Sciences of the United States of America*, 89: 3780–4. See also http://www.talkorigins.org/origins/postmonth/apr04.html; Blount, Z.D., *et al.* 2008. Historical contingency and the evolution of a key innovation in an experimental population of *Escherichia coli*. *Proceedings of the National Academy of Sciences of the United States of America*, 105(23): 7899–906.
21. Wellik, D.M. and Capecchi, M.R. 2003. Hox10 and Hox11 genes are required to globally pattern the mammalian skeleton. *Science*, 301: 363–7.
22. Resnick, D.M., *et al.* 2002. Independent origins and rapid evolution of the placenta in the fish genus *Poeciliopsis*. *Science*, 298: 1018.
23. Tada, T., *et al.* 2009. Evolutionary replacement of UV vision by violet vision in fish. *Proceedings of the National Academy of Sciences of the United States of America*, 106(41): 17457–62.
24. Carroll, S.B. 2006. *The Making of the Fittest: DNA and the Ultimate Forensic Record of Evolution*, New York: W.W. Norton; Carroll, S.B. 2007. God as genetic engineer. *Science*, 316: 1427–8.
25. In his 1984 book, *The Limits of Science*, Sir Peter Medawar makes the legitimate point that no amount of such verification logically compels acceptance of any given hypothesis. This is even true (or perhaps especially true), when we recognize that such verification can come from unexpected sources. For example, "all swans are white" entails the observation that all non-white things are non-swans (but does not require the converse). Therefore, as Medawar pointed out, we could in principle stop by our local landfill, pick up a decaying black boot, and cry out with vindication "Eureka, more support for my hypothesis that all swans are white!" Nevertheless, and despite this rather silly (but true) example, evidence that verifies still qualifies as evidence. In the real world, we expect such evidence to be a bit more immediately relevant to the hypothesis in question than a black boot is to a white swan.
26. Quoted from "Answers in Genesis" website, April 25, 2009. Available at: http://www.answersingenesis.org/get-answers#/topic/noahs-ark/v/qa.
27. Genesis 6: 17, New International version.
28. http://www.answersingenesis.org/articles/nab/really-a-flood-and-ark (accessed April 25, 2009).
29. ID advocates such as the Discovery Institute's Paul Nelson (e.g., his 2007 debate with Jerry Coyne on Canadian radio; available at: http://www.tvo.org/podcasts/theagenda/audio/TAWSPScienceAndReligion032707.mp3, accessed March 30,

2009), misleadingly portray evolution as requiring a single ancestor for all life, but as anticipated by Darwin in the very last sentence of the *Origin*, this is not true. Genuine scientific debate exists regarding the role of processes that are not typically regarded as a part of natural selection (e.g., lateral gene transfer) and on possible independent origins of single-celled organisms from abiotic predecessors. This does not decrease the importance of natural selection to the evolution of the vast majority of Earth's protists, plants, and animals. See Doolitte, W.F. and Bapteste, E. 2007. Pattern pluralism and the Tree of Life hypothesis. *Proceedings of the National Academy of Sciences of the United States of America*, 104(7): 2043–9.

30. Davis. P.W., *et al.* 1993. *Of Pandas and People: The Central Question of Biological Origins*, Dallas, TX: Haughton Publishing Co.
31. Meyer, S.C., *et al.* 2007. *Explore Evolution: The Arguments For and Against Neo-Darwinism*, Melbourne: Hill House.
32. See the response by Behe to Scott's Not (just) in Kansas anymore, available at http://www.sciencemag.org/content/288/5467/813.full/reply#sci_el_165 (accessed December 20, 2010). Additionally, in November 2010 I attended a lecture given by ID advocate Michael Behe, in which he claimed that common descent and even natural selection itself were, for his purposes, "trivial" components of Darwin's idea. In Behe's view, the role of a human-like intelligence guiding the evolutionary process is critical. While Darwin did reject a simplistic notion of a human-like, interventionist God, he did not reject all potential forms of divine "intelligence" (Weber, B.H. 2011. Design and its discontents. *Synthese*, 178, 271–89), as discussed in Chapters 1 and 12 in this book. More importantly, Darwin recognized that in order to be scientific, his contribution had to provide a mechanism by which biodiversity has appeared. Even though he acknowledged a potential creator, he knew that the tough part of understanding evolution was this mechanism, or the explanation of cause – not an attribution of agency. Recall the text from the *Origin* quoted in Chapter 1 of this book: "It is so easy to hide our ignorance under such expressions as the 'plan of creation,' 'unity of design,' etc., and to think that we give an explanation when we only restate a fact" (Darwin, C. 1859. *On the Origin of Species by Means of Natural Selection*, London: J. Murray, pp.481–2).
33. Cooke, J.P. 1864. *Religion and Chemistry; Proofs of God's Plan in the Atmosphere and its Elements*, New York: Charles Scribner. See also discussion in Weber, 2011, Design and its discontents: 276.
34. Gould, 1987, *Time's Arrow, Time's Cycle;* Clack, 2002, *Gaining Ground;* Carroll, 2006, *The Making of the Fittest*; Coyne, J.A. 2009. *Why Evolution is True*, Oxford and New York: Oxford University Press; Shubin, N. 2008. *Your Inner Fish: A Journey into the 3.5-Billion-Year History of the Human Body,* New York: Pantheon Books; Prothero, 2007, *Evolution*; Miller, K.R. 1999. *Finding Darwin's God: A Scientist's Search for Common Ground between God and Evolution*, New York: Cliff Street Books; Dawkins, R. 2009. *The Greatest Show on Earth: The Evidence for Evolution*, New York: Free Press; Kirschner, M.W. and Gerhart, J.C. 2005. *The Plausibility of Life: Resolving Darwin's Dilemma*, New Haven, CT: Yale University Press.

NOTES TO CHAPTER 3 (Pages 42 to 62)

1. Padian, K. and Angielczyk, K.D. 2007. Transitional forms versus transitional features, in Petto, A.J. and Godfrey, L.R. (eds.) *Scientists Confront Intelligent Design and Creationism*, New York: W.W. Norton, pp. 206–7. Importantly, some groups are defined by paleontologists arbitrarily on the basis of certain characters. For example, as we'll discuss further in Chapter 5, "mammals" are defined as vertebrates possessing a squamosal-dentary joint. This kind of character-based definition is arbitrary, and does not change the point made here that categories of animals recognized today are difficult to apply to the fossil record because of the mosaic, piece-by-piece nature of evolution.
2. Darwin, C. 1859. *On the Origin of Species by Means of Natural Selection*, London: J. Murray, p. 281: "So with natural species, if we look to forms very distinct ... we have no reason to suppose that links ever existed directly intermediate between them, but between each and an unknown common parent."
3. Norell, M.A. and Xu, X. 2005. Feathered dinosaurs. *Annual Review of Earth and Planetary Sciences*, 33: 277–99.
4. Lockwood, C.A. 2007. *The Human Story*, London: The Natural History Museum.
5. Padian and Angielczyk, 2007, Transitional forms versus transitional features.
6. Padian and Angielczyk, 2007, Transitional forms versus transitional features.
7. Paraphrased from an anonymous online review of Zimmer's 2001 book, *Evolution: Triumph of an Idea* at http://www.amazon.com/Evolution-Triumph-Idea-Carl-Zimmer/product-reviews/0061138401 (accessed May 21, 2010).
8. Elinson, R.P. and Beckham, Y. 2002. Development in frogs with large eggs and the origin of amniotes. *Zoology*, 105: 105–17.
9. Bagarinao, T. 1986. Yolk resorption, onset of feeding and survival potential of larvae of three tropical marine fish species reared in the hatchery. *Marine Biology*, 91: 449–59.
10. Elinson and Beckham, 2002, Development in frogs.
11. Kekalainen, J., *et al.* 2010. Genetic and potential non-genetic benefits increase offspring fitness of polyandrous females in non-resource based mating system. *BMC Evolutionary Biology*, 10: 20.
12. Wake, D.B. and Hanken, J. 1996. Direct development in the lungless salamanders: what are the consequences for developmental biology, evolution and phylogenesis? *International Journal of Development Biology*, 40(4): 859–69.
13. Manger, P.R. and Pettigrew, J.D. 1995. Electroreception and the feeding behaviour of the platypus. *Philosophical Transactions of the Royal Society of London B*, 347: 359–81.
14. Long, J., *et al.* 2002. *Prehistoric Mammals of Australia and New Guinea*. Baltimore, MD: Johns Hopkins University Press.
15. Rose, D.B. and Davis, R. 2005. *Dislocating the Frontier: Essaying the Mystique of the Outback*, Canberra: ANU Press. Available at: http://epress.anu.edu.au/dtf/html (accessed May 23, 2010).
16. Rose and Davis, 2005, *Dislocating the Frontier*.

17. http://www.youtube.com/watch?v=KeYySCRnt7E (accessed May 23, 2010).
18. http://www.iucnredlist.org/apps/redlist/details/23179/0 (accessed May 23, 2010).
19. Benton, M.J., *et al.* 2009. Calibrating and constraining molecular clocks, in Hedges, S.B. and Kumar, S. (eds.) *The Timetree of Life*, Oxford: Oxford University Press, pp.35–86.
20. Nilsson, M., *et al.* 2004. Marsupial relationships and a timeline for marsupial radiation in South Gondwana. *Gene*, 340: 189–96.
21. Long *et al.*, 2002, *Prehistoric Mammals of Australia and New Guinea.*
22. Tyndale-Biscoe, H. 2005. *Life of Marsupials*, Collingwood: CSIRO publishing (see table 5.1 and p.174 for data on marsupial gestation and lactation).
23. Tyndale-Biscoe, 2005, *Life of Marsupials.*
24. Tyndale-Biscoe, 2005, *Life of Marsupials*
25. Tyndale-Biscoe, H. 2001. Australasian marsupials: to cherish and to hold. *Reproduction, Fertility and Development*, 13: 477–85.
26. Tyndale-Biscoe, 2001, Australasian marsupials.
27. Tyndale-Biscoe, 2001, Australasian marsupials.
28. Fleagle, J.G. 1999. *Primate Adaptation and Evolution*, San Diego, CA: Academic Press; Seiffert, E.R., *et al.* 2009. Convergent evolution of anthropoid-like adaptations in Eocene adapiform primates. *Nature*, 461: 1118–22; Schmitz, J., *et al.* 2001. SINE insertions in cladistic analyses and the phylogenetic affiliations of *Tarsius bancanus* to other primates. *Genetics*, 157: 777–84; Hällstrom, B.M. and Janke, A. 2010. Mammalian evolution may not be strictly bifurcating. *Molecular Biology and Evolution*, 27(12): 2804–16.
29. Fleagle, 1999, *Primate Adaptation and Evolution.*
30. Ankel-Simons, F. and Rassmussen, D.T. 2008. Diurnality, nocturnality, and the evolution of primate visual systems. *Yearbook of Physical Anthropology*, 51: 100–17; Heesy, C.P. 2009. Seeing in stereo: the ecology and evolution of primate binocular vision and stereopsis. *Evolutionary Anthropology*, 18: 21–35.

NOTES TO CHAPTER 4 (Pages 63 to 88)

1. Milne, Edwards H. 1844. *Considerations… Annales des sciences naturelles* (third series) 1: 65–99.
2. Huxley, T.H. 1863. *Man's Place in Nature*, New York: D. Appleton; Haeckel, E. 1866. *Generelle Morphologie der Organismen: Allgemeine Grundzüge der organischen Formen-Wissenschaft, mechanisch begründet durch die von Charles Darwin reformirte Descendenz-Theorie*, Berlin: G. Reimer; Panchen, A.L. 1992. *Classification, Evolution, and the Nature of Biology*, Cambridge and New York: Cambridge University Press, pp.50–1; Gill, T. 1872. Arrangement of the families of fishes. *Smithsonian Miscellaneous Collections*, 247: 1–45.
3. Desmond, A. 1982. *Archetypes and Ancestors: Palaeontology in Victorian London, 1850–1875*, Chicago, IL: University of Chicago Press, p.60.
4. Desmond, 1982, *Archetypes and Ancestors*, p.61. See also Darwin's remarks on p.xviii of the *Origin* (sixth edition) to the effect that Owen at times sought credit for promulgating some form of natural selection prior to 1859.

5. Gould, S.J. 1986. The archaeopteryx flap: cardboard histories can be deceptive and destructive. *Natural History*, 95(9): 16–25.
6. Darwin, C. and Wallace, A.R. 1858. On the tendency of species to form varieties; and on the perpetuation of varieties and species by natural means of selection. *Journal of the Proceedings of the Linnean Society (Zoology)*, 3: 46–62.
7. Padian, K. and Angielczyk, K.D. 2007. Transitional forms versus transitional features, in Petto, A.J. and Godfrey, L.R. (eds.), *Scientists Confront Creationism: Intelligent Design and Beyond*, New York: W.W. Norton, pp.197–231; Prothero, D. 2007. *Evolution: What the Fossils Say and Why It Matters*, New York: Columbia University Press; Zimmer, C. 2009. *The Tangled Bank: An Introduction to Evolution*, Greenwood Village, CO: Roberts and Co.; Gould, S.J. 1996. *Full House: The Spread of Excellence from Plato to Darwin*, New York: Harmony Books.
8. Haeckel, 1866, *Generelle Morphologie der Organismen.*
9. Gill, 1872, Arrangement of the families of fishes.
10. Huxley, 1863, *Man's Place in Nature*, pp.80–1.
11. Meyer, A. and Zardoya, R. 2003. Recent advances in the (molecular) phylogeny of vertebrates. *Annual Review of Ecology, Evolution, and Systematics*, 34: 311–38; Delsuc, F. *et al.* 2008. Additional molecular support for the new chordate phylogeny. *Genesis*, 46: 592–604; Hugall, A.F., *et al.* 2007. Calibration choice, rate smoothing, and the pattern of tetrapod diversification according to the long nuclear gene RAG-1. *Systematic Biology*, 56: 543–63.
12. For example, see Asher, R.J., *et al.* 2005. Stem Lagomorpha and the antiquity of Glires. *Science*, 307(5712): 1091–4.
13. Benton, M.J., *et al.* 2009. Calibrating and constraining molecular clocks, in Hedges, S.B. and Kumar, S. (eds.), *The Timetree of Life*, Oxford: Oxford University Press, pp.35–86.
14. Niedzwiedzki, G., *et al.*, 2010. Tetrapod trackways from the early Middle Devonian period of Poland. *Nature*, 463(7277): 43–8.
15. Daeschler, E.B., *et al.* 2005. A Devonian tetrapod-like fish and the evolution of the tetrapod body plan. *Nature*, 440: 757–63.
16. http://www.colbertnation.com/the-colbert-report-videos/69314/may-18–2006/ted-daeschler (accessed April 28, 2010).
17. Clack, J.A. 1997. Devonian tetrapod trackways and trackmakers: a review of the fossils and footprints. *Palaeogeography, Palaeoclimatology, Palaeoecology*, 130: 227–50.
18. Norell, M.A. and Novacek, M.J. 1992. The fossil record and evolution: comparing cladistic and paleontologic evidence for vertebrate history. *Science*, 255(5052): 1690.
19. Wyss, A.D. 1987. Notes on Proteutheria, Insectivora, and Thomas Huxley's contribution to mammalian systematics. *Journal of Mammalogy*, 68: 135–8.
20. Clack, J.A. 2002. *Gaining Ground: The Origin and Evolution of Tetrapods*, Bloomington, IN: Indiana University Press.
21. See also http://evolution.berkeley.edu/evolibrary/article/0_0_0/evo_48 (accessed May 11, 2010).

22. http://www.answersingenesis.org (accessed June 10, 2010).
23. A very accessible discussion of several well-documented cases of evolution-in-action is found in chapter 5 of Dawkins, R. 2009. *The Greatest Show on Earth*, London: Bantam Press; Endler, J.A. 1986. *Natural Selection in the Wild*, Princeton, NJ: Princeton University Press.
24. Ohno, S. 1984. Birth of a unique enzyme from an alternative reading frame of the preexisted, internally repetitious coding sequence. *Proceedings of the National Academy of Sciences of the United States of America*, 81(8): 2421–5; Okamura, K., *et al.* 2006. Frequent appearance of novel protein-coding sequences by frame-shift translation. *Genomics*, 88(6): 690–7.
25. Blount, Z.D., *et al.* 2008. Historical contingency and the evolution of a key innovation in an experimental population of *Escherichia coli*. *Proceedings of the National Academy of Sciences of the United States of America*, 105(23): 7899–906.
26. Farmer, M.A. and Habura, A. 2010. Using protistan examples to dispel the myths of intelligent design. *Journal of Eukaryotic Microbiology*, 57(1): 3–10.
27. Reznick, D.N., *et al.* 1997. Evaluation of the rate of evolution in natural populations of guppies (*Poecilia reticulata*). *Science*, 275: 1934–7.
28. Kitano, J., *et al.* 2008. Reverse evolution of armor plates in the threespine stickleback. *Current Biology*, 18: 769–74.
29. Pulido, F. and Berthold, P. 2010. Current selection for lower migratory activity will drive the evolution of residency in a migratory bird population. *Proceedings of the National Academy of Sciences of the United States of America*, 107(16): 7341–6.
30. Pergams, O.R.W. and Lawler, J.J. 2009. Recent and widespread rapid morphological change in rodents. *PLoS ONE*, 4(7): e6452.
31. Eldredge, N. and Gould, S.J. 1972. Punctuated equilibria: an alternative to phyletic gradualism, in Schopf, T.J.M. (ed.), *Models in Paleobiology*, San Francisco, CA: Freeman & Cooper, pp. 82–115.
32. Coyne, J. and Charlesworth, B. 1996. Mechanisms of punctuated equilibrium. *Science*, 274(5293): 1748–9; Reeve, H.K. and Sherman, P.W. 1993. Adaptation and the goals of evolutionary research. *Quarterly Review of Biology*, 68(1): 3.
33. Gingerich, P.D. 1976. Paleontology and phylogeny: patterns of evolution at the species level in early Tertiary mammals. *American Journal of Science*, 276: 1–28; Bookstein, F.L., *et al.* 1978. Hierarchical linear modeling of the tempo and mode of evolution. *Paleobiology*, 4(2): 120–34.
34. Coyne and Charlesworth, 1996, Mechanisms of punctuated equilibrium; Reeve and Sherman, 1993, Adaptation and the goals of evolutionary research; Gingerich, 1976, Paleontology and phylogeny.
35. Gould, S.J. 2002. *The Structure of Evolutionary Theory*. Cambridge, MA: Belknap Press of Harvard University Press.
36. As of this writing (March 2009) this memorable performance is still available on the internet: http://www.youtube.com/watch?v=vnMYL8sF7bQ.
37. The first three editions were published each year between 1859 and 1861, the fourth in 1866, the fifth in 1869, and the sixth in 1872.

38. "[Creationists] claim that natural selection has the same chance at producing complex adaptations that a hurricane blowing through a junkyard has of assembling scattered pieces of metal into a functioning airplane. This analogy is fundamentally misleading. ... Natural selection is a *biased* process, not a *random* process. ... The misleading analogy between natural selection and a hurricane blowing through a junkyard should be junked" (Sober, E. 2008. *Evidence and Evolution: The Logic Behind the Science*, Cambridge and New York: Cambridge University Press, pp. 122–5).
39. Carroll, S. 2005. *The Making of the Fittest*, New York: Norton.
40. Gould, S.J. 1977. *Ontogeny and Phylogeny*, Cambridge, MA: Belknap Press of Harvard University Press.
41. Dawkins, R. 1986. *The Blind Watchmaker*. New York: W.W. Norton.
42. Sober, 2008, *Evidence and Evolution*, pp. 122–5.
43. Gould, S.J. 1987. *Time's Arrow, Time's Cycle: Myth and Metaphor in the Discovery of Geological Time*, Cambridge, MA: Harvard University Press, pp. 2–3.
44. Gould, 2002, *The Structure of Evolutionary Theory*, p. 986.
45. Eldredge and Gould, 1972, Punctuated equilibria.
46. Gould, S.J. 1980. Is a new and general theory of evolution emerging? *Paleobiology*, 6: 120.
47. Coyne and Charlesworth, 1996, Mechanisms of punctuated equilibrium.
48. Fisher, R.A. 1930. *The Genetical Theory of Natural Selection*, Oxford: Clarendon Press.
49. Dobzhansky, T. 1937. *Genetics and the Origin of Species*, New York: Columbia University Press.
50. Simpson, G.G. 1944. *Tempo and Mode in Evolution*, New York: Columbia University Press.
51. Simpson, G.G. 1953. *The Major Features of Evolution*, New York: Simon & Schuster, pp. 259–65.
52. Mayr, E. 1942. *Systematics and the Origin of Species*, New York: Columbia University Press.
53. Huxley, J.S. 1942. *Evolution: The Modern Synthesis*, London: Allen & Unwin.
54. Gould, S.J. and Lewontin, R. 1979. The spandrels of San-Marco and the Panglossian paradigm: a critique of the adaptationist program. *Proceedings of the Royal Society of London Series B: Biological Sciences*, 205(1161): 581–98.
55. Twain, M. 1903. Was the world made for man? (from a podcast by *Scientific American*, June 6, 2007: Mark Twain: fossil hunter and science writer. Available at: http://www.scientificamerican.com/podcast/episode.cfm?id=013429DB-E7F2–99DF-341402C912A40D31 (accessed November 2, 2010)).
56. Coyne and Charlesworth, 1996, Mechanisms of punctuated equilibrium; Dawkins, 1986, *The Blind Watchmaker*.
57. Simpson, 1944, *Tempo and Mode in Evolution*, pp. 117–18.
58. In *Variation of Animals and Plants under Domestication*, Darwin explicitly acknowledges group selection in cases such as the sterile worker caste in social insects and human altruism, as discussed by Sober, E. 2009. Did Darwin write the *Origin* backwards? *Proceedings of the National Academy of Sciences of the United States of America*, 106, 10048–55.

59. Alberch, P., *et al.* 1979. Size and shape in ontogeny and phylogeny. *Paleobiology*, 5(3): 296–317.
60. Gould, 2002, *The Structure of Evolutionary Theory*, p. 1003.
61. Gould, 1980, Is a new and general theory of evolution emerging?
62. Gould, 2002, *The Structure of Evolutionary Theory*, p. 715.
63. Sears, K.E. 2004. Constraints on the morphological evolution of marsupial shoulder girdles. *Evolution*, 58(10): 2353–70.
64. Gould, 2002, *The Structure of Evolutionary Theory*.
65. Coyne and Charlesworth, 1996, Mechanisms of punctuated equilibrium; Reeve and Sherman, 1993, Adaptation and the goals of evolutionary research; Dawkins, 1986, *The Blind Watchmaker*.

NOTES TO CHAPTER 5 (Pages 90 to 110)

1. Zhou, Z. 2004. The origin and early evolution of birds: discoveries, disputes, and perspectives from fossil evidence. *Naturwissenschaften*, 91: 455–71.
2. Darwin, C. 1859. *On the Origin of Species by Means of Natural Selection*, London: John Murray, p. 296.
3. Kemp, T.S. 2005. *The Origin and Evolution of Mammals*, Oxford and New York: Oxford University Press.
4. Carroll, R.L. 1982. Early evolution of reptiles. *Annual Review of Ecology and Systematics*, 13: 87–109.
5. Traditionally, turtles have been considered to be a third major division within amniotes, called "Anapsida." However, consensus among biologists is growing that they actually belong within the diapsid radiation (see, for example, Meyer and Zardoya, 2003. Recent advances in the (molecular) phylogeny of vertebrates. *Annual Review of Ecology, Evolution, and Systematics*, 34: 311–38; Hugall *et al.* 2007. Calibration choice, rate smoothing, and the pattern of tetrapod diversification according to the long nuclear gene RAG-1. *Systematic Biology*, 56(4): 543–63).
6. Angielczyk, K.D. 2009. *Dimetrodon* is not a dinosaur: using tree thinking to understand the ancient relatives of mammals and their evolution. *Evolution: Education Outreach*, 2: 257–71.
7. Benton, M.J. 2003. *When Life Nearly Died: The Greatest Mass Extinction Event of All Time*, London: Thames & Hudson.
8. Kemp, 2005, *The Origin and Evolution of Mammals*.
9. Darwin, 1859, *On the Origin of Species*, p. 282.
10. Clack, J.A. 2002. *Gaining Ground: The Origin and Evolution of Tetrapods*, Bloomington, IN: Indiana University Press.
11. Angielczyk, 2009, *Dimetrodon* is not a dinosaur.
12. Kielan-Jaworowska, Z., *et al.* 2004. *Mammals from the Age of Dinosaurs*, New York: Columbia University Press.
13. Kemp, 2005, *The Origin and Evolution of Mammals*; Carroll, 1982, Early evolution of reptiles; Angielczyk, 2009, *Dimetrodon* is not a dinosaur; Kielan-Jaworowska *et al.*, 2004, *Mammals from the Age of Dinosaurs*; Luo, Z.-X. and

Crompton, A.W. 1994. Transformation of the quadrate (incus) through the transition from non-mammalian cynodonts to mammals. *Journal of Vertebrate Paleontology*, 14: 341–74; Allin, E.F. 1975. Evolution of the mammalian middle ear. *Journal of Morphology*, 147: 403–38; Kemp, T.S. 2007. The origin of higher taxa: macroevolutionary processes and the case of the mammals. *Acta Zoologica (Stockholm)*, 88(1): 3–22; Sidor, C. and Hopson, J.A. 1998. Ghost lineages and "mammalness": assessing the temporal pattern of character acquisition in the Synapsida. *Paleobiology*, 24(2): 254–73.

14. Interestingly, some extinct crocodiles from the Cretaceous of Tanzania possess a dentition similar to that of mammals, with multi-cuspid molars and stabbing canines. Yet these fossils are clearly recognizable as crocodyliforms, based on the presence of, for example, multiple bones in the jaw (see O'Connor, P.M., *et al.*, 2010. The evolution of mammal-like crocodyliforms in the Cretaceous Period of Gondwana. *Nature*, 466(7307): 748–51).
15. Kemp, 2005, *The Origin and Evolution of Mammals*, figure 4.3; Kemp, T.S. 1972. The jaw articulation and musculature of the whaitsiid Therocephalia, in Joysey, K.A. and Kemp, T.S. (eds.), *Studies in Vertebrate Evolution*, Edinburgh: Oliver & Boyd, pp. 213–30.
16. Kemp, 2005, *The Origin and Evolution of Mammals*.
17. Carroll, 1982, *Early Evolution of Reptiles*.
18. Reichert, K.B. 1837. Über die Visceralbogen der Wirbelthiere im allgemeinen und deren Metamorphosen bei den Vögeln und Säugethieren. *Archiv für Anatomie, Physiologie, und wissenschaftliche Medizin*, 1837, 120–220.
19. Reichert, 1837, Über die Visceralbogen der Wirbelthiere im allgemeinen, 178. Translation by RJA.
20. Sedgwick, 1894. On the law of development commonly known as von Baer's law; and on the significance of ancestral rudiments in embryonic development. *Quarterly Journal of Microscopic Science*, 36, 35–52; Hopwood, N. 2006. Pictures of evolution and charges of fraud: Ernst Haeckel's embryological illustrations. *Isis*, 97(2): 260–301.
21. Richards, R.J. 2008. *The Tragic Sense of Life: Ernst Haeckel and the Struggle over Evolutionary Thought*, Chicago, IL: University of Chicago Press, p. 65.
22. Richardson, M.K., *et al.* 1998. Haeckel, embryos and evolution. *Science*, 280: 983–6; Pickett, K.M., *et al.* 2005. Iconoclasts of evolution: Haeckel, Behe, Wells and the ontogeny of a fraud. *The American Biology Teacher*, 67(5): 275–82.
23. Sánchez-Villagra, M.R., *et al.* 2002. Ontogenetic and phylogenetic transformations of the ear ossicles in marsupial mammals. *Journal of Morphology*, 251: 219–38.
24. Gaupp, E. 1912. Die Reichertsche Theorie, Hammer-, Amboss- und Kieferfrage. *Archiv für Anatomie und Entwicklungsgeschichte*, 1913: 1–416.
25. Sánchez-Villagra *et al.*, 2002, Ontogenetic and phylogenetic transformations of the ear ossicles in marsupial mammals; Takechi, M. and Kuratani, S. 2010. History of studies on mammalian middle ear evolution: a comparative morphological and developmental biology perspective. *Journal of Experimental Zoology Part B: Molecular and Developmental Evolution*, 314(6): 417–33; Maier, W. 1987.

Der Processus angularis bei *Monodelphis domestica* (Didelphidae; Marsupialia) und seine Beziehungen zum Mittelohr: Eine ontogenetische und evolutionsmorphologische Untersuchung. *Gegenbaurs Morphologisches Jahrbuch*, 133: 123–61.

26. Kim, Young-Ok. 2000. *Karl Bogislaus Reichert (1811–1883) Sein Leben und seine Forschungen zur Anatomie und Entwicklungsgeschichte*, Medizinische Doktorarbeit, Universität Mainz.
27. Kim, 2000, *Karl Bogislaus Reichert.*
28. Kim, 2000, *Karl Bogislaus Reichert.*
29. "...fast allgemein wurde zu jener Zeit Darwin's Lehre für vollständig verfehlt angesehen." Preyer, William Thierry. 1891. Briefe von Darwin. *Mit Erinnerungen und Erlaeuterungen. Deutsche Rundschau*, 17(9): 357. Available at: http://darwin-online.org.uk/content/frameset?viewtype=text&itemID=F6&pageseq=1 (accessed April 4, 2010).
30. Kim, 2000, *Karl Bogislaus Reichert.*
31. See the footnote on p. xiv of Darwin's *Origin*, sixth edition, in which he notes the foreshadowing of Lamarck's work by his own grandfather, and on the interesting fact that Erasmus Darwin, Goethe, and the French naturalist Geoffroy St. Hilaire articulated an evolution-like view of nature independently around 1794–5.
32. Pickett *et al.*, 2005, Iconoclasts of evolution; Gould, S.J. 1977. *Ever since Darwin: Reflections in Natural History*, New York: W.W. Norton. For an explicit claim that, except for selection itself, Goethe foresaw "alle wesentlichen Lehren des heutigen Darwinismus" (all essential teachings of today's Darwinism), see p. 159 of Lubosch, W. 1919. Was verdankt die vergleichend-anatomische Wissenschaft den Arbeiten Goethes? *Jahrbuch der Goethe Gesellschaft*, 6: 157–91. For this reference I thank Dr. Christian Mitgutsch of the Paläontologisches Institut, Zürich, who also points out that many German academics at the time were eager to give credit to Goethe for popular ideas, such as Darwinism. Lubosch would therefore not have been an impartial observer of Goethe's contributions to biology.
33. Gaupp, 1912, Die Reichertsche Theorie, Hammer-, Amboss- und Kieferfrage; Sánchez-Villagra *et al.*, 2002, Ontogenetic and phylogenetic transformations of the ear ossicles in marsupial mammals; Gaupp, E. 1899. Ontogenese und Phylogenese des Schallleitenden Apparates bei den Wirbeltieren. *Ergebnisse der Anatomie und Entwicklungsgeschichte*, 8: 990–1149.
34. Luo, Z.-X. 2007. Transformation and diversification in early mammal evolution. *Nature*, 450: 1011–19.
35. Luo, Z.-X., *et al.* 2004. Evolution of dental replacement in mammals. *Bulletin of Carnegie Museum of Natural History*, 36: 159–75.
36. Luo *et al.*, 2004, Evolution of dental replacement in mammals.
37. Gingerich, P.D. 1976. Paleontology and phylogeny: patterns of evolution at species level in early Tertiary mammals. *American Journal of Science*, 276(1): 1–28.
38. Eldredge, N. and Gould, S.J. 1972. Punctuated equilibria: an alternative to phyletic gradualism, in Schopf, T.J.M. (ed.), *Models in Paleobiology*, San Francisco, CA: Freeman & Cooper, pp. 82–115.

39. Kemp, 2009, *The Origin and Evolution of Mammals*; Angielczyk, 2009, *Dimetrodon* is not a dinosaur; Clack, 2002, *Gaining Ground;* Kielan-Jaworowska *et al.*, 2004, *Mammals from the Age of Dinosaurs;* Takechi and Kuratani, 2010, History of studies on mammalian middle ear evolution; Luo, 2007, Transformation and diversification in early mammal evolution.
40. Darwin, 1859, *On the Origin of Species*, p. 282.
41. Kemp, 2005, *The Origin and Evolution of Mammals*.

NOTES TO CHAPTER 6 (Pages 113 to 123)

1. Luckett, W.P. 1996. Ontogenetic evidence for incisor homologies in proboscideans, in Shoshani, J. and Tassy, P. (eds.), *The Proboscidea: Evolution and Palaeoecology of Elephants and their Relatives*, Oxford: Oxford University Press, pp. 26–31.
2. Prothero, D.R. and Schoch, R.M. 2002. *Horns, Tusks, and Flippers: The Evolution of Hoofed Mammals*, Baltimore, MD: Johns Hopkins University Press; Shoshani, J. and Tassy, P. (eds.) 1996. *The Proboscidea: Evolution and Palaeoecology of Elephants and their Relatives*. Oxford: Oxford University Press; Sanders, W.J., *et al.* 2010. Proboscidea, in Werdelin, L. and Sanders, W.J. (eds.), *Cenozoic Mammals of Africa*, Berkeley, CA: University of California Press, chapter 15; Prothero, D.R. 2007. *Evolution: What the Fossils Say and Why It Matters*, New York: Columbia University Press.
3. Maglio, V.J. 1973. Origin and evolution of the Elephantidae. *Transactions of the American Philosophical Society*, 63(3): 1–149; see figure 10 (p. 32) for an image of lower tusk evolution in *Loxodonta*.
4. Sanders *et al.*, 2010, Proboscidea.
5. Prothero and Schoch, 2002, *Horns, Tusks, and Flippers*.
6. Gould, S.J. 1996. *Full House: The Spread of Excellence from Plato to Darwin*, New York: Harmony Books.
7. Gheerbrant, E., *et al.* 2002. A new large mammal from the Ypresian of Morocco: evidence of surprising diversity of early proboscideans. *Acta Palaeontologica Polonica*, 47(3): 493–506.
8. Gheerbrant, E. 2009. Paleocene emergence of elephant relatives and the rapid radiation of African ungulates. *Proceedings of the National Academy of Sciences of the United States of America*, 106(26): 10717–21.
9. Sidor, C.A. and Hopson, J.A. 1998. Ghost lineages and "mammalness": assessing the temporal pattern of character acquisition in the Synapsida. *Paleobiology*, 24(2), 254–73.
10. Norell, M.A. and Novacek, M.J. 1992. The fossil record and evolution: comparing cladistic and paleontological evidence for vertebrate history. *Science*, 255(5052): 1690–3; Benton, M.J. 1995. Diversification and extinction in the history of life. *Science*, 268(5207): 52–8.
11. Fisher, D.C. 2008. Stratocladistics: integrating temporal data and character data in phylogenetic inference. *Annual Review of Ecology, Evolution, and Systematics*, 39: 365–85. It may be no accident that Professor Fisher, who has

pioneered the use of stratocladistics, is one of the world's foremost experts on the proboscidean fossil record.

12. Gheerbrant, 2009, Paleocene emergence of elephant relatives; Shoshani, J. 1996. Para- or monophyly of the gomphotheres and their position within Proboscidea, in Shoshani, J. and Tassy, P. (eds.), *The Proboscidea: Evolution and Palaeoecology of Elephants and their Relatives*, Oxford: Oxford University Press, pp. 149–77.
13. Gheerbrant, 2009, Paleocene emergence of elephant relatives; Tassy, P. 1996. Who is who among the Proboscidea, in Shoshani, J. and Tassy, P. (eds.), *The Proboscidea: Evolution and Palaeoecology of Elephants and their Relatives*, Oxford: Oxford University Press, pp. 39–48.
14. Shoshani, 1996, para- or monophyly of gomphotheres.
15. Tassy, 1996, Who is who among the Proboscidea.
16. Gheerbrant, 2009, Paleocene emergence of elephant relatives.
17. See also p. 422 in Darwin, C. 1859. *On the Origin of Species*, London: J. Murray.
18. "Wenn der Satzteil ein Nebensatz ist, muss ich das Verb an das Ende des Satzteils setzen."
19. "Mächtig ist er, aber auf die dunkle Seite könnte er gezogen werden!"
20. Vennemann, T. 2005. English: a German dialect? Available at: http://www.rotary-munich.de/2005–2006/theo-vennemann.pdf (accessed May 10, 2009).
21. Sweet, H. 1885. Alfred's *Orosius*. Available at: http://www.archive.org/details/extractsfromalfro2alfruoft (accessed May 10, 2009).
22. "Af him haefdon Pene pone weg forseten," or "the Carthaginians had blocked his way," quoted from Vennemann, English: a German dialect?: 8.
23. Vennemann, 2005, English: a German dialect?

NOTES TO CHAPTER 7 (Pages 125 to 138)

1. Gingerich, P.D., *et al.* 2001. Origin of whales from early artiodactyls: hands and feet of Eocene Protocetidae from Pakistan. *Science*, 239: 2239–42; Thewissen, J.G.M., *et al.* 2009. From land to water: the origin of whales, dolphins, and porpoises. *Evolution: Education Outreach*, 2: 272–88; Bajpai, S., *et al.* 2009. The origin and early evolution of whales: macroevolution documented on the Indian Subcontinent. *Journal of Biosciences*, 34: 673–86.
2. Zimmer, C. 1998. *At the Water's Edge*, New York: Touchstone; Gingerich, P.D. 2004. Whale evolution, in *McGraw-Hill Yearbook of Science and Technology*, New York: McGraw-Hill; Prothero, D.R. 2007. *Evolution: What the Fossils Say and Why It Matters*, New York: Columbia University Press.
3. Deméré, T.A., *et al.* 2008. Morphological and molecular evidence for a stepwise evolutionary transition from teeth to baleen in mysticete whales. *Systematic Biology*, 57(1): 15–37; Fitzgerald, E.M.G. 2010. The morphology and systematics of *Mammalodon colliveri* (Cetacea: Mysticeti), a toothed mysticete from the Oligocene of Australia. *Zoological Journal of the Linnean Society*, 158: 367–476.

4. Owen, R. 1841. Observations on the basilosaurus of Dr. Harlan (*Zeuglodon cetoides*, Owen). *Transactions of the Geological Society of London*, 6: 69–79.
5. Mossman, H.W. 1937. Comparative morphogenesis of the fetal membranes and accessory uterine structures. *Carnegie Institution of Washington Publication*, 482: 130–246.
6. Boyden, A. and Gemeroy, D. 1950. The relative position of the cetacea among the orders of Mammalia as indicated by precipitin tests. *Zoologica*, 35: 145–51.
7. Van Valen, L. 1966. Deltatheridia: a new order of mammals. *Bulletin of the American Museum of Natural History*, 132: 1–126.
8. Gingerich, P.D., *et al.* 1990. Hind limbs of Eocene *Basilosaurus*: evidence of feet in whales. *Science*, 249(4965): 154–7.
9. Novacek, M.J. 1992. Mammalian phylogeny: shaking the tree. *Nature*, 356: 121–5.
10. Gatesy, J., *et al.* 1996. Evidence from milk casein genes that cetaceans are close relatives of hippopotamid artiodactyls. *Molecular Biology and Evolution*, 13(7): 954–63.
11. Gingerich *et al.*, 1990, Hind limbs of Eocene *Basilosaurus*.
12. Gingerich *et al.*, 2001, Origin of whales from early artiodactyls.
13. Thewissen, J.G.F., *et al.* 2001. Skeletons of terrestrial cetaceans and the relationship of whales to artiodactyls. *Nature*, 413(6853): 277–81.
14. Gingerich, P.D. 2003. Land-to-sea transition in early whales: evolution of Eocene Archaeoceti (Cetacea) in relation to skeletal proportions and locomotion of living semiaquatic mammals. *Paleobiology*, 29(3): 429–54.
15. Gingerich, P.D., *et al.* 1994. New whale from the Eocene of Pakistan and the origin of cetacean swimming. *Nature*, 368: 844–7; Thewissen, J.G.M., *et al.* 1994. Fossil evidence for the origin of aquatic locomotion in archaeocete whales. *Science*, 263: 210–12.
16. Thewissen, J.G.M., *et al.* 2007. Whales originated from aquatic artiodactyls in the Eocene epoch of India. *Nature*, 450: 1190–5.
17. Geisler, J.H. and Uhen, M.D. 2005. Phylogenetic relationships of extinct cetartiodactyls: results of simultaneous analyses of molecular, morphological, and stratigraphic data. *Journal of Mammalian Evolution*, 12(1–2): 145–60; Geisler, J.H. and Theodor, J.M. 2009. Hippopotamus and whale phylogeny. *Nature*, 458: e1–4.
18. Geisler and Uhen, 2005, Phylogenetic relationships of extinct cetartiodactyls; Geisler and Theodor, 2009, Hippopotamus and whale phylogeny.
19. Geisler and Uhen, 2005, Phylogenetic relationships of extinct cetartiodactyls; Geisler and Theodor, 2009, Hippopotamus and whale phylogeny; Spaulding, M., *et al.* 2009. Relationships of Cetacea (Artiodactyla) among mammals: increased taxon sampling alters interpretations of key fossils and character evolution. *PLoS ONE*, 4(9): e7062.
20. Spaulding, M., *et al.* 2009. Relationships of Cetacea (Artiodactyla) among mammals: increased taxon sampling alters interpretations of key fossils and character evolution. *PLoS ONE*, 4(9), e7062.
21. Geisler and Uhen, 2005, Phylogenetic relationships of extinct cetartiodactyls; Geisler and Theodor, 2009, Hippopotamus and whale phylogeny.

22. Orliac, M., *et al.* 2010. Early Miocene hippopotamids (Cetartiodactyla) constrain the phylogenetic and spatiotemporal settings of hippopotamid origin. *Proceedings of the National Academy of Sciences of the United States of America*, 107(26): 11871–6.
23. Bajpai *et al.*, 2009, The origin and early evolution of whales; Zimmer, 1998, *At the Water's Edge*; Gingerich, 2004, Whale evolution.
24. Darwin, C. 1859. *On the Origin of Species by Means of Natural Selection*, London: J. Murray, p. 282.
25. Beatty, B.L. and Dooley, A.C. 2009. Injuries in a mysticete skeleton from the Miocene of Virginia, with a discussion of buoyancy and the primitive feeding mode in the chaeomysticeti. *Jeffersoniana*, 20: 1–27.
26. Thewissen *et al.*, 2009, From land to water.
27. Thewissen *et al.*, 2009, From land to water.
28. Fitzgerald, 2010, The morphology and systematics of *Mammalodon colliveri*.
29. Fitzgerald, 2010, The morphology and systematics of *Mammalodon colliveri*.
30. Fitzgerald, 2010, The morphology and systematics of *Mammalodon colliveri*.
31. Deméré *et al.*, 2008, Morphological and molecular evidence.
32. Fitzgerald (2010, The morphology and systematics of *Mammalodon colliveri*) expresses some uncertainty about the presence of baleen in *Aetiocetus*. However, regardless of the case for baleen in that particular fossil, evidence for the simultaneous presence of both teeth and baleen has also been suggested for *Llanocetus* (Fordyce, R.E. 2003. Early crown-group Cetacea in the southern ocean: the toothed archaic mysticete *Llanocetus*. *Journal of Vertebrate Paleontology*, 23(3): 50A).
33. Deméré *et al.*, 2008, Morphological and molecular evidence.
34. Deméré *et al.*, 2008, Morphological and molecular evidence.
35. Deméré *et al.*, 2008, Morphological and molecular evidence.
36. Deméré *et al.*, 2008, Morphological and molecular evidence.
37. Carroll, S.B. 2006. *The Making of the Fittest: DNA and the Ultimate Forensic Record of Evolution*, New York: W.W. Norton; Carroll, S.B. 2005. *Endless Forms Most Beautiful: The New Science of Evo Devo and the Making of the Animal Kingdom*, New York: W.W. Norton; Coyne, J. 2009. *Why Evolution is True*, Oxford and New York: Oxford University Press.

NOTES TO CHAPTER 8 (Pages 140 to 153)

1. Meyer, S.C., *et al.* 2007. *Explore Evolution*, Melbourne: Hill House. Three of the *Explore Evolution* authors (Meyer, Nelson, Minnich) are listed as "CSC Fellows" of the Discovery Institute (http://www.discovery.org/csc/fellows.php, accessed June 26, 2010). The publisher, Hill House, is affiliated with Australian Bernard d'Abrera. As of this writing (May 28, 2010), all of the books shown on www.hillhouse-publishers.com are authored by d'Abrera, and *Explore Evolution* is not among them. He was described in the pages of a June 2003 issue of *Creation* magazine as regarding evolution as "'viscid, asphyxiating baggage' that requires 'blind religious faith'." See http://en.wikipedia.org/wiki/Bernard_d'Abrera (accessed May 28, 2010).

2. Meyer *et al.*, 2007, *Explore Evolution*, pp.26–7.
3. Some mammals, such as soricine shrews, accumulate traces of iron in their teeth, giving them a reddish, pigmented appearance.
4. Gould, S.J. 2002. *The Structure of Evolutionary Theory*, Cambridge, MA: Belknap Press of Harvard University Press, p.986.
5. Meyer *et al.*, 2007, *Explore Evolution*, pp.26–7.
6. Carroll, R.L. 1982. Early evolution of reptiles. *Annual Review of Ecology and Systematics*, 13, 87–109.
7. Kemp, T.S. 2005. *The Origin and Evolution of Mammals*, Oxford and New York: Oxford University Press.
8. Norell, M.A. and Xu, X. 2005. Feathered dinosaurs. *Annual Review of Earth and Planetary Sciences*, 33, 277–99.
9. Novacek, M.J., *et al.* 1997. Epipubic bones in eutherian mammals from the late Cretaceous of Mongolia. *Nature*, 389: 483–6.
10. Norell, M.A. and Novacek, M.J. 1992. The fossil record and evolution: comparing cladistic and paleontological evidence for vertebrate history. *Science*, 255(5052): 1692.
11. Finarelli, J.A. and Clyde, W.C. 2004. Reassessing hominoid phylogeny: evaluating congruence in the morphological and temporal data. *Paleobiology*, 30: 614–51.
12. Benton, M.J. 1995. Diversification and extinction in the history of life. *Science*, 268(5207): 52–8; Sidor, C.A. and Hopson, J.A. 1998. Ghost lineages and "mammalness": assessing the temporal pattern of character acquisition in the Synapsida. *Paleobiology*, 24(2): 254–73.
13. Asher, R.J., *et al.* 2005. Stem Lagomorpha and the antiquity of Glires. *Science*, 307(5712): 1091–4.
14. Asher, R.J., *et al.* 2002. Morphology and relationships of *Apternodus* and other extinct, zalambdodont, placental mammals. *Bulletin of the American Museum of Natural History*, 243: 1–117.
15. Asher, R.J. and Avery, D.M. 2010. New golden moles (Afrotheria, Chrysochloridae) from the Pliocene of South Africa. *Paleontologica Electronica*, 13(1): 3A.
16. Morales, J., *et al.* 2008. El sistema de yacimientos de mamíferos miocenos del Cerro de los Batallones, Cuenca de Madrid: estado actual y perspectivas. *Palaeontologica Nova: Seminario de Paleontología de Zaragoza*, 8: 41–117; Anton, M. and Morales, J. (eds.) 2010. *Madrid antes del hombre*, Comunidad de Madrid: Dirreción General de Patrimonio Histórico.
17. Asher, R.J., *et al.* 2003. Relationships of endemic African mammals and their fossil relatives based on morphological and molecular evidence. *Journal of Mammalian Evolution*, 10: 131–94; Tabuce, R., *et al.* 2007. Early Tertiary mammals from North Africa reinforce the molecular Afrotheria clade. *Proceedings of the Royal Society B: Biological Sciences*, 274: 1159–66; Zack, S.P., *et al.* 2005. Affinities of "hyopsodontids" to elephant shrews and a Holarctic origin of Afrotheria. *Nature*, 434: 497–501.
18. Carroll, 1982, Early evolution of reptiles.

19. Kemp, 2005, *The Origin and Evolution of Mammals.*
20. Norell and Xu, 2005, Feathered dinosaurs; Norell and Novacek, 1992, The fossil record and evolution, 1692.
21. Novacek *et al.*, 1997, Epipubic bones in eutherian mammals.
22. Gould, 2002, *The Structure of Evolutionary Theory*; Pickett, M.A., *et al.* 2005. Iconoclasts of evolution: Haeckel, Behe, Wells & the ontogeny of a fraud. *American Biology Teacher*, 67(5), 275–82; Coyne, J. 2009. *Why Evolution is True*, Oxford and New York: Oxford University Press; Carroll, S.B. 2007. God as genetic engineer. *Science*, 316: 1427–8; Miller, K.R. 2007. Falling over the edge. *Nature*, 447, 1055–6.
23. See also National Center for Science Education website: "Critique: Exploring 'Explore Evolution'". Available at: http://ncse.com/creationism/analysis/explore-evolution (accessed March 25, 2011).
24. See discussion in Miller, K.R. 1996. A review of Darwin's *Black Box*. *Creation/Evolution*, 16, 36–40.
25. For example, in a 2007 interview, Richard Dawkins states "I've been told that prominent creationists love me because they can use me to say 'look you see ... science generally but especially evolution leads to atheism'. And were I to have gone into the court in Dover Pennsylvania, and the lawyer for the other side had said 'Mr Dawkins is it true that you were led to atheism because of your understanding of evolution?' I would have had to say yes." In fairness, in his next response to the interviewer, Dawkins says he does not equate science with atheism, but he does not explain in this interview how he can hold these apparently contradictory opinions. The entire interview is available at http://www.pointofinquiry.org/richard_dawkins_science_and_the_new_atheism/ (quote at *ca.* 17 minutes, accessed July 16, 2010).
26. Hinde, R.A. 2009. *Why Gods Persist: A Scientific Approach to Religion*, London and New York: Routledge.
27. This reflects a conclusion made by Nobel laureate Sir Peter Medawar in his 1984 book, *The Limits of Science*: "The existence of a limit to science is made clear by its inability to answer childlike elementary questions having to do with first and last things – questions such as 'how did everything begin?' ... Doctrinaire positivism dismisses all such questions as nonquestions or pseudoquestions – hardly an adequate rebuttal because the questions make sense to those who ask them, and the answers to those who try and give them" (pp. 59, 66).

NOTES TO CHAPTER 9 (Pages 154 to 177)

1. http://nobelprize.org/nobel_prizes/medicine/laureates/2001/press.html (accessed April 16, 2010).
2. Asher, R.J., *et al.* 2005. New material of *Centetodon* (Mammalia, Lipotyphla) and the importance of (missing) DNA sequences in systematic paleontology. *Journal of Vertebrate Paleontology*, 25(4): 911–23.
3. Felsenstein, J. 2004. *Inferring Phylogenies*, Sunderland, MA: Sinauer.
4. http://www.ncbi.nlm.nih.gov/nucleotide (accessed April 18, 2010).

5. Huxley, T.H. 1863. *Man's Place in Nature*, New York: Appleton, pp. 80–1.
6. Hull, D.L. 1967. Certainty and circularity in evolutionary taxonomy. *Evolution*, 21(1): 174–89.
7. http://www.ncbi.nlm.nih.gov/nucleotide (accessed April 18, 2010).
8. http://www.genomesize.com (accessed May 5, 2010).
9. Larkin, M.A., *et al.* 2007. Clustal W and Clustal X version 2.0. *Bioinformatics*, 23: 2947–8.
10. http://www.ebi.ac.uk/Tools/clustalw2/index.html (accessed May 8, 2010).
11. Hugall, A.F., *et al.* 2007. Calibration choice, rate smoothing, and the pattern of tetrapod diversification according to the long nuclear gene RAG-1. *Systematic Biology*, 56: 543–63.
12. Zardoya, R., *et al.* 2003. Complete nucleotide sequence of the mitochondrial genome of a salamander, *Mertensiella luschani*. *Gene*, 317(1–2): 17–27.
13. Ostrom, J.H. 1976. Archeopteryx and origin of birds. *Biological Journal of the Linnean Society*, 8(2): 91–182; Zhou, Z. 2004. The origin and early evolution of birds: discoveries, disputes, and perspectives from fossil evidence. *Naturwissenschaften*, 91: 455–71.
14. Hugall *et al.*, 2007, Calibration choice, rate smoothing; Zardoya *et al.*, 2003, Complete nucleotide sequence.
15. Penny, D., *et al.* 1982. Testing the theory of evolution by comparing phylogenetic trees constructed from five different protein sequences. *Nature*, 297: 197–200; Theobald, D.L. 2010. A formal test of the theory of universal common ancestry. *Nature*, 465: 219–23.
16. Hedges, S.B., *et al.* 1990. Tetrapod phylogeny inferred from 18s and 28s ribosomal RNA sequences and a review of the evidence for amniote relationships. *Molecular Biology and Evolution*, 7(6): 607–33.
17. Gardiner, B.G. 1993. Haematothermia: warm-blooded amniotes. *Cladistics*, 9(4): 369–95.
18. Arnason, U., *et al.* 2004. Mitogenomic analyses of deep gnathostome divergences: a fish is a fish. *Gene*, 333: 61–70.
19. D'Erchia, A.M., *et al.* 1996. The guinea pig is not a rodent. *Nature*, 381: 597–600.
20. Janke, A., *et al.* 1996. The mitochondrial genome of a monotreme: the platypus (*Ornithorhynchus anatinus*). *Journal of Molecular Evolution*, 42: 153–9.
21. Hugall *et al.*, 2007, Calibration choice, rate smoothing; Zardoya *et al.*, 2003, Complete nucleotide sequence; Hedges, S.B. 1994. Molecular evidence for the origin of birds. *Proceedings of the National Academy of Science of the United States of America*, 91: 2621–4.
22. Hällstrom, B.M. and Janke, A. 2009. Gnathostome phylogenomics utilizing lungfish EST sequences. *Molecular Biology and Evolution*, 26(2): 463–71.
23. Asher, R.J. 2007. A web-database of mammalian morphology and a reanalysis of placental phylogeny. *BMC Evolutionary Biology*, 7: 108.
24. Kullberg, M., *et al.* 2008. Phylogenetic analysis of 1.5 Mbp and platypus EST data refute the Marsupionta hypothesis and unequivocally support Monotremata as sister group to Marsupialia/Placentalia. *Zoologica Scripta*, 37: 115–27.

25. Hedges, 1994, Molecular evidence for the origin of birds; Hällstrom and Janke, 2009, Gnathostome phylogenomics; Kullberg *et al.*, 2008, Phylogenetic analysis of 1.5 Mbp.
26. Hugall *et al.*, 2007, Calibration choice, rate smoothing; Hällstrom and Janke, 2009, Gnathostome phylogenomics.
27. Penny *et al.*, 1982, Testing the theory of evolution; Theobald, 2010, A formal test of the theory of universal common ancestry.
28. For example, listen to Stephen C. Meyer's testimony before the Texas state school board in early 2009: "So you compare those sequences, and you produce those wonderful tree diagrams. The interesting thing ... is that's the conclusion of the analysis, the phylogenetic trees showing relatedness, showing a historical pattern of descent with modification. But that same conclusion is the assumption that the computer program uses to analyze the similarity data. ... Formally, it's a form of circular reasoning." Available at: http://www.idthefuture.com/2009/02/defending_critical_analysis_in.html (accessed August 17, 2009).
29. Penny *et al.*, 1982, Testing the theory of evolution; Theobald, 2010, A formal test of the theory of universal common ancestry.
30. Asher *et al.*, 2005, New material of *Centetodon*; Asher, 2007, A web-database of mammalian morphology; Asher, R.J. and Hofreiter, M. 2006. Tenrec phylogeny and the noninvasive extraction of nuclear DNA. *Systematic Biology*, 55(2): 181–94.
31. Swofford, D.L. 2002. *PAUP* Phylogenetic Analysis Using Parsimony (*and Other Methods) Version 4*, Sunderland, MA: Sinauer.
32. Goloboff, P. 1993. NONA version 1.9 computer program. Available at: http://www.cladistics.com.
33. Ronquist, F. and Huelsenbeck, J.P. 2003. MrBayes 3: Bayesian phylogenetic inference under mixed models. *Bioinformatics*, 19(12): 1572–4.
34. Page, R.D.M. and Charleston, M.A. 1999. Comments on Allard and Carpenter (1996), or the "Aquatic ape" hypothesis revisited. *Cladistics*, 15: 73–4.
35. Felsenstein, 2004, *Inferring Phylogenies*.
36. Hugall *et al.*, 2007, Calibration choice, rate smoothing.
37. The formula to calculate the number of rooted, bifurcating trees for n species (from Felsenstein, 2004, *Inferring Phylogenies*, p.23) is $(2n-3)! / 2^{n-2}(n-2)!$.
38. Zardoya *et al.*, 2003, Complete nucleotide sequence.
39. Hedges, 1994, Molecular evidence for the origin of birds.
40. Hällstrom and Janke, 2009, Gnathostome phylogenomics.
41. Davis, P., *et al.* 1993. *Of Pandas and People.* Richardson, TX: Foundation for Thought and Ethics.
42. Forrest, B. and Gross, P.R. 2007. *Creationism's Trojan Horse: The Wedge of Intelligent Design* (paperback edition with afterword on the Dover trial). Oxford and New York: Oxford University Press.
43. Lee, M.S.Y. 1999. Molecular phylogenies become functional. *Trends in Ecology & Evolution*, 14(5): 177–8.
44. Andrews, T.D., *et al.* 1998. Accelerated evolution of cytochrome *b* in simian primates: adaptive evolution in concert with other mitochondrial proteins? *Journal of Molecular Evolution*, 47: 249–57.

45. Schmitz, J., *et al.* 2002. The complete mitochondrial sequence of *Tarsius bancanus*: evidence for an extensive nucleotide compositional plasticity of primate mitochondrial DNA. *Molecular Biology and Evolution*, 19(4): 544–53.
46. Chatterjee, H.J., *et al.* 2009. Estimating the phylogeny and divergence times of primates using a supermatrix approach. *BMC Evolutionary Biology*, 9: 259.
47. Schmitz, J., *et al.* 2001. SINE insertions in cladistic analyses and the phylogenetic affiliations of *Tarsius bancanus* to other primates. *Genetics*, 157: 777–84; Hallström, B.M. and Janke, A. 2010. Mammalian evolution may not be strictly bifurcating. *Molecular Biology and Evolution*, 27(12): 2804–16.
48. Fleagle, J.G. 1999. *Primate Adaptation and Evolution*, San Diego, CA: Academic Press.
49. See also the National Center for Science Education website: "Critique: Exploring 'Explore Evolution'." Available at: http://ncse.com/creationism/analysis/explore-evolution (accessed March 25, 2011).
50. http://en.wikipedia.org/wiki/Yentl (accessed July 4, 2010).
51. Asher, R.J., *et al.* 2009. The new framework for understanding placental mammal evolution. *Bioessays*, 31(8): 853–64.
52. Asher and Hofreiter, 2006, Tenrec phylogeny and the noninvasive extraction of nuclear DNA.
53. Hallström and Janke, 2010, Mammalian evolution may not be strictly bifurcating; Asher *et al.*, 2009, The new framework for understanding placental mammal evolution.
54. LeGros Clark, W.E. and Sonntag, C.F. 1926. A monograph of *Orycteropus afer*: III. The skull, the skeleton of the trunk, and limbs. *Proceedings of the Zoological Society of London*, 30: 445–85.
55. Simpson, G.G. 1945. The principles of classification and a classification of mammals. *Bulletin of the American Museum of Natural History*, 85: 1–350.
56. Fischer, M.S. 1989. Hyracoids, the sister-group of perissodactyls, in Prothero, D.R. and Schoch, R.M. (eds.), *The Evolution of Perissodactyls*, New York: Oxford University Press, pp. 37–56.
57. Broom, R. 1916. On the structure of the skull in *Chrysochloris*. *Proceedings of the Zoological Society of London*, 1916: 449–59.
58. Stanhope, M.J., *et al.* 1998. Molecular evidence for multiple origins of Insectivora and for a new order of endemic African insectivore mammals. *Proceedings of the National Academy of Science of the United States of America*, 95(17): 9967–72.
59. Asher, R.J. 2001. Cranial anatomy in tenrecid insectivorans: character evolution across competing phylogenies. *American Museum Novitates*, 3352: 1–54; Asher, R.J., *et al.* 2003. Relationships of endemic African mammals and their fossil relatives based on morphological and molecular evidence. *Journal of Mammalian Evolution*, 10: 131–94.
60. Asher *et al.*, 2005, New material of *Centetodon*; Asher, 2007, A web-database of mammalian morphology; Asher *et al.*, 2003, Relationships of endemic African mammals and their fossil relatives.

61. Murphy, W.J., *et al.* 2001. Resolution of the early placental mammal radiation using Bayesian phylogenetics. *Science*, 294(5550): 2348–51.
62. Novacek, M.J. 1992. Mammalian phylogeny: shaking the tree. *Nature*, 356(6365): 121–5.
63. Beard, K.C. 1993. Phylogenetic systematics of the Primatomorpha, with special reference to Dermoptera, in Szalay, F.S., *et al.* (eds.), *Mammal Phylogeny Vol. 2: Placentals*, New York: Springer, pp. 129–50.
64. Fischer, 1989, Hyracoids, the sister-group of perissodactyls.
65. Lee, M.S.Y. and Camens, A.B. 2009. Strong morphological support for the molecular evolutionary tree of placental mammals. *Journal of Evolutionary Biology*, 22: 2243–57.
66. Asher *et al.*, 2009, The new framework for understanding placental mammal evolution.

NOTES TO CHAPTER 10 (Pages 179 to 200)

1. ID proponent Tom Woodward, quoted from the premier.org Christian radio program *Unbelievable*, podcast for August 30, 2008, around 16–17 minutes left in the show. See http://www.premier.org.uk/unbelievable (accessed August 1, 2010).
2. Surridge, A.K., *et al.* 2003. Evolution and selection of trichromatic vision in primates. *Trends in Ecology & Evolution*, 18(4): 198–205.
3. Wellik, D.M. 2007. Hox patterning of the vertebrate axial skeleton. *Developmental Dynamics*, 236: 2454–63.
4. Schienman, J.E., *et al.* 2006. Duplication and divergence of 2 distinct pancreatic ribonuclease genes in leaf-eating African and Asian colobine monkeys. *Molecular Biology and Evolution*, 23(8): 1465–79.
5. Kangas, A.T., *et al.* 2004. Nonindependence of mammalian dental characters. *Nature*, 432: 211–14; Jernvall, J., *et al.* 2000. Evolutionary modification of development in mammalian teeth: quantifying gene expression patterns and topography. *Proceedings of the National Academy of Sciences of the United States of America*, 97(26): 14444–8.
6. Avise, J. 2010. *Inside the Human Genome: The Case for Non-Intelligent Design*, Oxford: Oxford University Press.
7. Carroll, S.B. 2005. *Endless Forms Most Beautiful*, New York: W.W. Norton; Carroll, S.B. 2006. *The Making of the Fittest*, New York: W.W. Norton; Carroll, S.B., *et al.* 2005. *From DNA to Diversity*, 2nd edition, Oxford: Blackwell.
8. Coyne, J. 2009. *Why Evolution is True*, Oxford and New York: Oxford University Press.
9. Miller, K.R. 1999. *Finding Darwin's God*, New York: Cliff Street Books.
10. Shubin, N. 2008. *Your Inner Fish*, London: Penguin.
11. Kirschner, M.W. and Gerhart, J.C. 2005. *The Plausibility of Life: Resolving Darwin's Dilemma*, New Haven, CT: Yale University Press.
12. Weiss, K.M. and Buchanan, A.V. 2009. *The Mermaid's Tale*, Cambridge, MA: Harvard University Press.

13. Chandrasekaran, C. and Betrán, E. 2008. Origins of new genes and pseudogenes. *Nature Education*, 1(1). Available at: http://www.nature.com/scitable/topicpage/origins-of-new-genes-and-pseudogenes-835 (accessed August 15, 2010).
14. Al-Hashimi, N., *et al.* 2010. The enamelin genes in lizard, crocodile and frog, and the pseudogene in the chicken provide new insights on enamelin evolution in tetrapods. *Molecular Biology and Evolution*, 27(9), 2078–94; Meredith, R.W., *et al.* 2009. Molecular decay of the tooth gene enamelin (ENAM) mirrors the loss of enamel in the fossil record of placental mammals. *PLoS Genetics*, 5(9): e1000634.
15. Gerstein, M. and Zheng, D. 2006. The real life of pseudogenes. *Scientific American*, 295(2): 49–55; Karro, J., *et al.* 2007. Pseudogene.org: a comparison platform and comprehensive resource for pseudogene annotation. *Nucleic Acids Research*, 35: D55–60 (see http://www.pseudogene.org, accessed August 15, 2010).
16. Coyne, 2009, *Why Evolution is True.*
17. Gerstein and Zheng, 2006, The real life of pseudogenes.
18. Hayden, S., *et al.* 2010. Ecological adaptation determines functional mammalian olfactory subgenomes. *Genome Research*, 20: 1–9.
19. David-Gray, Z.K., *et al.* 2002. Adaptive loss of ultraviolet-sensitive/violet-sensitive (UVS/VS) cone opsin in the blind mole rat (*Spalax ehrenbergi*). *European Journal of Neuroscience*, 16(7): 1186–94.
20. Li, R., *et al.* 2010. The sequence and de novo assembly of the giant panda genome. *Nature*, 463(7279): 311–17.
21. Chandrasekaran and Betrán, 2008, Origins of new genes and pseudogenes; Gerstein and Zheng, 2006, The real life of pseudogenes; Ohno, S. 1970. *Evolution by Gene Duplication*, Berlin: Springer-Verlag.
22. Sasidharan, R. and Gerstein, M. 2008. Genomics: protein fossils live on as RNA. *Nature*, 453: 729–31.
23. Belshaw, R., *et al.* 2004. Long-term reinfection of the human genome by endogenous retroviruses. *Proceedings of the National Academy of Sciences of the United States of America*, 101(14): 4894–9; note that due to the capacity of viruses to infect hosts directly, independently of reproduction and descent, many cases of cross-species similarity exist that are not due to common ancestry. On the whole, however, retroviral DNA sequences are more similar in closely related species than they are among more distantly related species. Tarlinton, R.E., *et al.* 2006. Retroviral invasion of the koala genome. *Nature*, 442: 79–81.
24. Gifford, R. and Tristem, M. 2003. The evolution, distribution, and diversity of endogenous retroviruses. *Virus Genes*, 26(3): 291–315.
25. Belshaw *et al.*, 2004, Long-term reinfection of the human genome by endogenous retroviruses.
26. Gifford and Tristem, 2003, The evolution, distribution, and diversity; Britten, R. 1997. Mobile elements inserted in the distant past have taken on important functions. *Gene*, 205: 177–82; Volff, J.N. 2006. Turning junk into gold: domestication of transposable elements and the creation of new genes in eukaryotes. *Bioessays*, 28: 913–22.

27. Best, S., *et al.* 1996. Positional cloning of the mouse retrovirus restriction gene *Fv1*. *Nature*, 382: 826–9.
28. Yan, Y., *et al.* 2009. Origin, antiviral function and evidence for positive selection of the gammaretrovirus restriction gene *Fv1* in the genus *Mus*. *Proceedings of the National Academy of Sciences of the United States of America*, 106(9): 3529–63.
29. Gifford and Tristem, 2003, The evolution, distribution, and diversity; Volff, 2006, Turning junk into gold; Best *et al.*, 1996, Positional cloning of the mouse retrovirus; Mi, S. *et al.* 2000. Syncytin is a captive retroviral envelope protein involved in human placental morphogenesis. *Nature*, 403: 785–9; de Parseval, N. and Heidmann, T. 2005. Human endogenous retroviruses: from infectious elements to human genes. *Cytogenetic and Genome Research*, 110: 318–32.
30. Sturm, R.A., *et al.* 2008. A single SNP in an evolutionary conserved region within intron 86 of the HERC2 gene determines human blue–brown eye color. *American Journal of Human Genetics*, 82: 424–31.
31. Harding, R.M., *et al.* 2000. Evidence for variable selective pressures at MC1R. *American Journal of Human Genetics*, 66(4): 1351–61.
32. Surridge *et al.*, 2003, Evolution and selection of trichromatic vision; Neitz, M., *et al.* 1991. Spectral tuning of pigments underlying red–green color vision. *Science*, 252: 971–4; Bradley, B.J. and Mundy, N.I. 2008. The primate palette: the evolution of primate coloration. *Evolutionary Anthropology*, 17: 97–111.
33. Surridge *et al.*, 2003, Evolution and selection of trichromatic vision.
34. Surridge *et al.*, 2003, Evolution and selection of trichromatic vision; Dulai, K.S., *et al.* 1999. The evolution of trichromatic colour vision by opsin gene duplication in New World and Old World primates. *Genome Research*, 9: 629–38.
35. Dulai *et al.*, 1999, The evolution of trichromatic colour vision.
36. Ohno, 1970, *Evolution by Gene Duplication*; Caicedo, A.L., *et al.* 2009. Complex rearrangements lead to novel chimeric gene fusion polymorphisms at the *Arabidopsis thaliana* MAF2–5 flowering time gene cluster. *Molecular Biology and Evolution*, 26(3): 699–711; Han, M.V., *et al.* 2009. Adaptive evolution of young gene duplicates in mammals. *Genome Research*, 19: 859–67.
37. Hurles, M. 2004. Gene duplication: the genomic trade in spare parts. *PLoS Biology*, 2(7): 900–4; Zhang, J. 2003. Evolution by gene duplication: an update. *Trends in Ecology & Evolution*, 18(6): 292–8.
38. Ohno, 1970, *Evolution by Gene Duplication*; Hurles, 2004, Gene duplication; Zhang, 2003, Evolution by gene duplication.
39. Dulai *et al.*, 1999, The evolution of trichromatic colour vision.
40. Dulai *et al.*, 1999, The evolution of trichromatic colour vision.
41. Surridge *et al.*, 2003, Evolution and selection of trichromatic vision; Dulai *et al.*, 1999, The evolution of trichromatic colour vision.
42. Hoffmann, M., *et al.* 2007. Opsin gene duplication and diversification in the guppy, a model for sexual selection. *Proceedings of the Royal Society B: Biological Sciences*, 274(1606): 33–42.
43. Hoffmann *et al.*, 2007, Opsin gene duplication and diversification.

44. Hoffmann *et al.*, 2007, Opsin gene duplication and diversification; Elmer, K.R., *et al.* 2009. Color assortative mating contributes to sympatric divergence of neotropical crater lake cichlid fish. *Evolution*, 63: 2750–7; Salzburger, W. and Meyer, A. 2004. The species flocks of East African cichlid fishes: recent advances in molecular phylogenetics and population genetics. *Naturwissenschaften*, 91(6): 277–90.
45. Rube Goldberg (1883–1970) was a cartoonist who famously sketched overly complicated systems to effect simple tasks. See http://en.wikipedia.org/wiki/Rube_Goldberg_machine (accessed August 6, 2010).
46. Surridge *et al.*, 1999, Evolution and selection of trichromatic vision.
47. Kashi, Y. and King, D.G. 2006. Simple sequence repeats as advantageous mutators in evolution. *Trends in Ecology & Evolution*, 22(5): 253–9.
48. Kashi and King, 2006, Simple sequence repeats.
49. Fondon, J.W. and Garner, H.R. 2004. Molecular origins of rapid and continuous morphological evolution. *Proceedings of the National Academy of Sciences of the United States of America*, 101(52): 18058–63.
50. Sears, K.E., *et al.* 2007. The correlated evolution of Runx2 tandem repeats, transcriptional activity, and facial length in Carnivora. *Evolution and Development*, 9(6): 555–65.
51. Galant, R. and Carroll, S.B. 2002. Evolution of a transcriptional repression domain in an insect Hox protein. *Nature*, 415: 910–13.
52. Asher, R.J. and Lehmann, T. 2008. Dental eruption in afrotherian mammals. *BMC Biology*, 6: 14.
53. Fondon and Garner, 2004, Molecular origins.
54. Bradley, B.J., *et al.* 2009. Why the long face? Runx2 tandem repeats and the evolution of primate prognathism. *Journal of Vertebrate Paleontology*, 29(3): 69A.
55. Kashi and King, 2006, Simple sequence repeats; Fondon, J.W., *et al.* 2008. Simple sequence repeats: genetic modulators of brain function and behavior. *Trends in Neurosciences*, 31(8): 328–34.
56. Kashi and King, 2006, Simple sequence repeats.
57. Kashi and King, 2006, Simple sequence repeats; Chen, F., *et al.* 2010. Multiple genetic switches spontaneously modulating bacterial mutability. *BMC Evolutionary Biology*, 10: 277.
58. Wellik, 2007, Hox patterning of the vertebrate axial skeleton.
59. Pointer, M.A. and Mundy, N.I. 2008. Testing whether macroevolution follows microevolution: are colour differences among swans (*Cygnus*) attributable to variation at the MC1R locus? *BMC Evolutionary Biology*, 8: 249.
60. Tucker, A.S., *et al.* 2004. Bapx1 regulates patterning in the middle ear: altered regulatory role in the transition from the proximal jaw during vertebrate evolution. *Development*, 131(6): 1235–45; Bok, J., *et al.* 2007. Patterning and morphogenesis of the vertebrate inner ear. *International Journal of Developmental Biology*, 51: 521–33.
61. Bok *et al.*, 2007, Patterning and morphogenesis; Braunstein, E.M., *et al.* 2009. *Tbx1* and *Brn4* regulate retinoic acid metabolic genes during cochlear morphogenesis. *BMC Developmental Biology*, 9: 31.

62. Braunstein *et al.*, 2009, *Tbx1* and *Brn4*.
63. Tucker *et al.*, 2004, Bapx1 regulates patterning in the middle ear.
64. Tucker *et al.*, 2004, Bapx1 regulates patterning in the middle ear.
65. Braunstein *et al.*, 2009, *Tbx1* and *Brn4*.
66. Sears *et al.*, 2007, The correlated evolution of Runx2 tandem repeats.
67. Surridge *et al.*, 2003, Evolution and selection of trichromatic vision; Neitz, M., *et al.* 1991. Spectral tuning of pigments underlying red–green color vision. *Science*, 252(5008): 971–4.
68. Dulai *et al.*, 1999. The evolution of trichromatic colour vision.
69. Luo, Z.X., *et al.* 2010. Fossil evidence on evolution of inner ear cochlea in Jurassic mammals. *Proceedings of the Royal Society B: Biological Sciences*, 278(1702): 28–34.
70. Luo, Z.X., *et al.* 2007. A new eutriconodont mammal and evolutionary development in early mammals. *Nature*, 446: 288–93; Wang, Y., *et al.* 2001. An ossified Meckel's cartilage in two Cretaceous mammals and origin of the mammalian middle ear. *Science*, 294: 357–61.
71. See http://www.ensembl.org (accessed August 10, 2010).

NOTES TO CHAPTER 11 (Pages 204 to 219)

1. "The possibility of other laws and constants is fatal to the fine-tuning argument ... [Any] speculation about what form life might take in a different universe with a different electron mass, electromagnetic interaction strength, or different laws of physics is ... problematical. We simply do not have the knowledge to say whether life of *some* sort would not occur under different circumstances" (Stenger, V.J., 2007, *God: The Failed Hypothesis,* Amherst, NY: Prometheus, pp. 153–154, (italics original).
2. Conway-Morris, S. 2003. *Life's Solution: Inevitable Humans in a Lonely Universe*, Cambridge and New York: Cambridge University Press.
3. Dawkins, R. 1986 (1996 reprint). *The Blind Watchmaker*, New York: W.W. Norton.
4. Miller, K.R. 2004. The flagellum unspun: the collapse of "irreducible complexity," in Dembski, M.A. and Ruse, M. (eds.), *Debating Design: From Darwin to DNA*, Cambridge: Cambridge University Press, pp. 81–97; Padian, K. 2002. Waiting for the watchmaker. *Science*, 295: 2373.
5. Sober, E. 2008. *Evidence and Evolution: The Logic Behind the Science*, Cambridge and New York: Cambridge University Press.
6. Sober, 2008, *Evidence and Evolution.*
7. Meyer, S.C. 2009. *Signature in the Cell: DNA and the Evidence for Intelligent Design*, New York: HarperOne, pp. 281–4.
8. Bell, M.A. and Foster, S.A. 1994. *The Evolutionary Biology of the Threespine Stickleback*, Oxford and New York: Oxford University Press.
9. By the way, it may seem counterintuitive, but insects can indeed eat fish. Aquatic dragonfly larvae are one of the most voracious predators in small ponds and lakes, and feed on anything that's not too much bigger than their own size of

a few centimeters in length. Dramatic footage of a dragonfly larva feeding can be seen in David Attenborough's *Life in the Undergrowth*, e.g., http://www.youtube.com/watch?v=PVgJSGivl6o (accessed August 28, 2010).

10. Reimchen, T.E. 1994. Predators and evolution in threespine stickleback, in Bell, M.A. and Foster, S.A. (eds.), *The Evolutionary Biology of the Threespine Stickleback*, Oxford and New York: Oxford University Press pp. 240–73.
11. Bell and Foster, 1994, *The Evolutionary Biology of the Threespine Stickleback.*
12. Kitano, J., *et al.* 2008. Reverse evolution of armor plates in the threespine stickleback. *Current Biology*, 18: 769–74.
13. Colosimo, P.F., *et al.* 2005. Widespread parallel evolution in sticklebacks by repeated fixation of Ectodysplasin alleles. *Science*, 307: 1928–33.
14. Kitano *et al.*, 2008, Reverse evolution of armor plates.
15. Bell and Foster, 1994, *The Evolutionary Biology of the Threespine Stickleback*; Colosimo *et al.*, 2005, Widespread parallel evolution in sticklebacks; Bell, M.A., *et al.* 2004. Twelve years of contemporary armor evolution in a threespine stickleback population. *Evolution*, 58(4): 814–24; Cresko, W.A., *et al.* 2004. Parallel genetic basis for repeated evolution of armor loss in Alaskan threespine stickleback populations. *Proceedings of the National Academy of Sciences of the United States of America*, 101(16): 6050–5.
16. Kitano *et al.*, 2008, Reverse evolution of armor plates.
17. Bell and Foster, 1994, *The Evolutionary Biology of the Threespine Stickleback*; Colosimo *et al.*, 2005, Widespread parallel evolution in sticklebacks; Bell *et al.*, 2004, Twelve years of contemporary armor; Cresko *et al.*, 2004, Parallel genetic basis for repeated evolution of armor loss.
18. Shapiro, M.D., *et al.* 2004. Genetic and developmental basis of evolutionary pelvic reduction in threespine sticklebacks. *Nature*, 428: 717–23; Chan, Y.F., *et al.* 2010. Adaptive evolution of pelvic reduction in sticklebacks by recurrent deletion of a *Pitx1* enhancer. *Science*, 327(5963): 302–5; Shapiro, M.D., *et al.* 2006. Parallel genetic origins of pelvic reduction in vertebrates. *Proceedings of the National Academy of Sciences of the United States of America*, 103(37): 13753–8.
19. Shapiro *et al.*, 2004, Genetic and developmental basis of evolutionary.
20. Prager, E.M. and Wilson, A.C. 1975. Slow evolutionary loss of the potential for interspecific hybridization in birds: a manifestation of slow regulatory evolution. *Proceedings of the National Academy of Sciences of the United States of America*, 72: 200–4.
21. Chan *et al.*, 2010, Adaptive evolution of pelvic reduction in sticklebacks.
22. Shapiro *et al.*, 2006, Parallel genetic origins of pelvic reduction in vertebrates.
23. Chan *et al.*, 2010, Adaptive evolution of pelvic reduction in sticklebacks.
24. Chan *et al.*, 2010, Adaptive evolution of pelvic reduction in sticklebacks.
25. Coyne, J.A. and Hoekstra, H.E. 2007. Evolution of protein expression: new genes for a new diet. *Current Biology*, 17(23): R1014–16.
26. Shapiro *et al.*, 2006, Parallel genetic origins of pelvic reduction in vertebrates; Kangas, A.T., *et al.* 2004. Nonindependence of mammalian dental characters. *Nature*, 432: 211–14.
27. Bell and Foster, 1994, *The Evolutionary Biology of the Threespine Stickleback*; Reimchen, 1994, Predators and evolution in threespine stickleback; Kitano *et al.*,

2008, Reverse evolution of armor plates; Colosimo *et al.*, 2005, Widespread parallel evolution in sticklebacks; Bell *et al.*, 2004, Twelve years of contemporary armor evolution; Cresko *et al.*, 2004, Parallel genetic basis for repeated evolution of armor loss; Shapiro *et al.*, 2004, Genetic and developmental basis of evolutionary pelvic reduction; Chan *et al.*, 2010, Adaptive evolution of pelvic reduction in sticklebacks; Shapiro *et al.*, 2006, Parallel genetic origins of pelvic reduction in vertebrates; Walker, J.A. 1997. Ecological morphology of lacustrine threespine stickleback *Gasterosteus aculeatus* L. (Gasterosteidae) body shape. *Biological Journal of the Linnean Society*, 61: 3–50.

28. Mivart, S.G.J. 1871. *On the Genesis of Species*, London: Macmillan and Co., chapter 2.
29. Darwin, C. 1872. *On the Origin of Species by Means of Natural Selection* (sixth edition), London: J. Murray, p. 146.
30. Mivart, 1817, *On the Genesis of Species*, p. 52.
31. Behe, M. 1996. *Darwin's* Black Box*: The Biochemical Challenge to Evolution*, New York: Free Press, p. 36.
32. Mivart, 1871, *On the Genesis of Species*, p. 52.
33. Dawkins, 1986, *The Blind Watchmaker*; Miller, 2004, The flagellum unspun; Sober, 2008, *Evidence and Evolution*.
34. Behe, M., July 15, 2007 interview on the podcast "Things that matter most" by Rick Davis and Lael Arrington. Available at: http://www.thethingsthatmatter most.org/gallery07152007.htm (accessed August 28, 2010).
35. Meyer, S.C. 2004. The origin of biological information and the higher taxonomic categories. *Proceedings of the Biological Society of Washington*, 117(2): 213–39. Available at: http://www.discovery.org/a/2177 (accessed May 21, 2010).
36. Meyer, 2004, The origin of biological information.
37. Carroll, S.B. 2007. God as a genetic engineer. *Science*, 316: 1427; Miller, K.R. 2007. Falling over the edge. *Nature*, 447: 1055–6.
38. Romero, P.A. and Arnold, F.H. 2009. Exploring protein fitness landscapes by directed evolution. *Nature Reviews Molecular Cell Biology*, 10: 866–76.
39. Soskine, M. and Tawfik, D.S. 2010. Mutational effects and the evolution of new protein functions. *Nature Reviews Genetics*, 11: 572–82.
40. Dawkins, 1986, *The Blind Watchmaker*; Dean, A.W. and Thornton, J.W. 2007. Mechanistic approaches to the study of evolution: the functional synthesis. *Nature Reviews Genetics*, 8: 675–88; Bloom, J.D. and Arnold, F.H. 2009. In the light of directed evolution: pathways of adaptive protein evolution. *Proceedings of the National Academy of Sciences of the United States of America*, 106: 9995–10000.
41. Nelson, P., 2006. "What does natural selection really explain?" Calvary Baptist Church, September 6. Available at: http://www.calvaryabq.tv/live/?ServiceID=380 (accessed May 21, 2010).
42. Darwin, 1872, *Origin of Species*, sixth edition, p. 177.
43. Darwin, 1872, *Origin of Species*, sixth edition, p. 158.
44. Dial, K.P. 2003. Wing-assisted incline running and the evolution of flight. *Science*, 299: 402–4.

45. Gould, S.J. and Vrba, E. 1982. Exaptation: a missing term in the science of form. *Paleobiology*, 8(1): 4–15.
46. Miller, 2004, The flagellum unspun.
47. Behe, 1996, *Darwin's* Black Box.
48. Pallen, M.J. and Matzke, N.J. 2006. From *The Origin of Species* to the origin of bacterial flagella. *Nature Reviews Microbiology*, 4: 784–90.
49. Miller, 2004, The flagellum unspun; Pallen and Matzke, 2006 From, *The Origin of Species.*
50. Miller, 2004, The flagellum unspun; Miller, K.R. 2007. *Only a Theory: Evolution and the Battle for America's Soul*, New York: Viking.
51. Pallen and Matzke, 2006, From *The Origin of Species*; Liu, R. and Ochman, H. 2007. Stepwise formation of the bacterial flagellar system. *Proceedings of the National Academy of Sciences of the United States of America*, 104(17): 7116–21.
52. Miller, 2004, The flagellum unspun: 87.
53. Behe, M. 2004. Irreducible complexity: obstacle to Darwinian evolution, in Dembski, M.A. and Ruse, M. (eds.), *Debating Design: From Darwin to DNA*, Cambridge: Cambridge University Press, p. 359.
54. Pallen and Matzke, 2004, From *The Origin of Species*; Miller, *Only a Theory.*
55. Liu and Ochman, 2007, Stepwise formation of the bacterial flagellar system.
56. Minnich, S. and Meyer, S.C. 2004. Genetic analysis of coordinate flagellar and type III regulatory circuits, in Collins, M.W. and Brebbiam C.A. (eds.), *Proceedings of the Second International Conference on Design & Nature*, Rhodes: WIT Press, p. 8.
57. Miller, 2007, *Only a Theory.*
58. Doolittle, W.F. and Zhaxybayeva, O. 2007. Evolution: reducible complexity – the case for bacterial flagella. *Current Biology*, 17(13): R510–12.
59. Pallen and Matzke, 2004, From *The Origin of Species.*
60. Pallen and Matzke, 2004, From *The Origin of Species*; Miller, 2007, *Only a Theory*; Liu and Ochman, 2007, Stepwise formation of the bacterial flagellar system.
61. Liu and Ochman, 2007, Stepwise formation of the bacterial flagellar system; Doolittle and Zhaxybayeva, 2007, Evolution: reducible complexity.
62. Minnich and Meyer, 2004, Genetic analysis of coordinate flagellar.
63. Miller, 2004, The flagellum unspun; Dean and Thornton, 2007, Mechanistic approaches to the study of evolution; Pallen and Matzke, 2007, From *The Origin of Species*; Liu and Ochman, 2007, Stepwise formation of the bacterial flagellar system; Doolittle and Zhaxybayeva, 2007, Evolution: reducible complexity. Alexander, D. and White, R.S., 2004. Beyond Belief: Science, Faith, and Ethical Challenges, Oxford: Lion Books.

NOTES TO CHAPTER 12 (Pages 222 to 228)

1. District Court, Rice County Third Judicial District. Civil File No. CX-99–793, Rodney LeVake Plaintiff. Defendants' Memorandum of Law in Support of

Motion for Summary Judgment. Available at: http://ncse.com/creationism/legal/levake-v-independent-school-district-656 (accessed September 5, 2010).

2. Tyrangiel, J. 2000. History: Faribault, Minn. – the science of dissent. *Time Magazine* July 10, 2000. Available at: http://www.time.com/time/magazine/article/0,9171,997382,00.html (accessed September 5, 2010).
3. http://intelligentdesign.podomatic.com/entry/index/2010–01–11T17_00_16–08_00 (accessed September 5, 2010).
4. http://intelligentdesign.podomatic.com/entry/eg/2008–09–12T15_59_15–07_00 (accessed September 5, 2010).
5. http://intelligentdesign.podomatic.com/entry/eg/2008–09–12T15_59_15–07_00 (accessed September 5, 2010).
6. Tyrangiel, 2000, History: Faribault, Minn.; see also interviews with LeVake and one of his colleagues by the then-editor of *The American Biology Teacher*, Moore, R. 2004. When a biology teacher refuses to teach evolution: a talk with Rod LeVake. *The American Biology Teacher* 66(4): 246–50; Moore, R. 2004. Standing up for our profession: a talk with Ken Hubert. *The American Biology Teacher* 66(5): 325–7.
7. I would add that the Discovery Institute interviewer shares LeVake's ignorance of paleontology. Immediately after the quote cited above, following LeVake's statement that "the whole idea of change over time … is not borne out at all by the fossil record," the interviewer makes the following incredible statement: "So you have obviously studied biology quite a bit."
8. Moore, 2004, When a biology teacher refuses.
9. Attributed to Mariama Lowe, a student sympathetic to her former teacher and ID supporter Caroline Crocker, as quoted in the German newspaper *Mannheimer Morgen*, May 10, 2006. ("Menschen haben eine Seele, man kann sie nicht mit Tieren auf eine Stufe stellen. An die Evolution zu glauben, würde heissen, dass der Tod das letzte Wort ist.")
10. See also: Miller, K.R., 1999, *Finding Darwin's God: A Scientist's Search for Common Ground between God and Evolution*, New York: Cliff Street Books, p. 188: "To … opponents of evolution, the real risk is that evolution tells people that God is dead. And if people were to believe in that, they might indeed behave as if all is permitted. Social chaos would result – or *has resulted*, depending on the degree of pessimism with which one views the present state of American society."
11. Meyer, S.C. 2009. *Signature in the Cell: DNA and the Evidence for Intelligent Design*, New York: HarperOne, p. 547, note 24.
12. Meyer, 2009, *Signature in the Cell*, p. 35. In the Discovery Institute-supported movie *Unlocking the Mysteries of Life*, Meyer also implies a sense of inadequacy on Darwin's part for not explaining the origin of life: "Darwin … was trying to show how natural selection could have modified existing organisms to produce the great diversity of plant and animal life that fills the Earth today. But when it came to the base of the tree, which represented the origin of the first life, the first living cell, Darwin had very little to say. In fact in the *Origin of Species*, he didn't even address the question of how life might have originated from non-living

matter" (30 minutes, 20 seconds into the movie). Consider also the climax of the movie, which among its conclusions directed at a general, non-scientific audience, states "Charles Darwin transformed science with his theory of natural selection. Today, that theory faces a formidable challenge. Intelligent design has sparked both discovery and intense debate over the origin of life on Earth" (64 minutes, 15 seconds). This statement is a non-sequitur, not only because it falsely implies that Darwin's theory is about the "origin of life on Earth," but also because ID is an attribution of agency without a theory of cause, and to the extent that it is applied to the appearance of biodiversity after life appeared, offers nothing that could substitute for a real mechanism of biological change.

13. Peretó, J., *et al.* 2009. Charles Darwin and the origin of life. *Origins of Life and Evolution of Biospheres,* 39: 395–406.
14. Darwin, C. 1861. *On the Origin of Species by Means of Natural Selection* (third edition), London: J. Murray, p.514.
15. From a letter to Charles Lyell in 1860, quoted from Peretó *et al.*, Charles Darwin and the origin of life, p.397.
16. Aquinas, T. *ca.* 1260. *Summa Contra Gentiles*, book 3, chapters 69–70, translation by V.J. Bourke (1957). New York: Hanover House. Available at: http://www.op-stjoseph.org/Students/study/thomas/ContraGentiles3a.htm (accessed December 27, 2009).
17. From an October 2008 debate with Bart Ehrman, hosted by the group "Socrates in San Francisco." Available at: http://www.socratesinsf.com/NT-WRIGHT (accessed October 28, 2010).
18. Psalms 147: 9.
19. Cardinal Schönborn's previous misunderstanding was his view that neo-Darwinism is a concept "invented to avoid the overwhelming evidence for purpose and design found in modern science" (from his July 2005 op-ed piece for the *New York Times*; available at: http://www.nytimes.com/2005/07/07/opinion/07schonborn.html (accessed May 28, 2010).
20. http://www.bringyou.to/CardinalSchonbornChancePurpose.mp3 (accessed June 29, 2010).
21. See Miller, 1999, *Finding Darwin's God*, p.218: "Ordinary processes, rooted in the genuine materialism of science, ought to be sufficient to allow for God's work, yesterday, today, and tomorrow." See also Alexander and White (2004) p. 133: "Nature is 'what God does'. Therefore all scientific descriptions—without exception—represent descriptions of the sustaining activities of God in the world around us. ... In Christian theology there is no 'two-tier' universe that one can split into the 'designed' portion and the 'undesigned' portion. ... Every aspect of it ... has God as its author."
22. Darwin, C. 1859. *On the Origin of Species by Means of Natural Selection*, London: J. Murray, p.188.

BIBLIOGRAPHY

Ahlberg, P.E. and Systematics Association, 2001. *Major Events in Early Vertebrate Evolution: Palaeontology, Phylogeny, Genetics, and Development*, London and New York: Taylor & Francis.

Airoldi, C.A., S. Bergonzi, and B. Davies, 2010. Single amino acid change alters the ability to specify male or female organ identity. *Proceedings of the National Academy of Sciences of the United States of America*, 107(44), 18898–902.

Akey, J.M., A.L. Ruhe, D.T. Akey, A.K. Wong, C.F. Connelly, J. Madeoy, T.J. Nicholas, and M.W. Neff, 2010. Tracking footprints of artificial selection in the dog genome. *Proceedings of the National Academy of Sciences of the United States of America*, 107(3), 1160–5.

Alberch, P., S.J. Gould, G.F. Oster, and D.B. Wake, 1979. Size and shape in ontogeny and phylogeny. *Paleobiology*, 5(3), 296–317.

Albertson, R.C., Y.L. Yan, T.A. Titus, E. Pisano, M. Vacchi, P.C. Yelick, H.W. Detrich, and J.H. Postlethwait, 2010. Molecular pedomorphism underlies craniofacial skeletal evolution in Antarctic notothenioid fishes. *BMC Evolutionary Biology*, 10, 4.

Alexander, D., 2008. *Creation or Evolution: Do We Have to Choose?*, Oxford: Monarch Books.

Alexander, D. and White, R.S. 2004. *Beyond Belief: Science, Faith, and Ethical Challenges*, Oxford: Lion Books.

Al-Hashimi, N., A.G. Lafont, S. Delgado, K. Kawasaki, and J.Y. Sire, 2010. The enamelin genes in lizard, crocodile, and frog and the pseudogene in the chicken provide new insights on enamelin evolution in tetrapods. *Molecular Biology and Evolution*, 27(9), 2078–94.

Allin, E.F., 1975. Evolution of the mammalian middle ear. *Journal of Morphology*, 147(4), 403–37.

Anderson, J.S., R.R. Reisz, D. Scott, N.B. Frobisch, and S.S. Sumida, 2008. A stem batrachian from the Early Permian of Texas and the origin of frogs and salamanders. *Nature*, 453(7194), 515–18.

Andrews, T.D., L.S. Jermiin, and S. Easteal, 1998. Accelerated evolution of cytochrome b in simian primates: adaptive evolution in concert with other mitochondrial proteins? *Journal of Molecular Evolution*, 47(3), 249–57.

Angielczyk, K.D., 2009. *Dimetrodon* is not a dinosaur: using tree thinking to understand the ancient relatives of mammals and their evolution. *Evolution: Education and Outreach*, 2, 257–71.

Ankel-Simons, F. and D.T. Rasmussen, 2008. Diurnality, nocturnality, and the evolution of primate visual systems. *American Journal of Physical Anthropology*, 47 (Supp.), 100–17.

Anton, M. and J. Morales (eds.), 2010. *Madrid Antes del Hombre*, Madrid: Comunidad de Madrid: Dirreción General de Patrimonio Histórico.

Aquinas, T. and V.J. Bourke, 1957. *Summa Contra Gentiles*, New York: Hanover House.

Arnason, U., A. Gullberg, A. Janke, J. Joss, and C. Elmerot, 2004. Mitogenomic analyses of deep gnathostome divergences: a fish is a fish. *Gene*, 333, 61–70.

Arnegard, M.E., D.J. Zwickl, Y. Lu, and H.H. Zakon, 2010. From the cover: old gene duplication facilitates origin and diversification of an innovative communication system – twice. *Proceedings of the National Academy of Sciences of the United States of America*, 107(51), 22172–7.

Asher, R.J., 2001. Cranial anatomy in tenrecid insectivorans: character evolution across competing phylogenies. *American Museum Novitates*, 3352, 1–54.

2005. Insectivoran-grade placental mammals: character evolution and fossil history, in *The Rise of Placental Mammals*, ed. K.D. Rose and D. Archibald, Baltimore, MD: Johns Hopkins University Press, 50–70.

2007. A web-database of mammalian morphology and a reanalysis of placental phylogeny. *BMC Evolutionary Biology*, 7, 108.

Asher, R.J. and D.M. Avery, 2010. New golden moles (Afrotheria, Chrysochloridae) from the Early Pliocene of South Africa. *Palaeontologia Electronica*, 13(1), 1–12.

Asher, R.J. and M. Hofreiter, 2006. Tenrec phylogeny and the noninvasive extraction of nuclear DNA. *Systems Biology*, 55(2), 181–94.

Asher, R.J. and T. Lehmann, 2008. Dental eruption in afrotherian mammals. *BMC Biology*, 6, 14.

Asher, R.J., N. Bennett, and T. Lehmann, 2009. The new framework for understanding placental mammal evolution. *Bioessays*, 31(8), 853–64.

Asher, R.J., R.J. Emry, and M.C. McKenna, 2005. New material of *Centetodon* (Mammalia, Lipotyphla) and the importance of (missing) DNA sequences in systematic paleontology. *Journal of Vertebrate Paleontology*, 25(4), 911–23.

Asher, R.J., M.C. McKenna, R.J. Emry, A.R. Tabrum, and D.G. Kron, 2002. Morphology and relationships of *Apternodus* and other extinct, zalambdodont, placental mammals. *Bulletin of the American Museum of Natural History*, 273, 1–117.

Asher, R.J., J. Meng, J.R. Wible, M.C. McKenna, G.W. Rougier, D. Dashzeveg, and M.J. Novacek, 2005. Stem Lagomorpha and the antiquity of Glires. *Science,* 307(5712), 1091–4.

Asher, R.J., M.J. Novacek, and J. Geisler, 2003. Relationships of endemic African mammals and their fossil relatives based on morphological and molecular evidence. *Journal of Mammalian Evolution,* 10, 131–94.

Avise, J.C., 2010. *Inside the Human Genome: A Case for Non-Intelligent Design*, Oxford and New York: Oxford University Press.

Bagarinao, T., 1986. Yolk resorption, onset of feeding and survival potential of larvae of three tropical marine fish species reared in the hatchery. *Marine Biology,* 91, 449–59.

Bajpai, S., J.G.M. Thewissen, and A. Sahni, 2009. The origin and early evolution of whales: macroevolution documented on the Indian Subcontinent. *Journal of Biosciences,* 34(5), 673–86.

Baldi, C., S. Cho, and R.E. Ellis, 2009. Mutations in two independent pathways are sufficient to create hermaphroditic nematodes. *Science,* 326(5955), 1002–5.

Barrick, J.E., D.S. Yu, S.H. Yoon, H. Jeong, T.K. Oh, D. Schneider, R.E. Lenski, and J.F. Kim, 2009. Genome evolution and adaptation in a long-term experiment with *Escherichia coli. Nature,* 461(7268), U1243–74.

Beall, C.M., G.L. Cavalleri, L.B. Deng, R.C. Elston, Y. Gao, J. Knight, C.H. Li, J.C. Li, Y. Liang, M. McCormack, H.E. Montgomery, H. Pan, P.A. Robbins, K.V. Shianna, S.C. Tam, N. Tsering, K.R. Veeramah, W. Wang, P.C. Wangdui, M.E. Weale, Y.M. Xu, Z. Xu, L. Yang, M.J. Zaman, C.Q. Zeng, L. Zhang, X.L. Zhang, P.C. Zhaxi, and Y.T. Zheng, 2010. Natural selection on EPAS1 (HIF2 alpha) associated with low hemoglobin concentration in Tibetan highlanders. *Proceedings of the National Academy of Sciences of the United States of America,* 107(25), 11459–64.

Beard, K.C., 1993. Phylogenetic systematics of the Primatomorpha, with special reference to Dermoptera, in *Mammal Phylogeny Vol. 2: Placentals*, ed. F.S. Szalay, M.J. Novacek, and M.C. McKenna, New York: Springer, 129–50.

Beard, K.C., Y.S. Tong, M.R. Dawson, J.W. Wang, and X.S. Huang, 1996. Earliest complete dentition of an anthropoid primate from the late middle Eocene of Shanxi Province, China. *Science,* 272(5258), 82–5.

Beatty, B.L. and A.C. Dooley, 2009. Injuries in a Mysticete skeleton from the Miocene of Virginia, with a discussion of buoyancy and the primitive feeding mode in the Chaeomysticeti. *Jeffersoniana,* 20, 1–27.

Beck, R.M., H. Godthelp, V. Weisbecker, M. Archer, and S.J. Hand, 2008. Australia's oldest marsupial fossils and their biogeographical implications. *PLoS One,* 3(3), e1858.

Behe, M.J., 1996. *Darwin's* Black Box*: The Biochemical Challenge to Evolution,* New York: Free Press.

2004. Irreducible complexity: obstacle to Darwinian evolution, in *Debating Design: From Darwin to DNA*, ed. W.A. Dembski and M. Ruse, Cambridge: Cambridge University Press, 352–70.

2007. *The Edge of Evolution: The Search for the Limits of Darwinism,* New York: Free Press.

Bell, M.A. and S.A. Foster, 1994. *The Evolutionary Biology of the Threespine Stickleback*, Oxford and New York: Oxford University Press.

Bell, M.A., W.E. Aguirre, and N.J. Buck, 2004. Twelve years of contemporary armor evolution in a threespine stickleback population. *Evolution,* 58(4), 814–24.

Belshaw, R., V. Pereira, A. Katzourakis, G. Talbot, J. Paces, A. Burt, and M. Tristem, 2004. Long-term reinfection of the human genome by endogenous retroviruses. *Proceedings of the National Academy of Sciences of the United States of America,* 101(14), 4894–9.

Benton, M.J., 1995. Diversification and extinction in the history of life. *Science,* **268**(5207), 52–8.

2003. *When Life Nearly Died: The Greatest Mass Extinction of All Time,* New York: Thames & Hudson.

Benton, M.J., P.C.J. Donoghue, and R.J. Asher, 2009. Calibrating and constraining molecular clocks, in *The Timetree of Life*, ed. S.B. Hedges and S. Kumar, Oxford: Oxford University Press, 35–86.

Best, S., P. LeTissier, G. Towers, and J.P. Stoye, 1996. Positional cloning of the mouse retrovirus restriction gene *Fv1*. *Nature,* **382**(6594), 826–9.

Betti, L., F. Balloux, W. Amos, T. Hanihara, and A. Manica, 2009. Distance from Africa, not climate, explains within-population phenotypic diversity in humans. *Proceedings of the Royal Society B: Biological Sciences,* **276**(1658), 809–14.

Biedler, J.K. and Z.J. Tu, 2010. Evolutionary analysis of the kinesin light chain genes in the yellow fever mosquito *Aedes aegypti*: gene duplication as a source for novel early zygotic genes. *BMC Evolutionary Biology,* **10**.

Bloch, J.I. and D.M. Boyer, 2002. Grasping primate origins. *Science,* **298**(5598), 1606–10.

Bloom, J.D. and F.H. Arnold, 2009. In the light of directed evolution: pathways of adaptive protein evolution. *Proceedings of the National Academy of Sciences of the United States of America,* **106** (Supp. 1), 9995–10000.

Blount, Z.D., C.Z. Borland, and R.E. Lenski, 2008. Historical contingency and the evolution of a key innovation in an experimental population of *Escherichia coli. Proceedings of the National Academy of Sciences of the United States of America,* **105**(23), 7899–906.

Bok, J., W. Chang, and D.K. Wu, 2007. Patterning and morphogenesis of the vertebrate inner ear. *International Journal of Development Biology,* **51**(6–7), 521–33.

Bookstein, F.L., P.D. Gingerich, and A.G. Kluge, 1978. Hierarchical linear modeling of tempo and mode of evolution. *Paleobiology,* **4**(2), 120–34.

Boyden, A. and D. Gemeroy, 1950. The relative position of the Cetacea among the orders of mammalia as indicated by precipitin tests. *Zoologica,* **35**, 145–51.

Bradley, B.J. and N.I. Mundy, 2008. The primate palette: the evolution of primate coloration. *Evolutionary Anthropology,* **17**(2), 97–111.

Bradley, B.J., S. Chester, and R.J. Asher, 2009. Why the long face? Runx2 tandem repeats and the evolution of primate prognathism. *Journal of Vertebrate Paleontology,* **29**(3), 69A.

Braunstein, E.M., D.C. Monks, V.S. Aggarwal, J.S. Arnold, and B.E. Morrow, 2009. *Tbx1* and *Brn4* regulate retinoic acid metabolic genes during cochlear morphogenesis. *BMC Developmental Biology,* **9**, 31.

Britten, R.J., 1997. Mobile elements inserted in the distant past have taken on important functions. *Gene,* **205**(1–2), 177–82.

Broom, R., 1916. On the structure of the skull in *Chrysochloris. Proceedings of the Zoological Society of London*, 1916, 449–58.

Brunet, M., F. Guy, D. Pilbeam, H.T. Mackaye, A. Likius, D. Ahounta, A. Beauvilain, C. Blondel, H. Bocherens, J.R. Boisserie, L. De Bonis, Y. Coppens, J. Dejax, C. Denys, P. Duringer, V.R. Eisenmann, G. Fanone, P. Fronty, D. Geraads, T. Lehmann, F. Lihoreau, A. Louchart, A. Mahamat, G. Merceron, G. Mouchelin, O. Otero, P.P. Campomanes, M.P. De Leon, J.C. Rage, M. Sapanet, M. Schuster, J. Sudre, P. Tassy, X. Valentin, P. Vignaud, L. Viriot, A. Zazzo, and C. Zollikofer, 2002. A new hominid from the Upper Miocene of Chad, central Africa. *Nature,* **418**(6894), 145–51.

Burrell, D., 1993. *Freedom and Causation in Three Traditions*, Notre Dame, IN: University of Notre Dame Press.

Caicedo, A.L., C. Richards, I.M. Ehrenreich, and M.D. Purugganan, 2009. Complex rearrangements lead to novel chimeric gene fusion polymorphisms at the *Arabidopsis thaliana* MAF2–5 flowering time gene cluster. *Molecular Biology and Evolution,* **26**(3), 699–711.

Carrasco, M.A. and J.H. Wahlert, 1999. The cranial anatomy of *Cricetops dormitor*, an Oligocene fossil rodent from Mongolia. *American Museum Novitates*, **3275**, 1–14.

Carroll, R.L., 1982. Early evolution of reptiles. *Annual Review of Ecology and Systematics*, **13**, 87–109.

1988. *Vertebrate Paleontology and Evolution*, New York: Freeman.

1997. Limits to knowledge of the fossil record. *Zoology*, **100**, 221–31.

Carroll, S.B., 2005. *Endless Forms Most Beautiful: The New Science of Evo Devo and the Making of the Animal Kingdom*, New York: Norton.

2006. *The Making of the Fittest: DNA and the Ultimate Forensic Record of Evolution*, New York: W.W. Norton.

2007. God as genetic engineer. *Science,* **316**, 1427–8.

Carroll, S.B., J.K. Grenier, and S.D. Weatherbee, 2005. *From DNA to Diversity: Molecular Genetics and the Evolution of Animal Design*, Malden, MA: Blackwell Publishers.

Carroll, W.E., 2000. Creation, evolution, and Thomas Aquinas. *Revue des Questions Scientifiques,* **171**(4), 319–47.

Chan, Y.F., M.E. Marks, F.C. Jones, G. Villarreal, M.D. Shapiro, S.D. Brady, A.M. Southwick, D.M. Absher, J. Grimwood, J. Schmutz, R.M. Myers, D. Petrov, B. Jonsson, D. Schluter, M.A. Bell, and D.M. Kingsley, 2010. Adaptive evolution of pelvic reduction in sticklebacks by recurrent deletion of a *Pitx1* enhancer. *Science,* **327**(5963), 302–5.

Chandrasekaran, C. and E. Betrán, 2008. Origins of new genes and pseudogenes. *Nature Education,* **1**(1). Available at: http://www.nature.com/scitable/topicpage/origins-of-new-genes-and-pseudogenes-835.

Charles, C., V. Lazzari, P. Tafforeau, T. Schimmang, M. Tekin, O. Klein, and L. Viriot, 2009. Modulation of Fgf3 dosage in mouse and men mirrors evolution of mammalian dentition. *Proceedings of the National Academy of Sciences of the United States of America*, 106(52), 22364–8.

Chatterjee, H.J., S.Y.W. Ho, I. Barnes, and C. Groves, 2009. Estimating the phylogeny and divergence times of primates using a supermatrix approach. *BMC Evolutionary Biology*, 9, 259.

Chen, F., W.Q. Liu, A. Eisenstark, R.N. Johnston, G.R. Liu, and S.L. Liu, 2010. Multiple genetic switches spontaneously modulating bacterial mutability. *BMC Evolutionary Biology*, 10, 277.

Churcher, A.M. and J.S. Taylor, 2009. Amphioxus (*Branchiostoma floridae*) has orthologs of vertebrate odorant receptors. *BMC Evolutionary Biology*, 9, 242.

Clack, J.A., 1997. Devonian tetrapod trackways and trackmakers: a review of the fossils and footprints. *Palaeogeography, Palaeoclimatology, Palaeoecology*, 130(1–4), 227–50.

2002. *Gaining Ground: The Origin and Evolution of Tetrapods*, Bloomington, IN: Indiana University Press.

Clarke, J.A., D.T. Ksepka, R. Salas-Gismondi, A.J. Altamirano, M.D. Shawkey, L. D' Alba, J. Vinther, T.J. DeVries, and P. Baby, 2010. Fossil evidence for evolution of the shape and color of penguin feathers. *Science*, 330(6006), 954–7.

Clements, A., D. Bursac, X. Gatsos, A.J. Perry, S. Civciristov, N. Celik, V.A. Likic, S. Poggio, C. Jacobs-Wagner, R.A. Strugnell, and T. Lithgow, 2009. The reducible complexity of a mitochondrial molecular machine. *Proceedings of the National Academy of Sciences of the United States of America*, 106(37), 15791–5.

Collins, F.S., 2007. *The Language of God: A Scientist Presents Evidence for Belief*, Waterville, ME: Wheeler Publishers.

Colosimo, P.F., K.E. Hosemann, S. Balabhadra, G. Villarreal, M. Dickson, J. Grimwood, J. Schmutz, R.M. Myers, D. Schluter, and D.M. Kingsley, 2005. Widespread parallel evolution in sticklebacks by repeated fixation of Ectodysplasin alleles. *Science*, 307(5717), 1928–33.

Conway-Morris, S., 2003. *Life's Solution: Inevitable Humans in a Lonely Universe*, Cambridge and New York: Cambridge University Press.

Cooke, J.P. 1864. *Religion and Chemistry: Proofs of God's Plan in the Atmosphere and its Elements*, New York: Charles Scribner.

Cote, S., L. Werdelin, E.R. Seiffert, and J.C. Barry, 2007. Additional material of the enigmatic Early Miocene mammal Kelba and its relationship to the order Ptolemaiida. *Proceedings of the National Academy of Sciences of the United States of America*, 104(13), 5510–15.

Coyne, J.A., 2009. Seeing and believing: the never-ending attempt to reconcile science and religion, and why it is doomed to fail. *The New Republic*, (February 4).

2009. *Why Evolution is True*, Oxford and New York: Oxford University Press.

Coyne, J.A. and B. Charlesworth, 1996. Mechanisms of punctuated evolution. *Science*, 274(5293), 1748–9.

Coyne, J.A. and H.E. Hoekstra, 2007. Evolution of protein expression: new genes for a new diet. *Current Biology*, 17(23), R1014–16.

Cresko, W.A., A. Amores, C. Wilson, J. Murphy, M. Currey, P. Phillips, M.A. Bell, C.B. Kimmel, and J.H. Postlethwait, 2004. Parallel genetic basis for repeated evolution of armor loss in Alaskan threespine stickleback populations. *Proceedings of the National Academy of Sciences of the United States of America*, **101**(16), 6050–5.

Daeschler, E.B., N.H. Shubin, and F.A. Jenkins, Jr., 2006. A Devonian tetrapod-like fish and the evolution of the tetrapod body plan. *Nature*, **440**(7085), 757–63.

Darwin, C., 1859. *On the Origin of Species by Means of Natural Selection*, London: J. Murray.

1860. *On the Origin of Species by Means of Natural Selection* (second edition), London: J. Murray.

1861. *On the Origin of Species by Means of Natural Selection* (third edition), London: J. Murray.

1872. *On the Origin of Species by Means of Natural Selection* (sixth edition), London: J. Murray.

Darwin, C. and A.R. Wallace, 1858. On the tendency of species to form varieties; and on the perpetuation of varieties and species by natural means of selection. *Journal of the Proceedings of the Linnean Society (Zoology)*, **3**(1859), 46–62.

Datta, P.M. and S. Ray, 2006. Earliest lizard from the Late Triassic (Carnian) of India. *Journal of Vertebrate Paleontology*, **26**(4), 795–800.

David-Gray, Z.K., J. Bellingham, M. Munoz, A. Avivi, E. Nevo, and R.G. Foster, 2002. Adaptive loss of ultraviolet-sensitive/violet-sensitive (UVS/VS) cone opsin in the blind mole rat (*Spalax ehrenbergi*). *European Journal of Neuroscience*, **16**(7), 1186–94.

Davis, P.W., D.H. Kenyon, and C.B. Thaxton, 1993. *Of Pandas and People: The Central Question of Biological Origins*, Dallas, TX: Haughton Pub. Co.

Dawkins, R., 1986. *The Blind Watchmaker*, New York: W.W.Norton.

1996. *Climbing Mount Improbable*, New York: W.W. Norton.

2006. *The God Delusion*, Boston, MA: Houghton Mifflin Co.

2009. *The Greatest Show on Earth: The Evidence for Evolution*, New York: Free Press.

Dawson, M.R., L. Marivaux, C.K. Li, K.C. Beard, and G. Metais, 2006. Laonastes and the "Lazarus effect" in recent mammals. *Science*, **311**(5766), 1456–8.

de Parseval, N. and T. Heidmann, 2005. Human endogenous retroviruses: from infectious elements to human genes. *Cytogenetic and Genome Research*, **110**(1–4), 318–32.

Dean, A.M. and J.W. Thornton, 2007. Mechanistic approaches to the study of evolution: the functional synthesis. *Nature Reviews Genetics*, **8**(9), 675–88.

Delsuc, F., G. Tsagkogeorga, N. Lartillot, and H. Philippe, 2008. Additional molecular support for the new chordate phylogeny. *Genesis*, **46**(11), 592–604.

Dembski, W.A. and M. Ruse, 2004. *Debating Design: From Darwin to DNA*, New York: Cambridge University Press.

Deméré, T.A., M.R. Mcgowen, A. Berta, and J. Gatesy, 2008. Morphological and molecular evidence for a stepwise evolutionary transition from teeth to baleen in mysticete whales. *Systematic Biology*, **57**(1), 15–37.

DeMuizon, C., 1998. *Mayulestes ferox*, a borhyaenoid (Metatheria, Mammalia) from the early Palaeocene of Bolivia: phylogenetic and paleobiologic implications. *Geodiversitas*, **20**, 19–142.

Denlinger, R.P. and D.R.H. O' Connell, 2010. Simulations of cataclysmic outburst floods from Pleistocene glacial lake Missoula. *Geological Society of America Bulletin*, **122**(5–6), 678–89.

D'Erchia, A.M., C. Gissi, G. Pesole, C. Saccone, and U. Arnason, 1996. The guinea-pig is not a rodent. *Nature*, **381**(6583), 597–600.

Desmond, A.J., 1984. *Archetypes and Ancestors: Palaeontology in Victorian London, 1850–1875*, Chicago, IL: University of Chicago Press.

DeVries, P., 1986. Naturalism in the natural sciences: a Christian perspective. *Christian Scholars Review*, **15**, 388–9.

Dial, K.P. 2003. Wing-assisted incline running and the evolution of flight. *Science*, **299**, 402–4.

Di-Poi, N., J.I. Montoya-Burgos., H. Miller, O. Pourquie, M.C. Milinkovitch, and D. Duboule, 2010. Changes in Hox genes' structure and function during the evolution of the squamate body plan. *Nature*, **464**, 99–103.

Dobzhansky, T.G., 1937. *Genetics and the Origin of Species*, New York: Columbia University Press.

Domning, D.P., 2001. The earliest known fully quadrupedal sirenian. *Nature*, **413**(6856), 625–7.

Doolittle, W.F. and E. Bapteste, 2007. Pattern pluralism and the Tree of Life hypothesis. *Proceedings of the National Academy of Sciences of the United States of America*, **104**(7), 2043–9.

Doolittle, W.F. and O. Zhaxybayeva, 2007. Evolution: reducible complexity – the case for bacterial flagella. *Current Biology*, **17**(13), R510–12.

Dulai, K.S., M. von Dornum, J.D. Mollon, and D.M. Hunt, 1999. The evolution of trichromatic color vision by opsin gene duplication in New World and Old World primates. *Genome Research*, **9**(7), 629–38.

Eldredge, N. and S.J. Gould, 1972. Punctuated equilibria: an alternative to phyletic gradualism, in *Models in Paleobiology*, ed. T.J.M. Schopf, San Francisco: Freeman & Cooper, 82–115.

Elinson, R.P. and Y. Beckham, 2002. Development in frogs with large eggs and the origin of amniotes. *Zoology*, **105**(2), 105–17.

Elmer, K.R., T.K. Lehtonen, and A. Meyer, 2009. Color assortative mating contributes to sympatric divergence of neotropical cichlid fish. *Evolution*, **63**(10), 2750–7.

Emry, R.J., 1970. *A North American Oligocene Pangolin and Other Additions to the Pholidota*, New York: American Museum of Natural History.

Endler, J.A., 1986. *Natural Selection in the Wild*, Princeton, NJ: Princeton University Press.

Falkowski, P.G., M.E. Katz, A.J. Milligan, K. Fennel, B.S. Cramer, M.P. Aubry, R.A. Berner, M.J. Novacek, and W.M. Zapol, 2005. The rise of oxygen over the past 205 million years and the evolution of large placental mammals. *Science*, **309**(5744), 2202–4.

Farmer, M.A. and A. Habura, 2010. Using Protistan examples to dispel the myths of intelligent design. *Journal of Eukaryotic Microbiology*, 57(1), 3–10.
Felsenstein, J., 2004. *Inferring Phylogenies*, Sunderland, MA: Sinauer Associates.
Felsenstein, J. 2007. Has natural selection been refuted? The arguments of William Dembslzi. *Reports of the National Center for Science Education*, 27(3–4), 20–6.
Ferrada, E. and A. Wagner, 2010. Evolutionary innovations and the organization of protein functions in genotype space. *PLoS ONE*, 5(11), e14172.
Finarelli, J.A. and W.C. Clyde, 2004. Reassessing hominoid phylogeny: evaluating congruence in the morphological and temporal data. *Paleobiology*, 30(4), 614–51.
Fischer, M.S., 1989. Hyracoids, the sister-group of perissodactyls, in *The Evolution of Perissodactyls*, ed. D.R. Prothero and R.M. Schoch, New York: Oxford University Press, 37–56.
Fisher, D.C., 2008. Stratocladistics: integrating temporal data and character data in phylogenetic inference. *Annual Review of Ecology, Evolution and Systematics*, 39, 365–85.
Fisher, R.A., 1930. *The Genetical Theory of Natural Selection*, Oxford: Clarendon Press.
Fitzgerald, E.M.G., 2006. A bizarre new toothed mysticete (Cetacea) from Australia and the early evolution of baleen whales. *Proceedings of the Royal Society B: Biological Sciences*, 273(1604), 2955–63.
2010. The morphology and systematics of *Mammalodon colliveri* (Cetacea: Mysticeti), a toothed mysticete from the Oligocene of Australia. *Zoological Journal of the Linnean Society*, 158(2), 367–476.
Fleagle, J.G., 1999. *Primate Adaptation and Evolution*, San Diego, CA: Academic Press.
Fondon, J.W. and H.R. Garner, 2004. Molecular origins of rapid and continuous morphological evolution. *Proceedings of the National Academy of Sciences of the United States of America*, 101(52), 18058–63.
Fondon, J.W., E.A.D. Hammock, A.J. Hannan, and D.G. King, 2008. Simple sequence repeats: genetic modulators of brain function and behavior. *Trends in Neurosciences*, 31(7), 328–34.
Fordyce, R.E., 2003. Early crown-group Cetacea in the southern ocean: the toothed archaic mysticete *Llanocetus*. *Journal of Vertebrate Paleontology*, 23(3), 50A.
Forrest, B. and P.R. Gross, 2007. *Creationism's Trojan Horse: The Wedge of Intelligent Design*, Oxford and New York: Oxford University Press.
Fraser, G.J., C.D. Hulsey, R.F. Bloomquist, K. Uyesugi, N.R. Manley, and J.T. Streelman, 2009. An ancient gene network is co-opted for teeth on old and new jaws. *PLoS Biology*, 7(2), e1000031.
Fraser, N.C. and M.J. Benton, 1989. The Triassic reptiles brachyrhinodon and polysphenodon and the relationships of the sphenodontids. *Zoological Journal of the Linnean Society*, 96(4), 413–45.
Friedman, M., 2008. The evolutionary origin of flatfish asymmetry. *Nature*, 454(7201), 209–12.
Frobisch, N.B., R.L. Carroll, and R.R. Schoch, 2007. Limb ossification in the Paleozoic branchiosaurid Apateon (Temnospondyli) and the early evolution of preaxial dominance in tetrapod limb development. *Evolution & Development*, 9(1), 69–75.

Galant, R. and S.B. Carroll, 2002. Evolution of a transcriptional repression domain in an insect Hox protein. *Nature*, **415**(6874), 910–13.

Gardiner, B.G., 1993. Haematothermia: warm-blooded amniotes. *Cladistics: The International Journal of the Willi Hennig Society*, **9**(4), 369–95.

Gatesy, J., C. Hayashi, M.A. Cronin, and P. Arctander, 1996. Evidence from milk casein genes that cetaceans are close relatives of hippopotamid artiodactyls. *Molecular Biology and Evolution*, **13**(7), 954–63.

Gaupp, E., 1899. Ontogenese und phylogenese des schallleitenden apparates bei den wirbeltieren. *Ergebnisse der Anatomie und Entwicklungsgeschichte*, **8**, 990–1149.

1912. Die reichertsche theorie, hammer-, amboss- und kieferfrage. *Archiv für Anatomie und Entwicklungsgeschichte*, **1913**, 1–416.

Gee, H., 2001. *Rise of the Dragon: Readings from Nature on the Chinese Fossil Record [Teng fei zhi long]*, Chicago, IL: University of Chicago Press.

Geisler, J.H. and J.M. Theodor, 2009. Hippopotamus and whale phylogeny. *Nature*, **458**(7236), E1–4.

Geisler, J.H. and M.D. Uhen, 2005. Phylogenetic relationships of extinct cetartiodactyls: results of simultaneous analyses of molecular, morphological, and stratigraphic data. *Journal of Mammalian Evolution*, **12**(1–2), 145–60.

Gerstein, M. and D.Y. Zheng, 2006. The real life of pseudogenes. *Scientific American*, **295**(2), 48–55.

Gheerbrant, E., 2009. Paleocene emergence of elephant relatives and the rapid radiation of African ungulates. *Proceedings of the National Academy of Sciences of the United States of America*, **106**(26), 10717–21.

Gheerbrant, E. and P. Tassy, 2009. L'origine et l'évolution des éléphants. *Comptes Rendus Palevol*, **8**(2–3), 281–94.

Gheerbrant, E., S. Peigné, and H. Thomas, 2007. Première description du squelette d'un Hyracoïde paléogène: *Saghatherium antiquum* de l'Oligocène inférieur de Jebel al Hasawnah, Libye. *Palaeontographica Abt A*, **279**, 93–145.

Gheerbrant, E., J. Sudre, H. Cappetta, M. Iarochene, M. Amaghzaz, and B. Bouya, 2002. A new large mammal from the Ypresian of Morocco: evidence of surprising diversity of early proboscideans. *Acta Palaeontologica Polonica*, **47**(3), 493–506.

Gheerbrant, E., J. Sudre, P. Tassy, M. Amaghzaz, B. Bouya, and M. Iarochene, 2005. Nouvelles données sur *Phosphatherium escuilliei* (Mammalia, Proboscidea) de l'Éocène inférieur du Maroc, apports à la phylogénie des Proboscidea et des ongulés lophodontes. *Geodiversitas*, **27**, 239–333.

Gifford, R. and M. Tristem, 2003. The evolution, distribution and diversity of endogenous retroviruses. *Virus Genes*, **26**(3), 291–316.

Gill, T., 1872. *Arrangement of the Families of Fishes*, Washington, DC: Smithsonian Institution.

Gingerich, P.D., 1976. Paleontology and phylogeny: patterns of evolution at species level in early Tertiary mammals. *American Journal of Science*, **276**(1), 1–28.

2003. Land-to-sea transition in early whales: evolution of Eocene Archaeoceti (Cetacea) in relation to skeletal proportions and locomotion of living semiaquatic mammals. *Paleobiology*, **29**(3), 429–54.

2004. Whale evolution, in *McGraw-Hill Yearbook of Science and Technology*, New York: McGraw-Hill.

Gingerich, P.D., S.M. Raza, M. Arif, M. Anwar, and X.Y. Zhou, 1994. New whale from the Eocene of Pakistan and the origin of cetacean swimming. *Nature*, **368**(6474), 844–7.

Gingerich, P.D., B.H. Smith, and E.L. Simons, 1990. Hind limbs of Eocene *Basilosaurus*: evidence of feet in whales. *Science*, **249**(4965), 154–7.

Gingerich, P.D., M. ul Haq, I.S. Zalmout, I.H. Khan, and M.S. Malkani, 2001. Origin of whales from early artiodactyls: hands and feet of eocene Protocetidae from Pakistan. *Science*, **293**(5538), 2239–42.

Goloboff, P., 1993. NONA version 1.9. Available at: www.cladistics.com.

Gould, S.J., 1977. *Ever since Darwin: Reflections in Natural History*, New York: W.W. Norton.

1977. *Ontogeny and Phylogeny*, Cambridge, MA: Belknap Press of Harvard University Press.

1980. Is a new and general-theory of evolution emerging? *Paleobiology*, **6**(1), 119–30.

1982. In praise of Charles Darwin. *Discover*, **3**(2), 20–5.

1986. The archaeopteryx flap: cardboard histories can be deceptive and destructive. *Natural History*, **95**(9), 16–25.

1987. *Time's Arrow, Time's Cycle: Myth and Metaphor in the Discovery of Geological Time*, Cambridge, MA: Harvard University Press.

1991. *Bully for Brontosaurus: Reflections in Natural History*, New York: W.W. Norton.

1992. Impeaching a self-appointed judge. *Scientific American*, **267**(1), 118–21.

1996. *Full House: The Spread of Excellence from Plato to Darwin*, New York: Harmony Books.

2002. *The Structure of Evolutionary Theory*, Cambridge, MA: Belknap Press of Harvard University Press.

Gould, S.J. and R.C. Lewontin, 1979. Spandrels of San-Marco and the Panglossian paradigm: a critique of the adaptationist program. *Proceedings of the Royal Society of London Series B: Biological Sciences*, **205**(1161), 581–98.

Gould, S.J. and E.S. Vrba, 1982. Exaptation: a missing term in the science of form. *Paleobiology*, **8**(1), 4–15.

Gradstein, F.M., J.G. Ogg, and A.G. Smith (eds.) 2004. *A Geologic Time Scale 2004*, Cambridge: Cambridge University Press.

Gray, A., 1876. *Darwiniana: Essays and Reviews Pertaining to Darwinism*, New York: D. Appleton and Company.

Haeckel, E., 1866. *Generelle Morphologie der Organismen: Allgemeine Grundzüge der organischen Formen-Wissenschaft, mechanisch begründet durch die von Charles Darwin reformirte Descendenz-Theorie*, Berlin: Georg Reimer.

1873. *Natürliche Schöpfungsgeschichte*, Berlin: Georg Reimer.

1903. *Anthropogenie: Oder, Entwickelungsgeschichte des menschen*, Leipzig: W. Engelmann.

Hällstrom, B.M. and A. Janke, 2009. Gnathostome phylogenomics utilizing lungfish EST sequences. *Molecular Biology and Evolution*, **26**(2), 463–71.

2010. Mammalian evolution may not be strictly bifurcating. *Molecular Biology and Evolution,* **27**(12), 2804–16.

Han, M.V., J.P. Demuth, C.L. McGrath, C. Casola, and M.W. Hahn, 2009. Adaptive evolution of young gene duplicates in mammals. *Genome Research,* **19**(5), 859–67.

Harding, R.M., E. Healy, A.J. Ray, N.S. Ellis, N. Flanagan, C. Todd, C. Dixon, A. Sajantila, I.J. Jackson, M.A. Birch-Machin, and J.L. Rees, 2000. Evidence for variable selective pressures at MC1R. *American Journal of Human Genetics,* **66**(4), 1351–61.

Hayden, S., M. Bekaert, T.A. Crider, S. Mariani, W.J. Murphy, and E.C. Teeling, 2010. Ecological adaptation determines functional mammalian olfactory subgenomes. *Genome Research,* **20**(1), 1–9.

Hearst, N. and S. Chen, 2004. Condom promotion for AIDS prevention in the developing world: is it working? *Studies in Family Planning,* **35**(1), 39–47.

Hedges, S.B., 1994. Molecular evidence for the origin of birds. *Proceedings of the National Academy of Sciences of the United States of America,* **91**(7), 2621–4.

Hedges, S.B., K.D. Moberg, and L.R. Maxson, 1990. Tetrapod phylogeny inferred from 18s-ribosomal and 28s-ribosomal RNA sequences and a review of the evidence for amniote relationships. *Molecular Biology and Evolution,* **7**(6), 607–33.

Heesy, C.P., 2009. Seeing in stereo: the ecology and evolution of primate binocular vision and stereopsis. *Evolutionary Anthropology,* **18**(1), 21–35.

Hinde, R.A., 2009. *Why Gods Persist: A Scientific Approach to Religion,* London and New York: Routledge.

Hitchens, C., 2007. *God is Not Great: How Religion Poisons Everything,* New York: Twelve.

Hoffmann, M., N. Tripathi, S.R. Henz, A.K. Lindholm, D. Weigel, F. Breden, and C. Dreyer, 2007. Opsin gene duplication and diversification in the guppy, a model for sexual selection. *Proceedings of the Royal Society B: Biological Sciences,* **274**(1606), 33–42.

Hooper, W. 1994. *God in the Dock: Essays on Theology and Ethics by CS Lewis,* Grand Rapids: Wm B Eerdmans.

Hopwood, N., 2006. Pictures of evolution and charges of fraud: Ernst Haeckel's embryological illustrations. *Isis,* **97**(2), 260–301.

Horovitz, I., T. Martin, J. Bloch, S. Ladeveze, C. Kurz, and M.R. Sanchez-Villagra, 2009. Cranial anatomy of the earliest marsupials and the origin of opossums. *PLoS One,* **4**(12), e8278.

Hu, Y.M., J. Meng, Y.Q. Wang, and C.K. Li, 2005. Large Mesozoic mammals fed on young dinosaurs. *Nature,* **433**(7022), 149–52.

Hu, Y., Y. Wang, Z. Luo, and C. Li, 1997. A new symmetrodont mammal from China and its implications for mammalian evolution. *Nature,* **390**(6656), 137–42.

Huang, X.G., M.N.Y. Hui, Y. Liu, D.S.H. Yuen, Y. Zhang, W.Y. Chan, H.R. Lin, S.H. Cheng, and C.H.K. Cheng, 2009. Discovery of a novel prolactin in non-mammalian vertebrates: evolutionary perspectives and its involvement in teleost retina development. *PLoS One,* **4**(7), e6163.

Hugall, A.F., R. Foster, and M.S.Y. Lee, 2007. Calibration choice, rate smoothing, and the pattern of tetrapod diversification according to the long nuclear gene RAG-1. *Systematic Biology,* **56**(4), 543–63.

Hulbert, R.C., R.M. Petkewich, G.A. Bishop, and P. Aleshire 1998. A new Middle Eocene protocetid whale (Mammalia: Cetacea: Archaeoceti) from Georgia. *Journal of Paleontology*, **72**(5), 907–27.

Hull, D.L., 1967. Certainty and circularity in evolutionary taxonomy. *Evolution*, **21**(1), 174–89.

Hunter, G.W., 1914. *A Civic Biology: Presented in Problems,* New York: American Book Company.

Hurles, M., 2004. Gene duplication: the genomic trade in spare parts. *PLoS Biology*, **2**(7), 900–4.

Huxley, J., 1942. *Evolution, the Modern Synthesis*, London: G. Allen & Unwin Ltd.

Huxley, T.H., 1863. *Man's Place in Nature*. New York: D. Appleton.

1871. *Evidence as to Man's Place in Nature,* New York: D. Appleton and Company.

Janke, A., N.J. Gemmell, G. Feldmaier-Fuchs, A. von Haeseler, and S. Paabo, 1996. The mitochondrial genome of a monotreme: the platypus (*Ornithorhynchus anatinus*). *Journal of Molecular Evolution,* **42**(2), 153–9.

Jernvall, J., S.V.E. Keranen, and I. Thesleff, 2000. Evolutionary modification of development in mammalian teeth: quantifying gene expression patterns and topography. *Proceedings of the National Academy of Sciences of the United States of America,* **97**(26), 14444–8.

Ji, Q., Z.X. Luo, and S.A. Ji, 1999. A Chinese triconodont mammal and mosaic evolution of the mammalian skeleton. *Nature,* **398**(6725), 326–30.

Ji, Q., Z.X. Luo, C.X. Yuan, J.R. Wible, J.P. Zhang, and J.A. Georgi, 2002. The earliest known eutherian mammal. *Nature,* **416**(6883), 816–22.

Ji, Q., Z.X. Luo, X.L. Zhang, C.X. Yuan, and L. Xu, 2009. Evolutionary development of the middle ear in Mesozoic therian mammals. *Science,* **326**(5950), 278–81.

Johnson, P.E., 1991. *Darwin on Trial*, Washington, DC and Lanham, MD: Regnery Gateway.

Kangas, A.T., A.R. Evans, I. Thesleff, and J. Jernvall, 2004. Nonindependence of mammalian dental characters. *Nature,* **432**(7014), 211–14.

Kant, I. and John Davis Batchelder Collection (Library of Congress), 1788. *Critik der practischen Vernunft*, Riga: J.F. Hartknoch.

Karro, J.E., Y.P. Yan, D.Y. Zheng, Z.L. Zhang, N. Carriero, P. Cayting, P. Harrrison, and M. Gerstein, 2007. Pseudogene.org: a comprehensive database and comparison platform for pseudogene annotation. *Nucleic Acids Research,* **35**, D55–60.

Kashi, Y. and D.G. King, 2006. Simple sequence repeats as advantageous mutators in evolution. *Trends in Genetics,* **22**(5), 253–9.

Kekalainen, J., G. Rudolfsen, M. Janhunen, L. Figenschou, N. Peuhkuri, N. Tamper, and R. Kortet, 2010. Genetic and potential non-genetic benefits increase offspring fitness of polyandrous females in non-resource based mating system. *BMC Evolutionary Biology,* **10**, 20.

Kemp, T.S., 1972. The jaw articulation and musculature of the whaitsiid Therocephalia, in *Studies in Vertebrate Evolution*, ed. K.A. Joysey and T.S. Kemp, Edinburgh: Oliver & Boyd, 213–30.

1982. *Mammal Like Reptiles and the Origin of Mammals*, London: Academic Press.

2005. *The Origin and Evolution of Mammals*, Oxford and New York: Oxford University Press.

2007. The origin of higher taxa: macroevolutionary processes, and the case of the mammals. *Acta Zoologica*, **88**(1), 3–22.

Kielan-Jaworowska, Z., R. Cifelli, and Z.-X. Luo, 2004. *Mammals from the Age of Dinosaurs: Origins, Evolution, and Structure,* New York: Columbia University Press.

Kim, Y.-O., 2000. *Karl Bogislaus Reichert (1811–1883) Sein Leben und seine Forschungen zur Anatomie und Entwicklungsgeschichte*, Mainz: Universität Mainz.

Kirschner, M.W. and J.C. Gerhart, 2005. *The Plausibility of Life: Resolving Darwin's Dilemma*. New Haven, CT: Yale University Press.

Kitano, J., D.I. Bolnick, D.A. Beauchamp, M.M. Mazur, S. Mori, T. Nakano, and C.L. Peichel, 2008. Reverse evolution of armor plates in the threespine stickleback. *Current Biology,* **18**(10), 769–74.

Konopka, G., J.M. Bomar, K. Winden, G. Coppola, Z.O. Jonsson, F.Y. Gao, S. Peng, T.M. Preuss, J.A. Wohlschlegel, and D.H. Geschwind, 2009. Human-specific transcriptional regulation of CNS development genes by FOXP2. *Nature,* **462**(7270), U213–89.

Korth, W.W., 1980. Paradjidaumo (Eomyidae, Rodentia) from the Brule Formation, Nebraska. *Journal of Paleontology,* **54**(5), 933–41.

Krauss, L.M. and R. Dawkins, 2007. Should science speak to faith? *Scientific American,* **297**(1), 88–91.

Kullberg, M., B.M. Hällstrom, U. Arnason, and A. Janke, 2008. Phylogenetic analysis of 1.5 Mbp and platypus EST data refute the Marsupionta hypothesis and unequivocally support Monotremata as sister group to Marsupialia/Placentalia. *Zoologica Scripta,* **37**(2), 115–27.

Lambeck, K., 1980. *The Earth's Variable Rotation: Geophysical Causes and Consequences*, Cambridge and New York: Cambridge University Press.

Larkin, M.A., G. Blackshields, N.P. Brown, R. Chenna, P.A. McGettigan, H. McWilliam, F. Valentin, I.M. Wallace, A. Wilm, R. Lopez, J.D. Thompson, T.J. Gibson, and D.G. Higgins, 2007. Clustal W and clustal X version 2.0. *Bioinformatics,* **23**(21), 2947–8.

Le Gros Clark, W.E. and C.F. Sonntag, 1926. A monograph of *Orycteropus afer*: III. The skull. *Proceedings of the Zoological Society of London,* **96**, 445–85.

Lee, M.S.Y., 1999. Molecular phylogenies become functional. *Trends in Ecology & Evolution,* **14**(5), 177–8.

Lee, M.S.Y. and A.B. Camens, 2009. Strong morphological support for the molecular evolutionary tree of placental mammals. *Journal of Evolutionary Biology,* **22**(11), 2243–57.

Li, C., X.C. Wu, O. Rieppel, L.T. Wang, and L.J. Zhao, 2008. An ancestral turtle from the Late Triassic of southwestern China. *Nature,* **456**(7221), 497–501.

Li, G.A. and P.W.H. Holland, 2010. The origin and evolution of ARGFX homeobox loci in mammalian radiation. *BMC Evolutionary Biology,* **10**, 182.

Li, R.Q., W. Fan, G. Tian, H.M. Zhu, L. He, J. Cai, Q.F. Huang, Q.L. Cai, B. Li, Y.Q. Bai, Z.H. Zhang, Y.P. Zhang, W. Wang, J. Li, F.W. Wei, H. Li, M. Jian, J.W. Li, Z.L. Zhang, R. Nielsen, D.W. Li, W.J. Gu, Z.T. Yang, Z.L. Xuan, O.A. Ryder, F.C.C. Leung, Y. Zhou, J.J. Cao, X. Sun, Y.G. Fu, X.D. Fang, X.S. Guo, B. Wang, R. Hou, F.J. Shen, B. Mu, P.X. Ni, R.M. Lin, W.B. Qian, G.D. Wang, C. Yu, W.H. Nie, J.H. Wang, Z.G. Wu, H.Q. Liang, J.M. Min, Q. Wu, S.F. Cheng, J. Ruan, M.W. Wang, Z.B. Shi, M. Wen, B.H. Liu, X.L. Ren, H.S. Zheng, D. Dong, K. Cook, G. Shan, H. Zhang, C. Kosiol, X.Y. Xie, Z.H. Lu, H.C. Zheng, Y.R. Li, C.C. Steiner, T.T.Y. Lam, S.Y. Lin, Q.H. Zhang, G.Q. Li, J. Tian, T.M. Gong, H.D. Liu, D.J. Zhang, L. Fang, C. Ye, J.B. Zhang, W.B. Hu, A.L. Xu, Y.Y. Ren, G.J. Zhang, M.W. Bruford, Q.B. Li, L.J. Ma, Y.R. Guo, N. An, Y.J. Hu, Y. Zheng, Y.Y. Shi, Z.Q. Li, Q. Liu, Y.L. Chen, J. Zhao, N. Qu, S.C. Zhao, F. Tian, X.L. Wang, H.Y. Wang, L.Z. Xu, X. Liu, T. Vinar, Y.J. Wang, T.W. Lam, S.M. Yiu, S.P. Liu, H.M. Zhang, D.S. Li, Y. Huang, X. Wang, G.H. Yang, Z. Jiang, J.Y. Wang, N. Qin, L. Li, J.X. Li, L. Bolund, K. Kristiansen, G.K.S. Wong, M. Olson, X.Q. Zhang, S.G. Li, H.M. Yang, J. Wang, and J. Wang, 2010. The sequence and de novo assembly of the giant panda genome. *Nature,* 463(7279), 311–17.

Lindberg, D.C. and R.L. Numbers, 2003. *When Science & Christianity Meet,* Chicago, IL: University of Chicago Press.

Liu, R.Y. and H. Ochman, 2007. Stepwise formation of the bacterial flagellar system. *Proceedings of the National Academy of Sciences of the United States of America,* 104(17), 7116–21.

Lockwood, C.A., 2007. *The Human Story,* London: The Natural History Museum.

Long, J.A., M. Archer, T. Flannery, and S. Hand, 2002. *Prehistoric Mammals of Australia and New Guinea: One Hundred Million Years of Evolution,* Baltimore, MD: Johns Hopkins University Press.

Lubosch, W., 1919. Was verdankt die vergleichend-anatomische Wissenschaft den Arbeiten Goethes? *Jahrbuch der Goethe Gesellschaft,* 6, 157–91.

Luckett, W.P., 1996. Ontogenetic evidence for incisor homologies in proboscideans, in *The Proboscidea,* ed. J. Shoshani and P. Tassy, Oxford: Oxford University Press, 26–31.

Luís, C., C. Bastos-Silveira, E.G. Cothran, and M.D. Oom, 2006. Iberian origins of New World horse breeds. *Journal of Heredity,* 97(2), 107–13.

Luo, Z.X., 2007. Transformation and diversification in early mammal evolution. *Nature,* 450, 1011–19.

Luo, Z.X. and A.W. Crompton, 1994. Transformation of the quadrate (incus) through the transition from non-mammalian cynodonts to mammals. *Journal of Vertebrate Paleontology,* 14(3), 341–74.

Luo, Z.X., P.J. Chen, G. Li, and M. Chen, 2007. A new eutriconodont mammal and evolutionary development in early mammals. *Nature,* 446(7133), 288–93.

Luo, Z.X., Q. Ji, J.R. Wible, and C.X. Yuan, 2003. An early Cretaceous tribosphenic mammal and metatherian evolution. *Science,* 302(5652), 1934–40.

Luo, Z.X., Z. Kielan-Jaworowska, and R.L. Cifelli, 2004. Evolution of dental replacement in mammals. *Bulletin Carnegie Museum of Natural History,* 36, 159–75.

Luo, Z.X., I. Ruf, J.A. Schultz, and T. Martin, 2011. Fossil evidence on evolution of inner ear cochlea in Jurassic mammals. *Proceedings of the Royal Society B: Biological Sciences,* **278**(1702), 28–34.

MacFadden, B.J., 1992. *Fossil Horses: Systematics, Paleobiology, and Evolution of the Family Equidae*, Cambridge and New York: Cambridge University Press.

MacPhee, R.D.E. and I. Horovitz, 2004. New craniodental remains of the Quaternary Jamaican monkey *Xenothrix mcgregori* (Xenotrichini, Callicebinae, Pitheciidae), with a reconsideration of the *Aotus* hypothesis. *American Museum Novitates,* **3434**, 1–51.

Macqueen, D.J., M.L. Delbridge, S. Manthri, and I.A. Johnston, 2010. A newly classified vertebrate calpain protease, directly ancestral to capn1 and 2, episodically evolved a restricted physiological function in placental mammals. *Molecular Biology and Evolution,* **27**(8), 1886–902.

Macrini, T.E., C. deMuizon, R.L. Cifelli, and T. Rowe, 2007. Digital cranial endocast of *Pucadelphys andinus*, a Paleocene metatherian. *Journal of Vertebrate Paleontology*, **27**, 99–107.

Maglio, V.J., 1973. Origin and evolution of the Elephantidae. *Transactions of the American Philosophical Society*, **63**(3): 1–149.

Maier, W., 1987. Der Processus angularis bei *Monodelphis domestica* (Didelphidae; Marsupialia) und seine Beziehungen zum Mittelohr: eine ontogenetische und evolutionsmorphologische Untersuchung. *Gegenbaurs Morphologisches Jahrbuch*, **133**, 123–61.

Manger, P.R. and J.D. Pettigrew, 1995. Electroreception and the feeding behaviour of the platypus. *Philosophical Transactions of the Royal Society of London B,* **347**, 359–81.

Manica, A., W. Amos, F. Balloux, and T. Hanihara, 2007. The effect of ancient population bottlenecks on human phenotypic variation. *Nature*, **448**(7151), 346–8.

Mayr, E., 1942. *Systematics and the Origin of Species*, New York: Columbia University Press.

McElhinny, M.W. 1973. *Palaeomagnetism and Plate Tectonics*, Cambridge: Cambridge University Press.

Medawar, P.B., 1984. *The Limits of Science*, New York: Harper & Row.

Meng, J., Y. Hu, and C. Li, 2003. The osteology of *Rhombomylus* (Mammalia, Glires): implications for phylogeny and evolution of Glires. *Bulletin of the American Museum of Natural History*, **275**, 1–247.

Meredith, R.W., J. Gatesy, W.J. Murphy, O.A. Ryder, and M.S. Springer, 2009. Molecular decay of the tooth gene enamelin (ENAM) mirrors the loss of enamel in the fossil record of placental mammals. *PLoS Genetics,* **5**(9), e1000634.

Meyer, A. and R. Zardoya, 2003. Recent advances in the (molecular) phylogeny of vertebrates. *Annual Review of Ecology, Evolution, and Systematics,* **34**, 311–38.

Meyer, S.C., 2004. The origin of biological information and the higher taxonomic categories. *Proceedings of the Biological Society of Washington,* **117**(2), 213–39.

2009. *Signature in the Cell: DNA and the Evidence for Intelligent Design*, New York: HarperOne.

Meyer, S.C., S. Minnich, J. Moneymaker, P.A. Nelson, and R. Seelke, 2007. *Explore Evolution: The Arguments For and Against Neo-Darwinism,* Melbourne: Hill House.

Mi, S., X. Lee, X.P. Li, G.M. Veldman, H. Finnerty, L. Racie, E. LaVallie, X.Y. Tang, P. Edouard, S. Howes, J.C. Keith, and J.M. McCoy, 2000. Syncytin is a captive retroviral envelope protein involved in human placental morphogenesis. *Nature,* **403**(6771), 785–9.

Miller, K.B., 2009. The misguided attack on methodological naturalism, in *For the Rock Record*, ed. J.S. Schneidermann and W.D. Allmon, Berkeley, CA: University of California Press, 117–40.

Miller, K.R., 1996. A review of Darwin's *Black Box. Creation/Evolution,* **16**, 36–40.

1999. *Finding Darwin's God: A Scientist's Search for Common Ground between God and Evolution*, New York: Cliff Street Books.

2004. The flagellum unspun: the collapse of "irreducible complexity," in *Debating Design: from Darwin to DNA*, ed. M.A. Dembski and M. Ruse, Cambridge: Cambridge University Press, 81–97.

2007. Falling over the edge. *Nature,* **447**, 1055–6.

2008. *Only a Theory: Evolution and the Battle for America's Soul,* New York: Viking Penguin.

Milne-Edwards, H., 1844. Considérations sur quelques principes relatifs à la clasification naturelle des animaux, et plus particulièrement sur la distribution méthodique des mammifères. *Annales des Sciences Naturelles (Zoologie) 3d ser.,* **1**, 65–99.

Milner, A.C. and S.E.K. Sequeira, 1994. The temnospondyl amphibians from the Viséan of East Kirkton, West Lothian, Scotland. *Transactions of the Royal Society of Edinburgh, Earth Sciences,* **84**, 331–61.

Minguillon, C., J.J. Gibson-Brown, and M.P. Logan, 2009. Tbx4/5 gene duplication and the origin of vertebrate paired appendages. *Proceedings of the National Academy of Sciences of the United States of America,* **106**(51), 21726–30.

Minnich, S. and S.C. Meyer, 2004. Genetic analysis of coordinate flagellar and type III regulatory circuits, in *Proceedings of the Second International Conference on Design & Nature*, ed. M.W. Collins and C.A. Brebbia, Rhodes: WIT Press, 8.

Mitchell, C.E., 2009. It's not about the evidence: the role of metaphysics in the debate, in *For the Rock Record*, ed. J.S. Schneidermann and W.D. Allmon, Berkeley, CA: University of California Press, 93–116.

Mivart, S.G.J., 1871. *On the Genesis of Species,* London: Macmillan and Co.

Moore, R., 2004. Standing up for our profession: a talk with Ken Hubert. *American Biology Teacher,* **66**(5), 325–7.

2004. When a biology teacher refuses to teach evolution: a talk with Rod LeVake. *American Biology Teacher,* **66**(4), 246–50.

Morales, J., M. Pozo, P.G. Silva, M.S. Domingo, R. López-Antoñanzas, M.A. Álvarez Sierra, M. Antón, C. Martín Escorza, V. Quiralte, M.J. Salesa, I.M. Sánchez, B. Azanza, J.P. Calvo, P. Carrasco, I. García-Paredes, F. Knoll, M. Hernández Fernández, L. van den Hoek Ostende, L. Merino, A.J. van der Meulen, P. Montoya, S. Peigné, P. Peláez-Campomanes, A. Sánchez-Marco, A. Turner,

J. Abella, G.M. Alcalde, M. Andrés, D. DeMiguel, J.L. Cantalapiedra, S. Fraile, B.A. García Yelo, A.R. Gómez Cano, P. López Guerrero, A. Oliver Pérez, and G. Siliceo, 2008. El sistema de yacimientos de mamíferos miocenos del Cerro de los Batallones, Cuenca de Madrid: estado actual y perspectivas. *Palaeontologica Nova: Seminario de Paleontología de Zaragoza*, 8, 41–117.

Moreland, J.P. and J.M. Reynolds, 1999. *Three Views on Creation and Evolution*, Grand Rapids, MI: Zondervan Publishing House.

Morris, T. and D.N. Petcher, 2006. *Science & Grace: Gods Reign in the Natural Sciences*, Wheaton, IL: Crossway Books.

Mossman, H.W., 1937. *Comparative Morphogenesis of the Fetal Membranes and Accessory Uterine Structures*, Washington, DC.

Murphy, W.J., E. Eizirik, S.J. O'Brien, O. Madsen, M. Scally, C.J. Douady, E. Teeling, O.A. Ryder, M.J. Stanhope, W.W. de Jong, and M.S. Springer, 2001. Resolution of the early placental mammal radiation using Bayesian phylogenetics. *Science*, **294**(5550), 2348–51.

Murphy, W.J., T.H. Pringle, T.A. Crider, M.S. Springer, and W. Miller. 2007. Using genomic data to unravel the root of the placental mammal phylogeny. *Genome Research*, 17(4), 413–21.

Negoro, S., T. Ohki, N. Shibata, N. Mizuno, Y. Wakitani, J. Tsurukame, K. Matsumoto, I. Kawamoto, M. Takeo, and Y. Higuchi, 2005. X-ray crystallographic analysis of 6-aminohexanoate-dimer hydrolase: molecular basis for the birth of a nylon oligomer-degrading enzyme. *Journal of Biological Chemistry*, **280**(47), 39644–52.

Neitz, M., J. Neitz, and G.H. Jacobs, 1991. Spectral tuning of pigments underlying red–green color vision. *Science*, **252**(5008), 971–4.

Ni, X., Y. Wang, Y. Hu, and C. Li, 2004. A euprimate skull from the early Eocene of China. *Nature*, **427**(6969), 65–8.

Niedzwiedzki, G., P. Szrek, K. Narkiewicz, M. Narkiewicz, and P.E. Ahlberg, 2010. Tetrapod trackways from the early Middle Devonian period of Poland. *Nature*, **463**(7277), 43–8.

Nilsson, M.A., U. Arnason, P.B.S. Spencer, and A. Janke, 2004. Marsupial relationships and a timeline for marsupial radiation in South Gondwana. *Gene*, **340**(2), 189–96.

Norell, M.A. and M.J. Novacek, 1992. The fossil record and evolution: comparing cladistic and paleontological evidence for vertebrate history. *Science*, **255**(5052), 1690–3.

Norell, M.A. and X. Xu, 2005. Feathered dinosaurs. *Annual Review of Earth and Planetary Sciences*, 33, 277–99.

Novacek, M.J., 1986. The skull of leptictid insectivorans and the higher-level classification of eutherian mammals. *Bulletin of the American Museum of Natural History*, 183(1), 1–111.

1992. Mammalian phylogeny: shaking the tree. *Nature*, **356**(6365), 121–5.

Novacek, M.J., G.W. Rougier, J.R. Wible, M.C. McKenna, D. Dashzeveg, and I. Horovitz, 1997. Epipubic bones in eutherian mammals from the late Cretaceous of Mongolia. *Nature*, **389**(6650), 483–6.

Numbers, R.L., 2003. Science without God: natural laws and Christian beliefs, in *When Science and Christianity Meet*, ed. D.C. Lindberg and R.L. Numbers, Chicago, IL: University of Chicago Press, 265–85.

2006. *The Creationists*, Cambridge, MA: Harvard.

O'Connor, P.M. and L.P.A.M. Claessens, 2005. Basic avian pulmonary design and flow-through ventilation in non-avian theropod dinosaurs. *Nature*, **436**(7048), 253–6.

O'Connor, P.M., J.J. Sertich, N.J. Stevens, E.M. Roberts, M.D. Gottfried, T.L. Hieronymus, Z.A. Jinnah, R. Ridgely, S.E. Ngasala, and J. Temba. 2010. The evolution of mammal-like crocodyliforms in the Cretaceous Period of Gondwana. *Nature*, **466**(7307), 748–51.

Ohno, S., 1970. *Evolution by Gene Duplication*, Berlin and New York: Springer-Verlag.

1984. Birth of a unique enzyme from an alternative reading frame of the preexisted, internally repetitious coding sequence. *Proceedings of the National Academy of Sciences of the United States of America – Biological Sciences*, **81**(8), 2421–5.

Okamura, K., L. Feuk, T. Marques-Bonet, A. Navarro, and S.W. Scherer, 2006. Frequent appearance of novel protein-coding sequences by frameshift translation. *Genomics*, **88**(6), 690–7.

O'Leary, M.A., S.G. Lucas, and T.E. Williamson, 2000. A new specimen of ankalagon (Mammalia, Mesonychia) and evidence of sexual dimorphism in mesonychians. *Journal of Vertebrate Paleontology*, **20**(2), 387–93.

Orliac, M., J.R. Boisserie, L. MacLatchy, and F. Lihoreau, 2010. Early Miocene hippopotamids (Cetartiodactyla) constrain the phylogenetic and spatiotemporal settings of hippopotamid origin. *Proceedings of the National Academy of Sciences of the United States of America*, **107**(26), 11871–6.

Ostrom, J.H., 1976. Archeopteryx and origin of birds. *Biological Journal of the Linnean Society*, **8**(2), 91–182.

Owen, R., 1841. Observations on the basilosaurus of Dr. Harlan (*Zeuglodon cetoides*, Owen). *Transactions of the Geological Society of London*, **6**(2), 69–79.

2009. Vatican says evolution does not prove the non-existence of God. *Times Online*, March 6. Available at: http://www.timesonline.co.uk/tol/comment/faith/article5859797.ece (accessed November 7, 2009).

Padian, K., 2002. Waiting for the watchmaker. *Science*, **295**, 2373.

2008. Trickle-down evolution: an approach to getting major evolutionary adaptive changes into textbooks and curricula. *Integrative and Comparative Biology*, **48**(2), 175–88.

2009. Truth or consequences? Engaging the "truth" of evolution. *PLoS Biology*, **7**(3), e1000077.

Padian, K. and K.D. Angielczyk, 2007. Transitional forms versus transitional features, in *Scientists Confront Creationism: Intelligent Design and Beyond*, ed. A.J. Petto and L.R. Godfrey, New York: W.W. Norton, 197–231.

Page, R.D.M. and M.A. Charleston, 1999. Comments on Allard and Carpenter (1996), or the "aquatic ape" hypothesis revisited. *Cladistics: The International Journal of the Willi Hennig Society*, **15**(1), 73–4.

Paley, W., 1802. *Natural Theology*, London: Printed for R. Faulder by Wilks and Taylor.

Pallen, M.J. and N.J. Matzke, 2006. From *The Origin of Species* to the origin of bacterial flagella. *Nature Reviews Microbiology*, **4**(10), 784–90.

Panchen, A.L., 1992. *Classification, Evolution, and the Nature of Biology*, Cambridge and New York: Cambridge University Press.

Parker, H.G., B.M. VonHoldt, P. Quignon, E.H. Margulies, S. Shao, D.S. Mosher, T.C. Spady, A. Elkahloun, M. Cargill, P.G. Jones, C.L. Maslen, G.M. Acland, N.B. Sutter, K. Kuroki, C.D. Bustamante, R.K. Wayne, and E.A. Ostrander, 2009. An expressed Fgf4 retrogene is associated with breed-defining chondrodysplasia in domestic dogs. *Science*, **325**(5943), 995–8.

Pennock, R.T., 1999. *Tower of Babel: The Evidence Against the new Creationism*, Cambridge, MA: MIT Press.

2007. God of the gaps: the argument from ignorance and the limits of methodological naturalism, in *Scientists Confront Intelligent Design and Creationism*, ed. A.J. Petto and L.R. Godfrey, New York: W.W. Norton, 309–38.

Penny, D., L.R. Foulds, and M.D. Hendy, 1982. Testing the theory of evolution by comparing phylogenetic trees constructed from 5 different protein sequences. *Nature*, **297**(5863), 197–200.

Peretó, J., J.L. Bada, and A. Lazcano, 2009. Charles Darwin and the origin of life. *Origins of Life and Evolution of Biospheres*, **39**(5), 395–406.

Pergams, O.R.W. and J.J. Lawler, 2009. Recent and widespread rapid morphological change in rodents. *PLoS One*, **4**(7), e6452.

Petto, A.J. and L.R. Godfrey, 2007. *Scientists Confront Intelligent Design and Creationism*, New York: W.W. Norton.

Phillips, M.J., T.H. Bennett, and M.S.Y. Lee, 2009. Molecules, morphology, and ecology indicate a recent, amphibious ancestry for echidnas. *Proceedings of the National Academy of Sciences of the United States of America*, **106**(40), 17089–94.

Phipps, W.E., 1983. Darwin, the scientific creationist. *Christian Century*, **1983**, 809–11.

Pickett, K.M., J.W. Wenzel, and S.W. Rissing, 2005. Iconoclasts of evolution: Haeckel, Behe, Wells & the ontogeny of a fraud. *American Biology Teacher*, **67**(5), 275–82.

Pointer, M.A. and N.I. Mundy, 2008. Testing whether macroevolution follows microevolution: are colour differences among swans (*Cygnus*) attributable to variation at the MC1R locus? *BMC Evolutionary Biology*, **8**, 249.

Pool, C.A., 2007. *Olmec Archaeology and Early Mesoamerica*, Cambridge and New York: Cambridge University Press.

Prager, E.M. and A.C. Wilson, 1975. Slow evolutionary loss of potential for interspecific hybridization in birds: manifestation of slow regulatory evolution. *Proceedings of the National Academy of Sciences of the United States of America*, **72**(1), 200–4.

Preyer, William Thierry, 1891. Briefe von Darwin. Mit Erinnerungen und Erlaeuterungen. *Deutsche Rundschau*, **17**(9): 357. Available at: http://darwin-online.org.uk/content/frameset?viewtype=text&itemID=F6&pageseq=1 (accessed April 4, 2010).

Prothero, D.R. 2007. *Evolution: What the Fossils Say and Why It Matters*, New York: Columbia University Press.

Prothero, D.R. and R.M. Schoch, 1989. *The Evolution of Perissodactyls*, New York: Clarendon Press and Oxford University Press.

2002. *Horns, Tusks, and Flippers: The Evolution of Hoofed Mammals*, Baltimore, MD: Johns Hopkins University Press.

Pulido, F. and P. Berthold, 2010. Current selection for lower migratory activity will drive the evolution of residency in a migratory bird population. *Proceedings of the National Academy of Sciences of the United States of America*, 107(16), 7341–6.

Radin, M. 1931. *The Trial of Jesus of Nazareth*, Chicago, IL: University of Chicago Press.

Rage, J.C. and Z. Rocek, 1989. Redescription of *Triadobatrachus massinoti* (Piveteau, 1936) an anuran amphibian from the early Triassic. *Palaeontographica Abt. A*, **206**, 1–16.

Ravi, V., K. Lam, B.H. Tay, A. Tay, S. Brenner, and B. Venkatesh, 2009. Elephant shark (*Callorhinchus milii*) provides insights into the evolution of Hox gene clusters in gnathostomes. *Proceedings of the National Academy of Sciences of the United States of America*, 106(38), 16327–32.

Reeve, H.K. and P.W. Sherman, 1993. Adaptation and the goals of evolutionary research. *Quarterly Review of Biology*, **68**(1), 1–32.

Reichert, K.B., 1837. Über die Visceralbogen der Wirbelthiere im allgemeinen und deren Metamorphosen bei den Vögeln und Säugethieren. *Archiv für Anatomie, Physiologie, und wissenschaftliche Medizin*, **1837**, 120–220.

Reimchen, T.E., 1994. Predators and evolution in threespine stickleback, in *The Evolutionary Biology of the Threespine Stickleback*, ed. M.A. Bell and S.A. Foster, Oxford and New York: Oxford University Press.

Reznick, D.N., M. Mateos, and M.S. Springer, 2002. Independent origins and rapid evolution of the placenta in the fish genus *Poeciliopsis*. *Science*, **298**(5595), 1018–20.

Reznick, D.N., F.H. Shaw, F.H. Rodd, and R.G. Shaw, 1997. Evaluation of the rate of evolution in natural populations of guppies (*Poecilia reticulata*). *Science*, 275(5308), 1934–7.

Richards, R.J., 2008. *The Tragic Sense of Life: Ernst Haeckel and the Struggle over Evolutionary Thought*, Chicago, IL: University of Chicago Press.

Richardson, M.K., J. Hanken, L. Selwood, G.M. Wright, R.J. Richards, C. Pieau, and A. Raynaud, 1998. Haeckel, embryos, and evolution. *Science*, 280(5366), 985–6.

Rivera, A.S., M.S. Pankey, D.C. Plachetzki, C. Villacorta, A.E. Syme, J.M. Serb, A.R. Omilian, and T.H. Oakley, 2010. Gene duplication and the origins of morphological complexity in pancrustacean eyes, a genomic approach. *BMC Evolutionary Biology*, 10, 123.

Roberts, M.D., 2007. *Can We Trust the Gospels? Investigating the Reliability of Matthew, Mark, Luke, and John*, Wheaton, IL: Crossway Books.

Rogers, R.L., T. Bedford, A.M. Lyons, and D.L. Hartl, 2010. Adaptive impact of the chimeric gene Quetzalcoatl in *Drosophila melanogaster*. *Proceedings of the National Academy of Sciences of the United States of America*, 107(24), 10943–8.

Rohner, N., M. Bercsenyi, L. Orban, M.E. Kolanczyk, D. Linke, M. Brand, C. Nusslein-Volhard, and M.P. Harris, 2009. Duplication of Fgfr1 permits Fgf signaling to serve as a target for selection during domestication. *Current Biology,* 19(19), 1642–7.

Romero, P.A. and F.H. Arnold, 2009. Exploring protein fitness landscapes by directed evolution. *Nature Reviews Molecular Cell Biology,* 10(12), 866–76.

Ronquist, F. and J.P. Huelsenbeck, 2003. MrBayes 3: Bayesian phylogenetic inference under mixed models. *Bioinformatics,* 19(12), 1572–4.

Rose, D.B. and R. Davis, 2005. *Dislocating the Frontier: Essaying the Mystique of the Outback*, Canberra: ANU E Press.

Rose, K.D., 2006. *The Beginning of the Age of Mammals,* Baltimore, MD: Johns Hopkins University Press.

Rosenbaum, J.E., 2009. Patient teenagers? A comparison of the sexual behavior of Virginity pledgers and matched nonpledgers. *Pediatrics,* 123(1), E110–20.

Rougier, G.W., J.R. Wible, and M.J. Novacek, 1998. Implications of Deltatheridium specimens for early marsupial history. *Nature,* 396(6710), 459–63.

Rybczynski, N., M.R. Dawson, and R.H. Tedford, 2009. A semi-aquatic Arctic mammalian carnivore from the Miocene epoch and origin of Pinnipedia. *Nature,* 458(7241), 1021–4.

Salazar-Ciudad, I. and J. Jernvall, 2010. A computational model of teeth and the developmental origins of morphological variation. *Nature,* 464(7288), 583–6.

Salzburger, W. and A. Meyer, 2004. The species flocks of East African cichlid fishes: recent advances in molecular phylogenetics and population genetics. *Naturwissenschaften,* 91(6), 277–90.

Sánchez-Villagra, M.R., O. Aguilera, and I. Horovitz, 2003. The anatomy of the world's largest extinct rodent. *Science,* 301(5640), 1708–10.

Sánchez-Villagra, M.R., S. Gemballa, S. Nummela, K.K. Smith, and W. Maier, 2002. Ontogenetic and phylogenetic transformations of the ear ossicles in marsupial mammals. *Journal of Morphology,* 251(3), 219–38.

Sanders, W.J., E. Gheerbrant, J.M. Harris, H. Saegusa, and C. Delmer, 2010. Proboscidea, in *Cenozoic Mammals of Africa,* Berkeley, CA: University of California Press.

Sasidharan, R. and M. Gerstein, 2008. Genomics: protein fossils live on as RNA. *Nature,* 453(7196), 729–31.

Schmid, L. and M.R. Sánchez-Villagra, 2010. Potential genetic bases of morphological evolution in the Triassic fish saurichthys. *Journal of Experimental Zoology Part B – Molecular and Developmental Evolution,* 314B(7), 519–26.

Schienman, J.E., R.A. Holt, M.R. Auerbach, and C.B. Stewart. 2006. Duplication and divergence of 2 distinct pancreatic ribonuclease genes in leaf-eating African and Asian colobine monkeys. *Molecular Biology and Evolution,* 23(8), 1465–79.

Schmitz, J., M. Ohme, and H. Zischler, 2001. SINE insertions in cladistic analyses and the phylogenetic affiliations of *Tarsius bancanus* to other primates. *Genetics,* 157(2), 777–84.

2002. The complete mitochondrial sequence of *Tarsius bancanus*: evidence for an extensive nucleotide compositional plasticity of primate mitochondrial DNA. *Molecular Biology and Evolution,* **19**(4), 544–53.

Schneiderman, J.S. and W.D. Allmon (eds.), 2009. *For the Rock Record: Geologists on Intelligent Design,* Berkeley, CA: University of California Press.

Schultze, H.P. and S.L. Cumbaa, 2001. *Dialipina* and the characters of basal actinopterygians, in *Major Events in Early Vertebrate Evolution: Palaeontology, Phylogeny and Development*, ed. P.E. Ahlberg, London: Taylor & Francis, 315–32.

Scott, E.C., 2000. Not (just) in Kansas anymore. *Science,* **288**(5467), 813–15.

2009. *Evolution vs. Creationism: An Introduction,* Berkeley, CA: University of California Press.

Sears, K.E., 2004. Constraints on the morphological evolution of marsupial shoulder girdles. *Evolution,* **58**(10), 2353–70.

Sears, K.E., A. Goswami, J.J. Flynn, and L.A. Niswander, 2007. The correlated evolution of Runx2 tandem repeats, transcriptional activity, and facial length in Carnivora. *Evolution & Development,* **9**(6), 555–65.

Sedgwick, A., 1894. On the law of development commonly known as von Baer's law; and on the significance of ancestral rudiments in embryonic development. *Quarterly Journal of Microscopic Science,* **36**, 35–52.

Seiffert, E.R. and E.L. Simons, 2000. *Widanelfarasia*, a diminutive placental from the late Eocene of Egypt. *Proceedings of the National Academy of Sciences of the United States of America,* **97**(6), 2646–51.

Seiffert, E.R., J.M. Perry, E.L. Simons, and D.M. Boyer, 2009. Convergent evolution of anthropoid-like adaptations in Eocene adapiform primates. *Nature,* **461**(7267), 1118–21.

Seiffert, E.R., E.L. Simons, and Y. Attia, 2003. Fossil evidence for an ancient divergence of lorises and galagos. *Nature,* **422**(6930), 421–4.

Seiffert, E.R., E.L. Simons, W.C. Clyde, J.B. Rossie, Y. Attia, T.M. Bown, P. Chatrath, and M.E. Mathison, 2005. Basal anthropoids from Egypt and the antiquity of Africa's higher primate radiation. *Science,* **310**(5746), 300–4.

Seiffert, E.R., E.L. Simons, T.M. Ryan, T.M. Bown, and Y. Attia, 2007. New remains of Eocene and Oligocene Afrosoricida (Afrotheria) from Egypt, with implications for the origin(s) of afrosoricid zalambdodonty. *Journal of Vertebrate Paleontology,* **27**(4), 963–72.

Shanks, N. and K. Green, 2011. Intelligent design in theological perspective. *Synthese,* **178**, 307–30.

Shapiro, M.D., M.A. Bell, and D.M. Kingsley, 2006. Parallel genetic origins of pelvic reduction in vertebrates. *Proceedings of the National Academy of Sciences of the United States of America,* **103**(37), 13753–8.

Shapiro, M.D., M.E. Marks, C.L. Peichel, B.K. Blackman, K.S. Nereng, B. Jonsson, D. Schluter, and D.M. Kingsley, 2004. Genetic and developmental basis of evolutionary pelvic reduction in threespine sticklebacks. *Nature,* **428**(6984), 717–23.

Shoshani, J., 1996. Para- or monophyly of the gomphotheres and their position within Proboscidea, in *The Proboscidea: Evolution and Palaeoecology of Elephants and their Relatives*, ed. J. Shoshani and P. Tassy, Oxford: Oxford University Press, 149–77.

Shoshani, J. and P. Tassy, 1996. *The Proboscidea: Evolution and Palaeoecology of Elephants and their Relatives*, Oxford and New York: Oxford University Press.

Shu, D.G., H.L. Luo, S.C. Morris, X.L. Zhang, S.X. Hu, L. Chen, J. Han, M. Zhu, Y. Li, and L.Z. Chen, 1999. Lower Cambrian vertebrates from South China. *Nature*, **402**(6757), 42–6.

Shubin, N., 2008. *Your Inner Fish: A Journey into the 3.5-Billion-Year History of the Human Body*, New York: Pantheon Books.

Shubin, N.H. and F.A. Jenkins, 1995. An early Jurassic jumping frog. *Nature*, **377**(6544), 49–52.

Shubin, N., C. Tabin, and S. Carroll, 2009. Deep homology and the origins of evolutionary novelty. *Nature*, **457**(7231), 818–23.

Sidor, C.A. and J.A. Hopson, 1998. Ghost lineages and "mammalness": assessing the temporal pattern of character acquisition in the Synapsida. *Paleobiology*, **24**(2), 254–73.

Simmons, N.B., K.L. Seymour, J. Habersetzer, and G.F. Gunnell, 2008. Primitive early Eocene bat from Wyoming and the evolution of flight and echolocation. *Nature*, **451**(7180), 818–21.

Simons, E.L., 1995. Skulls and anterior teeth of *Catopithecus* (Primates, Anthropoidea) from the Eocene and anthropoid origins. *Science*, **268**(5219), 1885–8.

Simons, E.L. and R.F. Kay, 1988. New material of *Qatrania* from Egypt with comments on the phylogenetic position of the Parapithecidae (Primates, Anthropoidea). *American Journal of Primatology*, **15**(4), 337–47.

Simons, E.L., P.A. Holroyd, and T.M. Bown, 1991. Early Tertiary elephant-shrews from Egypt and the origin of the Macroscelidea. *Proceedings of the National Academy of Sciences of the United States of America*, **88**(21), 9734–7.

Simpson, G.G., 1944. *Tempo and Mode in Evolution*, New York: Columbia University Press.

1945. The principles of classification and a classification of mammals. *Bulletin of the American Museum of Natural History*, **85**, 1–307.

1953. *The Major Features of Evolution*, New York: Columbia University Press.

1961. 100 years without Darwin are enough. *Teachers College Record*, **62**(8), 617–28.

1967. *The Meaning of Evolution*, New Haven, CT: Yale University Press.

Sober, E., 2008. *Evidence and Evolution: The Logic Behind the Science*, Cambridge and New York: Cambridge University Press.

2009. Did Darwin write the *Origin* backwards? *Proceedings of the National Academy of Sciences of the United States of America*, **106**, 10048–55.

Sokal, R.R. and F.J. Rohlf, 1995. *Biometry: The Principles and Practice of Statistics in Biological Research*, New York: W.H. Freeman.

Soskine, M. and D.S. Tawfik, 2010. Mutational effects and the evolution of new protein functions. *Nature Reviews Genetics*, **11**(8), 572–82.

Spaulding, M., M.A. O'Leary, and J. Gatesy, 2009. Relationships of Cetacea (Artiodactyla) among mammals: increased taxon sampling alters interpretations of key fossils and character evolution. *PLoS One*, **4**(9), e7062.

Srivastava, M., C. Larroux, D.R. Lu, K. Mohanty, J. Chapman, B.M. Degnan, and D.S. Rokhsar, 2010. Early evolution of the LIM homeobox gene family. *BMC Biology*, **8**, 4.

Stanhope, M.J., V.G. Waddell, O. Madsen, W. de Jong, S.B. Hedges, G.C. Cleven, D. Kao, and M.S. Springer, 1998. Molecular evidence for multiple origins of Insectivora and for a new order of endemic African insectivore mammals. *Proceedings of the National Academy of Sciences of the United States of America*, **95**(17), 9967–72.

Stenger, V.J., 2007, *God: The Foiled Hypothesis*, Amherst, NY: Prometheus.

Stolfi, A., T.B. Gainous, J.J. Young, A. Mori, M. Levine, and L. Christiaen, 2010. Early chordate origins of the vertebrate second heart field. *Science*, **329**(5991), 565–8.

Sturm, R.A., D.L. Duffy, Z.Z. Zhao, F.P.N. Leite, M.S. Stark, N.K. Hayward, N.G. Martin, and G.W. Montgomery, 2008. A single SNP in an evolutionary conserved region within intron 86 of the HERC2 gene determines human blue-brown eye color. *American Journal of Human Genetics*, **82**(2), 424–31.

Surridge, A.K., D. Osorio, and N.I. Mundy, 2003. Evolution and selection of trichromatic vision in primates. *Trends in Ecology & Evolution*, **18**(4), 198–205.

Suwa, G., R.T. Kono, S. Katoh, B. Asfaw, and Y. Beyene, 2007. A new species of great ape from the late Miocene epoch in Ethiopia. *Nature*, **448**(7156), 921–4.

Sweet, H., 1885. Alfred's Orosius. Available at: http://www.archive.org/details/extractsfromalfro2alfruoft (accessed May 10, 2009).

Swofford, D.L., 2002. *PAUP* Phylogenetic Analysis Using Parsimony (*and Other Methods)*, Sunderland, MA: Sinauer.

Szalay, F.S. and B.A. Trofimov, 1996. The Mongolian late Cretaceous Asiatherium, and the early phylogeny and paleobiogeography of Metatheria. *Journal of Vertebrate Paleontology*, **16**(3), 474–509.

Szalay, F.S., M.J. Novacek, and M.C. McKenna, 1993. *Mammal Phylogeny*, New York: Springer-Verlag.

Tabuce, R., L. Marivaux, M. Adaci, M. Bensalah, J.L. Hartenberger, M. Mahboubi, F. Mebrouk, P. Tafforeau, and J.J. Jaeger, 2007. Early tertiary mammals from north Africa reinforce the molecular afrotheria clade. *Proceedings of the Royal Society B: Biological Sciences*, **274**(1614), 1159–66.

Tabuce, R., L. Marivaux, R. Lebrun, M. Adaci, M. Bensalah, P.H. Fabre, E. Fara, H.G. Rodrigues, L. Hautier, J.J. Jaeger, V. Lazzari, F. Mebrouk, S. Peigne, J. Sudre, P. Tafforeau, X. Valentin, and M. Mahboubi, 2009. Anthropoid versus strepsirhine status of the African Eocene primates Algeripithecus and Azibius: craniodental evidence. *Proceedings of the Royal Society B: Biological Sciences*, **276**(1676), 4087–94.

Tada, T., A. Altun, and S. Yokoyama, 2009. Evolutionary replacement of UV vision by violet vision in fish. *Proceedings of the National Academy of Sciences of the United States of America*, **106**(41), 17457–62.

Takechi, M. and S. Kuratani, 2010. History of studies on mammalian middle ear evolution: a comparative morphological and developmental biology perspective. *Journal of Experimental Zoology Part B: Molecular and Developmental Evolution,* **314B**(6), 417–33.

Tarlinton, R.E., J. Meers, and P.R. Young, 2006. Retroviral invasion of the koala genome. *Nature,* **442**(7098), 79–81.

Tassy, P., 1996. Who is who among the Proboscidea, in *The Proboscidea*, ed. J. Shoshani and P. Tassy, Oxford: Oxford University Press, 39–48.

Theobald, D.L., 2010. A formal test of the theory of universal common ancestry. *Nature,* **465**(7295), 219–22.

Thewissen, J.G.M., L.N. Cooper, M.T. Clementz, S. Bajpai, and B.N. Tiwari, 2007. Whales originated from aquatic artiodactyls in the Eocene epoch of India. *Nature,* **450**(7173), 1190–4.

Thewissen, J.G.M., L.N. Cooper, J.C. George, and S. Bajpai, 2009. From land to water: the origin of whales, dolphins, and porpoises. *Evolution: Education and Outreach,* **2**, 272–88.

Thewissen, J.G.M., S.T. Hussain, and M. Arif, 1994. Fossil evidence for the origin of aquatic locomotion in archaeocete whales. *Science,* **263**(5144), 210–12.

Thewissen, J.G.M., E.M. Williams, L.J. Roe, and S.T. Hussain, 2001. Skeletons of terrestrial cetaceans and the relationship of whales to artiodactyls. *Nature,* **413**(6853), 277–81.

Tucker, A.S., R.P. Watson, L.A. Lettice, G. Yamada, and R.E. Hill, 2004. Bapx1 regulates patterning in the middle ear: altered regulatory role in the transition from the proximal jaw during vertebrate evolution. *Development,* **131**(6), 1235–45.

Tummers, M. and I. Thesleff, 2009. The importance of signal pathway modulation in all aspects of tooth development. *Journal of Experimental Zoology Part B: Molecular and Developmental Evolution,* **312B**(4), 309–19.

Tyndale-Biscoe, C.H., 2001. Australasian marsupials: to cherish and to hold. *Reproduction, Fertility and Development,* **13**(7–8), 477–85.

2005. *Life of Marsupials*, Collingwood, VIC: CSIRO Publishing.

Tyrangiel, J., 2000. History: Faribault, Minn – the science of dissent. *Time Magazine*, July 10.

Van Biema, D., 2006. God vs. science. *Time Magazine*, November 5.

Van Valen, L.M., 1966. *Deltatheridia, a New Order of Mammals,* New York: American Museum of Natural History.

Vargas, A.O. and G.P. Wagner, 2009. Frame-shifts of digit identity in bird evolution and cyclopamine-treated wings. *Evolution & Development,* **11**(2), 163–9.

Vennemann, T., 2005. English: a German dialect? Available at: http://www.rotary-munich.de/2005–2006/theo-vennemann.pdf (accessed May 10, 2009).

Volff, J.N., 2006. Turning junk into gold: domestication of transposable elements and the creation of new genes in eukaryotes. *Bioessays,* **28**(9), 913–22.

Wake, D.B. and J. Hanken, 1996. Direct development in the lungless salamanders: what are the consequences for developmental biology, evolution and phylogenesis? *International Journal of Developmental Biology,* **40**(4), 859–69.

Walker, J.A., 1997. Ecological morphology of lacustrine threespine stickleback *Gasterosteus aculeatus* L. (Gasterosteidae) body shape. *Biological Journal of the Linnean Society*, **61**, 3–50.

Wang, B., 2001. On Tsaganomyidae (Rodentia, Mammalia) of Asia. *American Museum Novitates,* 3317, 1–50.

Wang, Y., Y. Hu, J. Meng, and C. Li, 2001. An ossified Meckel's cartilage in two Cretaceous mammals and origin of the mammalian middle ear. *Science,* **294**(5541), 357–61.

Wapinski, I., J. Pfiffner, C. French, A. Socha, D.A. Thompson, and A. Regev, 2010. Gene duplication and the evolution of ribosomal protein gene regulation in yeast. *Proceedings of the National Academy of Sciences of the United States of America,* **107**(12), 5505–10.

Weber, B.H., 2011. Design and its discontents. *Synthese,* **178**, 271–89.

Weinstock, J., E. Willerslev, A. Sher, W.F. Tong, S.Y.W. Ho, D. Rubenstein, J. Storer, J. Burns, L. Martin, C. Bravi, A. Prieto, D. Froese, E. Scott, X.L. Lai, and A. Cooper, 2005. Evolution, systematics, and phylogeography of Pleistocene horses in the New World: a molecular perspective. *PLoS Biology,* **3**(8), 1373–9.

Weiss, K.M., 2007. The Scopes trial. *Evolutionary Anthropology,* **16**(4), 126–31.

Weiss, K.M. and A.V. Buchanan, 2009. *The Mermaid's Tale: Four Billion Years of Cooperation in the Making of Living Things*, Cambridge, MA: Harvard University Press.

Wellik, D.M., 2007. Hox patterning of the vertebrate axial skeleton. *Developmental Dynamics,* **236**(9), 2454–63.

Wellik, D.M. and M.R. Capecchi, 2003. Hox10 and Hox11 genes are required to globally pattern the mammalian skeleton. *Science,* **301**(5631), 363–7.

Werdelin, L. and W.J. Sanders, 2010. *Cenozoic Mammals of Africa,* Berkeley, CA: University of California Press.

Wesley-Hunt, G.D. and J.J. Flynn, 2005. Phylogeny of the Carnivora: basal relationships among the Carnivoramorphans, and assessment of the position of "Miacoidea" relative to Carnivora. *Journal of Systematic Palaeontology,* **3**(1), 1–28.

White, T.D., B. Asfaw, Y. Beyene, Y. Haile-Selassie, C.O. Lovejoy, G. Suwa, and G. Woldegabriel, 2009. *Ardipithecus ramidus* and the paleobiology of early hominids. *Science,* **326**(5949), 75–86.

Wible, J.R., M.J. Novacek, and G.W. Rougier, 2004. New data on the skull and dentition in the Mongolian Late Cretaceous eutherian mammal zalambdalestes. *Bulletin of the American Museum of Natural History,* **281**, 1–144.

Wible, J.R., G.W. Rougier, M.J. Novacek, and R.J. Asher, 2009. The eutherian mammal *Maelestes Gobiensis* from the Late Cretaceous of Mongolia and the phylogeny of Cretaceous eutheria. *Bulletin of the American Museum of Natural History,* **327**, 1–123.

Wieland, C., 1992. Darwin's real message: have you missed it? *Creation,* **14**(4), 16–19.

Wyss, A.R., 1987. Notes on Prototheria, Insectivora, and Thomas Huxley's contribution to mammalian systematics. *Journal of Mammalogy,* **68**(1), 135–8.

Xu, X., X.T. Zheng, and H.L. You, 2009. A new feather type in a nonavian theropod and the early evolution of feathers. *Proceedings of the National Academy of Sciences of the United States of America,* **106**(3), 832–4.

Yan, Y., A. Buckler-White, K. Wollenberg, and C.A. Kozak, 2009. Origin, antiviral function and evidence for positive selection of the gammaretrovirus restriction gene *Fv1* in the genus *Mus*. *Proceedings of the National Academy of Sciences of the United States of America,* **106**(9), 3259–63.

Yomo, T., I. Urabe, and H. Okada, 1992. No stop codons in the antisense strands of the genes for nylon oligomer degradation. *Proceedings of the National Academy of Sciences of the United States of America,* **89**(9), 3780–4.

Young, R.L., V. Caputo, M. Giovannotti, T. Kohlsdorf, A.O. Vargas, G.E. May, and G.P. Wagner, 2009. Evolution of digit identity in the three-toed Italian skink *Chalcides chalcides*: a new case of digit identity frame shift. *Evolution & Development,* **11**(6), 647–58.

Zack, S.P., T.A. Penkrot, J.I. Bloch, and K.D. Rose, 2005. Affinities of "hyopsodontids" to elephant shrews and a Holarctic origin of Afrotheria. *Nature,* **434**(7032), 497–501.

Zalmout, I.S., W.J. Sanders, L.M. MacLatchy, G.F. Gunnell, Y.A. Al-Mufarreh, M.A. Ali, A.A.H. Nasser, A.M. Al-Masari, S.A. Al-Sobhi, A.O. Nadhra, A.H. Matari, J.A. Wilson, and P.D. Gingerich, 2010. New Oligocene primate from Saudi Arabia and the divergence of apes and Old World monkeys. *Nature,* **466**(7304), 360–4.

Zardoya, R., E. Malaga-Trillo, M. Veith, and A. Meyer, 2003. Complete nucleotide sequence of the mitochondrial genome of a salamander, *Mertensiella luschani*. *Gene,* **317**(1–2), 17–27.

Zhang, G.R., S.T. Wang, J.Q. Wang, N.Z. Wang, and M. Zhua, 2010. A basal antiarch (placoderm fish) from the Silurian of Qujing, Yunnan, China. *Palaeoworld,* **19**, 129–35.

Zhang, J.Z., 2003. Evolution by gene duplication: an update. *Trends in Ecology & Evolution,* **18**(6), 292–8.

Zhou, Z.H., 2004. The origin and early evolution of birds: discoveries, disputes, and perspectives from fossil evidence. *Naturwissenschaften,* **91**(10), 455–71.

Zhu, M., W. Zhao, L. Jia, J. Lu, T. Qiao, and Q. Qu, 2009. The oldest articulated osteichthyan reveals mosaic gnathostome characters. *Nature,* **458**, 469–74.

Zimmer, C., 1998. *At the Water's Edge: Macroevolution and the Transformation of Life,* New York: Free Press.

2001. *Evolution: The Triumph of an Idea,* New York: HarperCollins.

2009. *The Tangled Bank: An Introduction to Evolution*, Greenwood Village, CO: Roberts and Co. Publishers.

INDEX